Petroleum and Natural Gas Engineering

Petroleum and Natural Gas Engineering

Guest Editors

Jiaxue Li
Pengfei Zhao

MDPI

Basel • Beijing • Wuhan • Barcelona • Belgrade • Novi Sad • Cluj • Manchester

Guest Editors
Jiaxue Li
Engineering Institute
China University of
Petroleum (Beijing)
Beijing
China

Pengfei Zhao
School of Geoscience
and Technology
Southwest Petroleum
University
Chengdu
China

Editorial Office
MDPI AG
Grosspeteranlage 5
4052 Basel, Switzerland

This is a reprint of the Special Issue, published open access by the journal *Energies* (ISSN 1996-1073), freely accessible at: https://www.mdpi.com/journal/energies/special_issues/S14Z410XRG.

For citation purposes, cite each article independently as indicated on the article page online and as indicated below:

Lastname, A.A.; Lastname, B.B. Article Title. *Journal Name* **Year**, *Volume Number*, Page Range.

ISBN 978-3-7258-5093-8 (Hbk)
ISBN 978-3-7258-5094-5 (PDF)
https://doi.org/10.3390/books978-3-7258-5094-5

Contents

energies

MDPI

Editorial

Overview of the Progress in Practice and Experimental Research in Petroleum and Natural Gas Engineering

Xiaobin Li [1], Jiaxue Li [2], Pengfei Zhao [3], Gan Feng [1,4,*] and Lei Zhao [5]

1 State Key Laboratory of Water Resource Protection and Utilization in Coal Mining, National Institute of Low Carbon and Clean Energy, Beijing 102209, China; lxb162197@126.com
2 Engineering Institute, China University of Petroleum (Beijing), Karamay 834000, China; 20192146@cupk.edu.cn
3 School of Geoscience and Technology, Southwest Petroleum University, Chengdu 610500, China; 202199010067@swpu.edu.cn
4 State Key Laboratory of Hydraulics and Mountain River Engineering, College of Water Resource & Hydropower, Sichuan University, Chengdu 610065, China
5 Shanxi Coal Transportation and Sales Group Yechuan Coal Industry Co., Ltd., Jingcheng 048000, China; zhaolei_7@126.com
* Correspondence: fenggan@whu.edu.cn

1. Introduction

The global energy system is undergoing unprecedented changes. Despite the urgent need for the development of existing renewable energy systems, the promotion of carbon neutrality goals, and enhanced energy security, oil and natural gas—traditional fossil fuels—continue to provide most of humanity's primary energy needs [1–3]. Our reserves of oil are substantial; however, complex geological and geographical environments, which are the result of different mechanisms behind the formation of oil in terrestrial and marine settings, pose significant challenges to their efficient extraction. Simultaneously, there are considerable shale gas reserves within the ground [4–6], and other natural gases, such as tight sandstone gas and coalbed methane, are not only abundant but also widely distributed. The geological conditions under which these resources are found are similarly complex [7–10], characterized by high ground stress, fault structures, low porosity, and fractures in gas reservoirs. These factors complicate the efficient extraction of natural gas. However, the efficient extraction of oil and natural gas is crucial to minimizing resource waste. For instance, in regions where extraction is challenging, abandonment can lead to the permanent loss of unexploited resources due to geological disturbances, complicating future extraction efforts. On the other hand, efficient mining introduces new challenges related to mining equipment, processes, and technological advancements, necessitating the development of advanced methodologies and predictive techniques.

Efficient oil and gas recovery methods, enhanced oil recovery, fracture propagation prediction, and reservoir management continue to pose significant challenges in the field of oil and gas extraction. The development of unconventional resources, such as shale oil and gas, tight gas, and natural gas hydrates, is reshaping the global energy supply landscape. Meanwhile, the advancement of conventional oil and gas resources is undergoing profound transformation through the integration of digital technology. The real-time data analysis of intelligent drilling systems, completion systems, and nanoscale fracturing fluid technology exemplifies this change [11–13]. Intelligent drilling systems enhance drilling efficiency and mitigate the risks associated with mountain springs during construction by enabling real-time data analysis. The use of nanoparticle fracturing fluid technology in unconventional oil and gas development increases the efficiency of shale gas recovery. The integration of new composite-material pipelines into smart pipeline monitoring systems, which utilize IoT technology, significantly advances the storage, transportation, and extraction of oil and gas. Methane capture technology effectively reduces carbon emissions. Photothermal-driven geological storage systems facilitate negative carbon cycling in the CO_2-enhanced oil recovery (EOR) process. Hybrid power supply systems that combine offshore wind power, photovoltaic power generation, and oil and gas platforms play a crucial role in reducing carbon emissions. Digital models of deep-sea oil fields substantially improve the accuracy of development plans. Artificial intelligence algorithms are vital for predicting oil and gas reservoirs, enhancing prediction accuracy. Moreover, the wear resistance of graphene-reinforced drill bits has been significantly improved. Some of the innovations in oil and gas storage and transportation technologies are also noteworthy. In response to carbon neutrality targets, the oil and gas industry is establishing a new environmental technology framework. Digital twin technology is redefining the oil and gas development model, while the application of artificial intelligence, big data, and Internet of Things (IoT) technology is accelerating the digital transformation of the industry. Over the next decade, several changes are expected, including the upgrade of carbon management technology from auxiliary processes to core business units, the promotion of intelligent transitions across the entire industry chain through digital twin systems, and the collaborative design of engineering systems that include wind, solar, and hydrogen technologies. These changes not only extend the economic lifecycle of fossil fuels but also strive to create a strategic buffer period during which new energy systems can mature.

This Special Issue aims to highlight recent developments and innovations in oil and gas engineering, with a particular emphasis on hydraulic fracturing and enhanced oil recovery (EOR) technologies, as well as the application of advanced artificial intelligence in oil and gas extraction practices and experimental research. These methods are essential to optimizing extraction processes and enhancing production efficiency. We also seek to address advancements in drilling technology, which play a critical role in the exploration and extraction of oil and gas resources. Research topics of interest included innovations in hydraulic fracturing fluids and proppants, the modeling and simulation of fracture propagation, environmental impacts and mitigation strategies related to hydraulic fracturing, real-time monitoring and adaptive fracturing techniques, chemical and thermal EOR methods, successful case studies of EOR implementations, the integration of EOR with reservoir management, advances in drilling technology and equipment, intelligent construction in oil and gas extraction, drilling optimization and cost reduction strategies, wellbore stability and control, and new materials and technologies for drilling in challenging environments. This collective effort aims to further our understanding of oil and gas engineering. A total of 11 high-quality papers have been published in this Special Issue, all of which have undergone rigorous review and screening.

2. Recent Developments in Petroleum and Natural Gas Engineering

Scholars have conducted extensive research on topics related to petroleum and natural gas engineering. The papers published in our Special Issue cover key aspects of this engineering, and we have elaborated on them below.

Li et al. utilized artificial intelligence and GRN-VSN neural networks to predict oilfield indicators. First, they input highly relevant parameters into that serve as predictors of the key indicators driven by artificial intelligence into their model. Subsequently, the Shapley Additive Explanation (SHAP) was employed to interpret the artificial model and evaluate its predicted results. Additionally, its performance was compared to that of the ResNet-50 neural network, long short-term memory (LSTM) neural network, and backpropagation (BP) neural network in terms of oil extraction efficiency. Among these networks, LSTM excels in continuous sequence prediction and demonstrates a superior performance. Artificial intelligence algorithms have significantly enhanced the prediction accuracy of key production indicators in offshore oil fields, achieving an accuracy rate of at least 92%.

Aiming at improving the extraction efficiency of petroleum and natural gas, Chen et al. investigated the characteristics of a reservoir within a specific block of the Ordos Basin, analyzing the flow lines and horizontal principal stresses present to identify the conditions necessary for the formation of waterflood-induced fractures (WIFs). Subsequently, they constructed a permeability evolution equation for the fractured reservoir and cap rock areas between oil wells. Finally, numerical simulation methods were employed to examine the characteristics of WIFs in horizontal and vertical well networks by applying different injection modes. Notably, the study revealed that WIFs formed at locations where the maximum principal stress and flow were aligned, influencing the distribution of permeability. To effectively adjust the extension range of WIFs, the authors recommended controlling the injection rate of the vertical well along the flow line and at the maximum principal stress point, thereby cyclically optimizing oil production. Additionally, Wei et al. provided a comprehensive introduction to the principles of and technology behind oxygen-reducing agent-assisted gravity drainage (OAGD). They conducted a detailed analysis of the factors affecting oil displacement, which primarily include the reservoir dip angle, layer characteristics, injection rate, and low-temperature oxidation reactions. Low-temperature oxidation significantly enhances oil recovery (EOR) due to the dynamic balance between fuel deposition and light hydrocarbon volatilization, as well as the synergistic optimization of the concentration, temperature, and pressure of oxygen. To expand the sweeping volume and delay gas breakthroughs, the injection rate can be appropriately controlled to stabilize the oil–gas interface. The enhanced gravity separation effect results in highly efficient oil displacement in steeply dipping reservoirs.

The following studies focus on fracture networks in petroleum and natural gas reservoirs. Shale oil reservoirs are characterized by low permeability and porosity, necessitating the use of horizontal wells with multiple fractures for their extraction. Multiphase flow characteristics complicate fluid movement during shale oil extraction. Liu et al. propose a productivity model for multi-fractured horizontal wells in shale oil reservoirs, which they establish using the principles of pressure superposition, conformal transformation, and fractal theory, which are solved simultaneously. The effectiveness of this model has been validated using on-site experimental data. Another group of authors conducted an analysis of various complex factors affecting the productivity of shale oil wells. Their research findings indicate that phase transition behavior significantly reduces oil production, while fluid desorption markedly increases production. Gu et al. employed a Pearson correlation analysis to investigate the relationships between natural fractures, mineral composition and content, horizontal stress differences, and yield parameters, aiming to evaluate the impact of fracture network complexity on yield. Furthermore, they proposed the fracture network index (FNI) model, which is based on the support vector machine (SVM) algorithm and an improved particle swarm optimization (IPSO) algorithm, to assess the complexity of fracture networks. Finally, the correlation between the fracture network index and gas produced from various hydraulic fracturing operations was analyzed and quantified. Their research findings indicate that the Pearson correlation coefficient is 0.39, revealing that natural fracture density has a dominant controlling effect on gas production, while other factors exert relatively minor effects. The coefficient of determination (R^2) for the IPSO-SVM-FNI model, when deployed on the training set, was better than that of traditional models, demonstrating its superior data fitting. The IPSO-SVM-FNI model also exhibits high prediction accuracy. The fracture network index (FNI) predominantly falls within the range of [0.2, 0.8]. For hydraulic fracturing operations with a high fracture network index (FNI), oil and gas production is comparatively high, indicating a positive correlation between reservoir fracture complexity and shale gas production.

The following studies focus on oil and gas extraction. Chen et al. utilized a surfactant polymer (SP) system combined with a viscosity reducer to enhance oil recovery efficiency. The combination of OAB (a beta surfactant) and LPS-3 (an anionic surfactant) significantly reduced interfacial tension and enhanced lotion stability, with optimal results achieved when their ratio was 1:9. Their research indicates that the BRH-325 polymer possesses multiple desirable characteristics, including enhanced viscosity, resistance to high temperatures, and long-term stability. The viscosity reducer contains graphene nanowedges, which can decrease the viscosity of heavy oil by approximately 97%. Indoor core flooding simulation experiments were conducted to verify the effectiveness of this method, resulting in an increase in the recovery rate of about 33%. Microbial-enhanced oil recovery remains a compelling area of research. Zhao et al. collected outcrop rock samples and conducted simulation experiments on microbial-enhanced oil recovery. Concurrently, they studied the changes in biochemical parameters, including Bacillus subtilis concentration, nutrient concentration, displacement pressure, and surface tension, seen throughout the process. Their research findings indicate that after injecting microorganisms into reservoirs, cells and nutrients tend to be distributed along the primary pathways of the injection wells and fluid flow. Bacteria exhibit adsorption and retention abilities which are greater than those of nutrients. The combined effects of microbial reproduction and metabolic products increase the pressure within the model. However, from the injection well to the production well, pressure gradually decreases, with high-pressure areas primarily concentrated near the injection well. The fermentation mixture of Bacillus subtilis can enhance oil recovery by 6.5%.

Regarding other research areas, two-dimensional nuclear magnetic resonance (NMR) provides rapid measurements in petroleum and natural gas engineering. Zhang et al. conducted research on high-porosity and high-permeability heavy-oil loose-sandstone reservoirs using two-dimensional NMR testing technology. They obtained the distribution patterns of crude oils with varying viscosities from the NMR spectra. Additionally, a model relating NMR parameters to oil viscosity was established using T1 and T2 spectra, thereby creating a novel method for estimating oil viscosity. This technology has been applied in practical engineering, and results indicate that the error between the actual viscosity and the theoretically calculated viscosity of the oil is 15%. The reliability of the method was further validated by analyzing the consistency between the oil discrimination chart and the oil type. This research meets the accuracy requirements of well logging interpretation. Zhang et al. employed the pressure transient analysis (PTA) method to examine the shut-in pressure data of shale gas wells following fracturing. The results indicated that the pressure derivative gradually dispersed after one day of shut-in. Two wells exhibited zero-slope pressure derivatives over one week of fluid immersion, suggesting that the duration of fluid immersion was appropriate. In contrast, the other four wells demonstrated an increase in their pressure derivative after one week of fluid immersion, indicating that a longer immersion period is necessary to fully achieve the desired effect. Xu et al. established a heat transfer model for a composite insulation structure featuring multi-layer insulation and liquid nitrogen screen (LNCS) insulation, as well as a numerical model. They investigated the changes in the natural convection characteristics, thermal stratification, pressure distribution, and self-pressurization characteristics of LHe-4 storage tanks. Additionally, a self-pressurization thermodynamic model for LHe-4 storage tanks was developed. Their research findings indicate that the mLee model significantly improves the prediction of self-enhancement characteristics compared to the Lee model. An increase in operational time is associated with a rise in the thermal stratification degree (TSD) of the storage tank, and gradual increases in operational duration increase the self-pressurization of the tank. The graph of the interface mass transfer rate reveals a pattern of low values in the middle and high values at both ends, which is attributed to the strong evaporation point on the wall in contact with the phase interface. Yang et al. investigated the fixed tooth strength of roller drill bits, testing the maximum fastening force of the fixed teeth under various conditions and analyzing the change in this force. Their research findings indicate that high temperatures can weaken the strength of the fixed teeth. The maximum fastening force decreases with increasing temperature, becoming approximately 49–65% lower than that in a normal-temperature environment. Under consistent temperature conditions, the maximum fastening force occurs at a perforation distance of 10 mm. An increase in tooth diameter is associated with a rise in the maximum fastening force, indicating an improvement in the fixing effect. The relationship between maximum fastening force and interference fit is non-linear, with the maximum tightening force occurring at an interference fit of 0.095 mm.

Despite the significant achievements of this Special Issue, oil and gas engineering continues to face numerous challenges. For instance, technological bottlenecks in deepwater oil and gas development and ultra-deep resource exploration urgently need to be addressed. Cross-disciplinary technologies, such as hydrogen energy coupling and carbon capture, utilization, and storage (CCUS), require enhanced system integration. Additionally, the use of artificial intelligence in oil and gas extraction engineering needs to be further developed, and innovations in high-temperature drill bit design and production processes for deep oil and gas extraction are required. The advancement of efficiency and environmental adaptability in oil and gas extraction will inevitably lead to the establishment of a diversified, intelligent, and sustainable new energy system.

3. Closing Remarks

The papers published in this Special Issue addresses various aspects of oil and gas extraction, including improvements in extraction methods, reservoir modifications, and fracture network prediction. Although petroleum and natural gas engineering is a broad field, the findings presented in this Special Issue will stimulate and enhance research in the industry. Understanding these topics is crucial for recognizing the latest advancements in oil and gas engineering. We would like to express our gratitude to all the authors who contributed to this Special Issue. Given the broad scope of oil and gas engineering, we will continue to explore new research directions with these scholars and engage in in-depth investigations.

Author Contributions: All authors contributed in similar ways to all sections of this paper. All authors have read and agreed to the published version of the manuscript.

Funding: This research was funded by the open fund of State Key Laboratory of Water Resource Protection and Utilization in Coal Mining (Grant No. GJNY-21-41-01), the National Natural Science Foundation of China (Grant Nos. 52374099; 52404109), and the Natural Science Foundation of Sichuan Province, China (Grant No. 2025YFHZ0323).

Conflicts of Interest: Author Lei Zhao was employed by the company Shanxi Coal Transportation and Sales Group Yechuan Coal Industry Co., Ltd. The authors declare no conflicts of interest.

List of Contributions

1. Li, K.; Wang, K.; Tang, C.; Pan, Y.; He, Y.; Cai, S.; Chen, S.; Zhou, Y. Prediction of Key Development Indicators for Offshore Oilfields Based on Artificial Intelligence. *Energies* **2024**, *17*, 4594. https://doi.org/10.3390/en17184594
2. Chen, J.; Cai, D.; Zhang, T.; Yu, L.; Zhou, D.; Cheng, S. Asynchronous Injection–Production Method in the High Water Cut Stage of Tight Oil Reservoirs. *Energies* **2024**, *17*, 4838. https://doi.org/10.3390/en17194838
3. Wei, J.; Yu, H.; Gao, M.; Yan, P.; Tan, K.; Yan, Y.; Wei, K.; Sun, M.; Yu, X.; Chen, Z.; et al. Research and Application of Oxygen-ReducedAir-Assisted Gravity Drainage for Enhanced Oil Recovery. *Energies* **2025**, *18*, 557. https://doi.org/10.3390/en18030557
4. Liu, X.; Guo, P.; Ren, J.; Wang, Z.; Tu, H. Productivity Model for Multi-Fractured Horizontal Wells with Complex Fracture Networks in Shale Oil Reservoirs Considering Fluid Desorption and Two-Phase Behavior. *Energies* **2024**, *17*, 6012. https://doi.org/10.3390/en17236012
5. Gu, Z.; Liu, B.; Liu, W.; Liu, L.; Wei, H.; Yu, B.; Dong, L.; Zhong, P.; Lin, H. Analysis on Correlation Model Between Fracture Network Complexity and Gas-Well Production: A Case in the Y214 Block of Changning, China. *Energies* **2024**, *17*, 6026. https://doi.org/10.3390/en17236026
6. Chen, X.; Hou, Q.; Liu, Y.; Liu, G.; Zhang, H.; Sun, H.; Zhu, Z.; Liu, W. Experimental Study on Surfactant–Polymer Flooding After Viscosity Reduction for Heavy Oil in Matured Reservoir. *Energies* **2025**, *18*, 756. https://doi.org/10.3390/en18030756
7. Zhao, Y.; Xiu, J.; Huang, L.; Yi, L.; Ma, Y. Two-Dimensional Physical Simulation of the Seepage Law of Microbial Flooding. *Energies* **2025**, *18*, 1246. https://doi.org/10.3390/en18051246
8. Zhang, W.; Li, S.; Wang, S.; Sun, J.; Cai, W.; Yu, W.; Dai, H.; Yang, W. Application of Two-Dimensional NMR for Quantitative Analysis of Viscosity in Medium–High-Porosityand-Permeability Sandstones in North China Oilfields. *Energies* **2024**, *17*, 5257. https://doi.org/10.3390/en17215257
9. Zhang, J.; Guo, B.; Hussain, M. Use of Pressure Transient Analysis Method to Assess Fluid Soaking in Multi-Fractured Shale Gas Wells. *Energies* **2025**, *18*, 549. https://doi.org/10.3390/en18030549
10. Xu, J.; Liu, F.; Zhang, J.; Li, C.; Liu, Q.; Li, C.; Jia, W.; Fu, S.; Li, L. Numerical Study on Heat Leakage, Thermal Stratification, and Self-Pressurization Characteristics in Liquid Helium Storage Tanks. *Energies* **2024**, *17*, 6254. https://doi.org/10.3390/en17246254
11. Yang, Y.; Song, D.; Chen, L.; Yang, Y.; Ren, H.; Qiu, S.; Huang, Z. Research on the Fixation Strength of High-Temperature Geothermal Drilling Cone Bit Teeth. *Energies* **2025**, *18*, 2469. https://doi.org/10.3390/en18102469

References

1. Alabido, Z.; Alabido, D. Purifying Seawater from Petroleum Derivatives Resulting from Oil Tankers using Autonomous Robot: Seaswarm Robot. *Transp. Res. Procedia* **2025**, *84*, 43–50. [CrossRef]
2. Alhammadi, A.; Soar, J.; Yusaf, T.; Ali, B.M.; Kadirgama, K. Redefining procurement paradigms: A critical review of buyer-supplier dynamics in the global petroleum and natural gas industry. *Extr. Ind. Soc.* **2023**, *16*, 101351. [CrossRef]
3. Karim, M.A.; Abdullah, M.Z.; Deifalla, A.F.; Azab, M.; Waqar, A. An assessment of the processing parameters and application of fibre-reinforced polymers (FRPs) in the petroleum and natural gas industries: A review. *Results Eng.* **2023**, *18*, 101091. [CrossRef]
4. Feng, G.; Kang, Y.; Wang, X.C.; Hu, Y.Q.; Li, X.H. Investigation on the failure characteristics and fracture classification of shale under Brazilian test conditions. *Rock Mech. Rock Eng.* **2020**, *53*, 3325–3340. [CrossRef]
5. Xie, H.Q.; Feng, G.; Liu, H.Z.; He, Q.; Xiao, M.L.; Pei, J.L.; Reza, T. Study on the characteristics of crack initiation in deep dense shale containing circular hole under varying stress conditions. *J. Cent. South Univ.* **2025**, *32*, 244–261. [CrossRef]
6. Feng, G.; Kang, Y.; Sun, Z.D.; Wang, X.C.; Hu, Y.Q. Effects of Supercritical CO_2 adsorption on the mechanical characteristics and failure mechanisms of shale. *Energy* **2019**, *173*, 870–882. [CrossRef]
7. Wang, E.; Zhu, H.; Yi, X.; Li, Q.; Zhao, P.; Micheal, M.; Tan, H.; Zhang, Z. Numerical simulation of fracture propagation in high-energy gas fracturing of shale reservoir. *Geoenergy Sci. Eng.* **2025**, *252*, 213915. [CrossRef]

. Shui, H.; Wang, Y.; Li, Q.; Fan, C.; Li, Y.; Zeng, Y.; Guo, W. Analysis of heat transfer performance and system energy efficiency of catalytic combustion heaters for low calorific value waste gas application to oil shale in-situ conversion. *Energy* **2024**, *294*, 130690. [CrossRef]

. Gao, S.; Ye, L.; Liu, H.; Zhang, J.; Zhu, W.; An, W. Research on the basic theory and application of enhanced recovery in tight sandstone gas reservoirs. *Heliyon* **2025**, *11*, e41306. [CrossRef]

0. Jiu, H.; Li, D.; Zhang, Y.; Xin, G.; Ren, S. Microscopic adsorption study from coal rank comparison on the feasibility of CO_2-rich industrial waste gas injection enhanced coal bed methane recovery. *J. Mol. Liq.* **2024**, *415 Pt 1*, 126337. [CrossRef]

1. Xia, W.; Tang, Y.; Li, G.; Yue, C.; Han, Y.; Wu, X.; Fan, S. Wellbore stability prediction method based on intelligent analysis model of drilling cuttings logging images. *Geoenergy Sci. Eng.* **2025**, *252*, 213961. [CrossRef]

2. Eren, T.; Polat, C. Numerical investigation of the application of intelligent horizontal well completion. *J. Nat. Gas Sci. Eng.* **2020**, *83*, 103599. [CrossRef]

3. Pan, J.; Mou, P.; Ju, Y.; Wang, K.; Zhu, Q.; Ge, T.; Yu, K. Micro-nano-scale pore stimulation of coalbed methane reservoirs caused by hydraulic fracturing experiments. *J. Pet. Sci. Eng.* **2022**, *214*, 110512. [CrossRef]

energies

MDPI

Review

Research and Application of Oxygen-Reduced-Air-Assisted Gravity Drainage for Enhanced Oil Recovery

Jiangfei Wei [1,2], Hongwei Yu [3,4,*], Ming Gao [1,2,3,4], Peifeng Yan [5], Kesheng Tan [1,2], Yutong Yan [1,2], Keqiang Wei [1,2], Mingyan Sun [1,2], Xianglong Yu [1,2], Zhihua Chen [1,2] and Qiang Chen [1,2]

1 University of Chinese Academy of Sciences, Beijing 101400, China; weijiangfei23@mails.ucas.ac.cn (J.W.)
2 Institute of Porous Flow and Fluid Mechanics, Chinese Academy of Sciences, Langfang 065007, China
3 National Key Laboratory of Enhanced Oil and Gas Recovery, Beijing 100083, China
4 China National Petroleum Corporation, Beijing 100007, China
5 PetroChina Huabei Oilfield Company, Renqiu 062550, China
* Correspondence: yhongwei@petrochina.com.cn

Abstract: This paper summarizes the research progress and applications of oxygen-reduced-air-assisted gravity drainage (OAGD) in enhanced oil recovery (EOR). The fundamental principles and key technologies of OAGD are introduced, along with a review of domestic and international field trials. Factors influencing displacement performance, including low-temperature oxidation reactions, injection rates, and reservoir dip angles, are discussed in detail. The findings reveal that low-temperature oxidation significantly improves the recovery efficiency through the dynamic balance of light hydrocarbon volatilization and fuel deposition, coupled with the synergistic optimization of the reservoir temperature, pressure, and oxygen concentration. Proper control of the injection rate stabilizes the oil–gas interface, expands the swept volume, and delays gas channeling. High-dip reservoirs, benefiting from enhanced gravity segregation, demonstrate superior displacement efficiency. Finally, the paper highlights future directions, including the optimization of injection parameters, deepening studies on reservoir chemical reaction mechanisms, and integrating intelligent gas injection technologies to enhance the effectiveness and economic viability of OAGD in complex reservoirs.

Keywords: oxygen-reduced-air-assisted gravity drainage; low-temperature oxidation; gas injection rate; reservoir dip angle

1. Introduction

With the continuous growth of the global energy demand, conventional oilfields are progressively transitioning into high water-cut development stages, significantly increasing the difficulty of recovering residual oil. Consequently, oilfield development is facing increasingly severe challenges [1–3]. Against this backdrop, enhanced oil recovery (EOR) technologies have become a critical research focus in the field of oilfield development [4–6]. Gas-assisted gravity drainage (GAGD), as an advanced EOR technology, leveraging gravity segregation effects, has gained significant attention due to its ability to utilize the density difference between gas and crude oil to expand the swept volume and enhance the displacement efficiency. Among these, OAGD technology has demonstrated promising potential, particularly in reducing oxidation risks, delaying gas channeling, and stabilizing gas drive fronts. This technology shows broad application prospects, especially in low-permeability, ultra-low-permeability, and complex reservoirs, where it exhibits remarkable oil recovery potential [7].

Compared to conventional water flooding, oxygen-reduced air gravity drainage demonstrates significant advantages in expanding sweep efficiency and enhancing gas flooding stability. However, the diverse and complex nature of reservoir conditions presents numerous challenges for its practical application. Key issues include optimizing the injection rates and oxygen concentrations to delay gas channeling and maximize the recovery efficiency, clarifying the mechanisms of chemical interactions between oxygen-reduced air, reservoir minerals, and crude oil and their impact on reservoir stability, and developing effective injection production strategies tailored to heterogeneous reservoirs. Addressing these challenges remains critical for advancing this technology.

In recent years, researchers both domestically and internationally have conducted extensive studies on OAGD technology. These studies have focused on various aspects, including the mechanisms of LTO reactions and the optimization of gas injection parameters, gas migration patterns, and oil–gas interface stability. Laboratory-scale physical simulations and numerical modeling have unveiled the dynamic changes of the oil–gas interface during gas injection and identified key influencing factors. Field trials have further validated the feasibility of this technology in low-permeability reservoirs, high-dip reservoirs, and heavy oil reservoirs [8]. However, these studies still exhibit notable limitations, particularly in the investigation of multi-factor synergistic effects under complex reservoir conditions, the impact of reservoir heterogeneity on oil recovery efficiency, and the optimization of injection production strategies for engineering applications. Further exploration is required in these areas.

This paper provides a systematic review of the research status of this technology, focusing on its fundamental principles, key techniques, and influencing factors. It further addresses existing scientific issues and engineering challenges, proposing future research directions and development recommendations to offer theoretical insights and technical guidance for the advancement of oxygen-reduced air gravity drainage technology and its application in practical reservoir development (Figure 1).

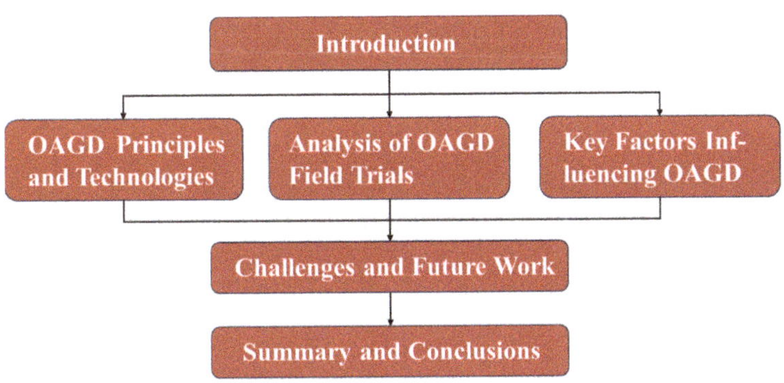

Figure 1. Structural flow chart of the OAGD review.

2. OAGD

2.1. Fundamental Principles of Oxygen-Reduced Air Gravity Drainage

OAGD combines the principles of oxygen-reduced air injection with gas-assisted gravity drainage (GAGD). Compared to traditional water flooding techniques, GAGD leverages the density difference between gas and crude oil to effectively mobilize residual oil under the influence of gravity. This method is particularly suitable for high-angle, thick, or complex heterogeneous reservoirs where conventional technologies struggle to achieve sufficient coverage and recovery [9–11]. The core advantage of this technology lies in its ability to fully utilize gravity segregation effects. The injected gas diffuses upward within the reservoir, causing crude oil to flow downward under gravitational forces toward the production well, thereby significantly enhancing the vertical displacement efficiency (Figure 2). This displacement method prevents water breakthrough and channeling issues commonly encountered in water flooding processes, effectively expanding the swept volume and mobilizing a greater proportion of residual oil [12].

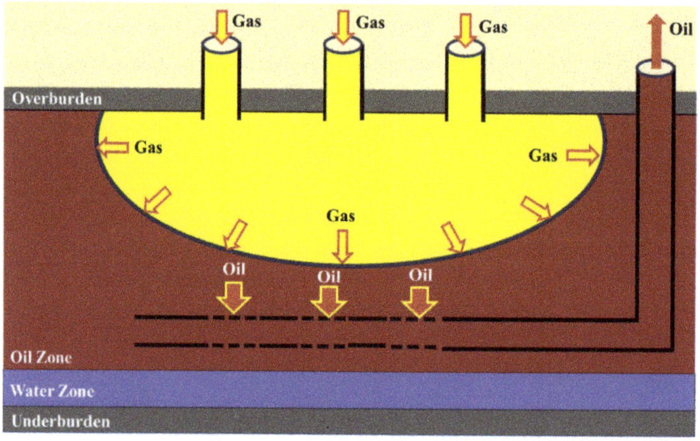

Figure 2. Schematic diagram of the GAGD process.

The selection of oxygen-reduced air as the medium for oxygen-reduced-air-assisted gravity drainage is primarily based on its safety, economic feasibility, and EOR efficiency. Compared to conventional air injection techniques, oxygen-reduced air reduces the oxygen content, effectively mitigating the explosion risks associated with oxidation reactions while retaining the critical displacement efficiency characteristics of air injection [13,14]. This refinement not only enhances the controllability of the technology but also expands its applicability in complex reservoir conditions. Its core advantage lies in combining LTO reactions with gravity segregation effects to improve oil recovery. During the displacement process, LTO generates flue gases (such as CO_2 and water vapor), which further increase the swept volume, reduce crude oil viscosity, and elevate the reservoir temperature, thereby enhancing the oil recovery efficiency [15–17]. Compared to conventional water flooding or other gas displacement methods, oxygen-reduced air flooding demonstrates exceptional performance in high-dip, low-permeability reservoirs. It effectively mobilizes residual oil and delays gas breakthrough, significantly improving the recovery efficiency [18–20].

2.2. Oxygen-Reduced Air Preparation Technologies

In terms of resources and cost effectiveness, oxygen-reduced air offers inherent advantages. Its production relies on well-established technologies, such as pressure swing adsorption (PSA), membrane separation, and cryogenic separation, ensuring low costs and abundant gas supply. This makes it particularly suitable for oilfield development projects with high economic requirements. Compared to nitrogen or carbon dioxide flooding, the acquisition and preparation of oxygen-reduced air are more convenient while maintaining significant oil recovery efficiency and operational safety [21]. Therefore, oxygen-reduced air has emerged as an ideal displacement medium for gas-assisted gravity drainage, making it a preferred choice for EOR during the later stages of oilfield development.

2.2.1. Cryogenic Separation

Cryogenic separation involves a series of purification, compression, and cooling processes, where air is passed through a primary heat exchanger to achieve liquefaction, forming liquid air. The liquid air is then introduced into a distillation column for separation, utilizing the differing boiling points of oxygen and nitrogen. Oxygen, with its higher boiling point, gradually accumulates at the lower section of the distillation column, forming oxygen-enriched liquid air. Meanwhile, oxygen-reduced air can be obtained from the upper section of the column as required [22]. The typical process of cryogenic separation is illustrated in Figure 3.

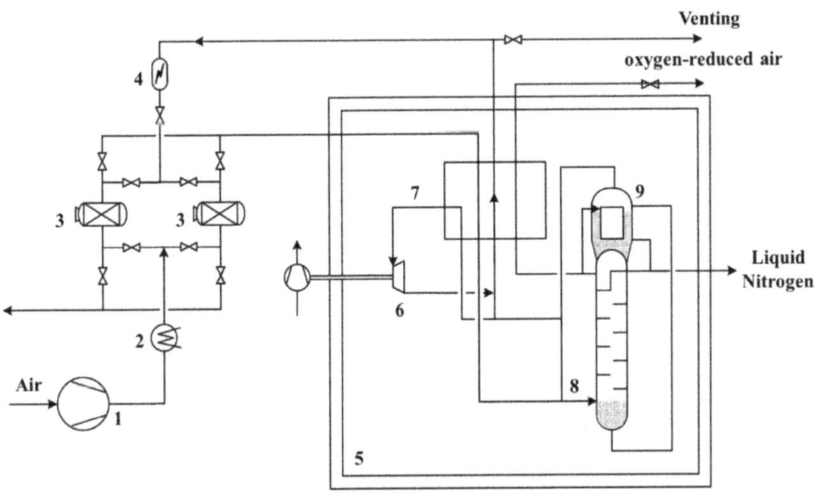

Figure 3. Cryogenic separation process flow. 1—air compressor, 2—pre-cooling unit, 3—molecular sieve adsorber, 4—electric heater, 5—cold box, 6—turbo expander, 7—main heat exchanger, 8—rectification column, and 9—condenser evaporator.

2.2.2. Membrane Separation Method

The core of membrane separation technology lies in the selection of membrane materials. For specific membrane materials, nitrogen and oxygen in the air exhibit significant differences in solubility and diffusion rates, particularly under higher pressure conditions. By utilizing the pressure differential across the membrane, faster-diffusing gases such as water vapor and oxygen preferentially pass through to the low-pressure side, forming oxygen-enriched gas, while slower-diffusing nitrogen accumulates on the high-pressure side, thereby achieving the separation of oxygen-reduced air [23]. The process flow of the membrane separation method is shown in Figure 4.

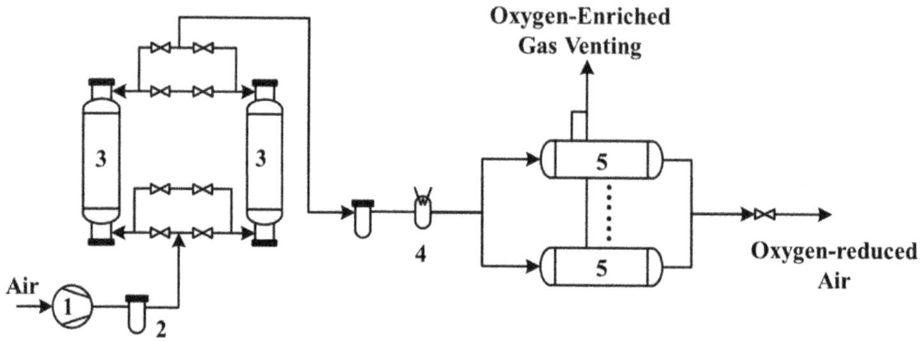

Figure 4. Process flow of membrane separation method. 1—air compressor, 2—filter, 3—dryer, 4—electric heater, and 5—membrane module.

2.2.3. Pressure Swing Adsorption (PSA) Technology

PSA technology is a separation process based on the adsorption phenomenon. It utilizes porous solid media in contact with gas or liquid phases to selectively adsorb specific components from the fluid on the solid surface, thereby achieving separation and enrichment of the desired components [24]. This technology adjusts the pressure during the adsorption and desorption processes, significantly altering the composition and concentration of the fluid's components [25]. The entire PSA nitrogen production process can be divided into four sub-processes: pressurized adsorption, pressure maintenance, depressurization regeneration, and purging [22,26,27]. The classic PSA process flow is shown in Figure 5.

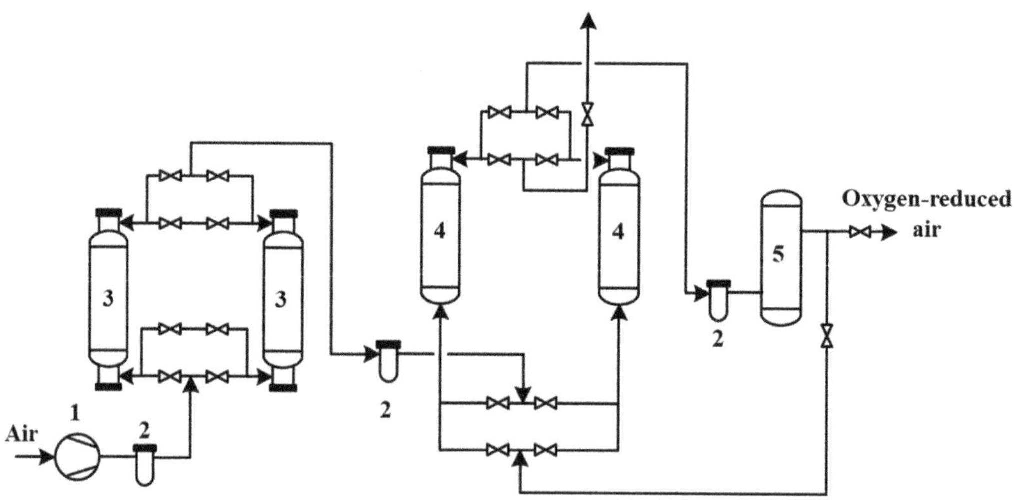

Figure 5. Pressure swing adsorption process flow. 1—air compressor, 2—filter, 3—dryer, 4—PSA tower, and 5—buffer tank.

2.2.4. Comparison of Oxygen-Reduced Air Preparation Processes

In oilfield applications requiring nitrogen, the nitrogen purity in oxygen-reduced air must be maintained at 90–95% to ensure safety. The PSA (pressure swing adsorption) process offers significant economic advantages with lower equipment investment costs, being particularly suited for medium-purity nitrogen requirements. Compared to cryogenic separation, PSA does not require operation at low temperatures, simplifying the process flow, reducing equipment investment and operating costs, and greatly enhancing economic efficiency (Table 1).

Furthermore, PSA technology's rapid start-up and shutdown capabilities make it well suited to fluctuating demands for oxygen-reduced air in oilfields, accommodating frequent start–stop operations. In contrast, cryogenic separation, with longer start-up and shutdown times, is less flexible and unsuitable for the dynamic needs of oilfields. Additionally, PSA is simple to operate, requires minimal maintenance, and can achieve long-term, low-cost, and stable operation in the complex environments of oilfields.

Table 1. Comparison of oxygen-reduced air preparation processes.

Comparison Criteria	Cryogenic Separation	Membrane Separation	PSA
Process complexity	Complex process, more equipment, and long flow paths	Simpler than PSA, with no switching valves	Simple process with less equipment
Start/stop flexibility	Low flexibility, with 12 h to start and 24 h to shut down	High flexibility and short start-up time	Flexible, with rapid start-up/shutdown
Nitrogen purity efficiency	Highest efficiency for high-purity nitrogen; suitable for >99% purity	Similar to PSA, efficiency decreases above 99%	Higher efficiency below 97%; efficiency decreases above 99%
Air compression requirement	Medium pressure requirements	Higher pressure requirements	Medium pressure requirements
Product pressure stability	Stable output pressure	Stable output pressure	Requires buffer tank for pressure stabilization
Investment cost	High equipment and land requirements; high investment cost	Membrane components are expensive; high investment	Low initial investment cost

2.3. *Field Trials in China and Internationally*

2.3.1. International Field Trials

Air injection technology has progressed internationally since its first trial in 1963, demonstrating significant production increases. By 1996, it had achieved large-scale application, with recovery rates improving from 6% to over 30% (Table 2). This evolution validated its enhanced oil recovery potential, refined the approach through optimization, and facilitated its global adoption.

Table 2. Field trials of air injection/oxygen-reduced air injection abroad.

Year	Field	Trial Results	Significance
1963–1966	Nebraska Sloss [28]	Increased oil production by over 1 million barrels.	Demonstrated the effectiveness of air injection in EOR in water-flooded reservoirs.
1971–1982	W. Heidelberg [29]	Recovery factor improved from 6% to 30%.	Validated the feasibility of air and flue gas injection in high-temperature deep reservoirs.
1977	BRRU [30]	Recovery factor improved to 21%; cumulative production increased by over 15%.	Highlighted the potential of high-pressure air injection in low-permeability, high-pressure reservoirs.
1987–1994	MPHU [31]	Recovery factor increased from 15% to 28.2%; gas-to-oil ratio reached 1182.62 m^3/t.	Demonstrated the significant enhancement in recovery for low-yield reservoirs.
1996	Horse Creek [32]	Increased production by 1 million tons; recovery factor improved by over 10%.	Showcased excellent economic and recovery performance of high-pressure air injection for further promotion.

In 1965, gravity-stable displacement trials in the United States laid the foundation for subsequent developments. By 1975, gas injection enhanced the recovery efficiency in the Hawkins Field. In 1992, gravity-stable displacement effectiveness was validated in Hungarian oilfields. In 1997, Indonesia's oilfields achieved a recovery rate of 59.2%. By 2000, the Cantarell Field successfully stabilized the oil–gas interface and boosted production, marking the maturity and widespread adoption of the technology (Table 3).

Table 3. Research progress on gravity-stable gas injection field trials abroad.

Year	Field	Injection Method	Significance
1965	America [33]	Top–down gas injection	Conducted the first vertical gravity-stable gas injection field trial, establishing a foundation for subsequent studies.
1995–1997	Handil Main Zone [34]	Top–down non-miscible dry gas injection	Doubled oil recovery compared to water flooding, reaching 59.2%. Laboratory studies showed a 24% increase in displacement efficiency.
1981–1992	Nagy Lengyel [35]	Gas injection	Over four years, 39.6 billion m^3 of gas was injected, producing an additional 1.402 million barrels of oil. The oil–gas interface remained stable, with no gas channeling observed.
2000	Cantarell [36]	Top–down nitrogen injection	Increased oil recovery by over 5%, effectively controlling the water cut and increasing oil output. This was the first nitrogen non-miscible gas injection field trial in the region.

2.3.2. Field Trials in China

In recent years, air injection technology has seen significant advancements in China. Trials conducted in oilfields such as Zhejiang and Jilin have demonstrated great potential. The Baise Oilfield achieved a cumulative production increase of 14,800 tons, and the Tuha

Oilfield improved recovery rates by 10–20%. Moreover, oilfields in Zhongyuan, Liaohe, Zhejiang, and Jilin reported increased production and reduced water cuts, underscoring the potential for broader application of this technology (Table 4).

Table 4. Field trials of air injection/oxygen-reduced air injection in China.

Year	Field	Trial Results	Significance
1996	Baise Field [37]	Cumulative production increased by 14,800 tons, with significant economic benefits.	Validated the effectiveness of air/foam-assisted water injection in controlling water and enhancing oil production.
2003	Tuha Field [38]	Oil recovery efficiency improved by 10–20% compared to water flooding under LTO.	Provided theoretical and practical support for applying air injection in complex reservoirs in Tuha Field.
2007	Zhongyuan Field [39]	Oil production increased by 12%; water cut reduced by 4%, with no gas channeling observed.	Demonstrated the effectiveness of air/foam injection in high-temperature, high-salinity heterogeneous reservoirs.
2012	Liaohe Field [40]	Annual decline rate reduced from 22% to 14.5%; cumulative oil production increased by 110,000 tons.	Successfully applied oxygen-reduced air injection technology in buried hill reservoirs, laying the groundwork for large-scale implementation.
2016	Zhejiang Field	Daily oil production increased to 2.5 tons/day; water cut decreased by 20%.	Addressed water injection challenges and enhanced recovery efficiency and output.
2017	Jilin Field [41]	Daily oil production increased by 2.2 times; water cut reduced by 3.7 percentage points.	Provided a successful case study of oxygen-reduced air injection for high water-cut, low-permeability reservoirs.

Top gas injection technology in China is still in its exploratory phase. In 1994, the Yanling Oilfield conducted the first trial; in 2007, the Weizhou Oilfield optimized injection production well patterns; in 2015, the Huabei Oilfield achieved a recovery increase of over 10%; and in 2024, the Qinghai Oilfield enhanced the stability of gas drive, further improving the gravity-stable gas injection efficiency (Table 5).

Table 5. Research status of gravity-stable gas injection field trials in China.

Year	Oilfield	Injection Method	Significance
1994	Yanling Oilfield [42]	Top–down nitrogen injection	Enhanced recovery by over 5%, with significant water control and oil increment effects. Conducted China's first top–down non-miscible nitrogen injection field trial.
2007	Weizhou Oilfield [43]	Top–down gas injection	Laboratory and simulation studies confirmed the effectiveness of top–down gas injection in improving recovery. Clarified the principles for well placement of injectors and producers.
2016	Huabei Oilfield [44]	Top–down air injection	Predicted recovery improvement of over 10%, with cumulative oil production of 1.789 million tons. Ensured safe production without gas explosion risks.
2024	Qinghai Oilfield	Top–down oxygen-reduced air injection	Research confirmed that injection production coordination and pressure-controlled zonal production significantly stabilized the oil–gas interface, enhancing the gravity-stable gas injection efficiency.

3. Key Factors Influencing OAGD

3.1. LTO

The LTO process of oxygen-reduced air gravity drainage is a critical mechanism for EOR. It is influenced by multiple factors, including the segmentation of oxidation stages, regulation of oxygen concentration, characteristics of oxidation product formation, and experimental methodologies. Simultaneously, experimental studies and numerical modeling provide theoretical support for elucidating the reaction pathways and oil recovery mechanisms of LTO. Future research should focus on the coupling effects of the reservoir temperature, pressure, and oxygen concentration on LTO efficiency. An in-depth exploration of its dynamic effects and application potential is essential to optimize technical parameters and improve the stability and feasibility of oil recovery.

The LTO process of crude oil can be divided into four stages: light hydrocarbon evaporation, low-temperature oxidation, fuel deposition, and high-temperature oxidation [45,46]. Among these, the LTO stage, due to its lowest activation energy, is the most likely phase for crude oil oxidation reactions to occur. During this stage, the oxidation process generates heat and flue gases (such as CO_2 and water vapor), which help reduce crude oil viscosity and increase the reservoir temperature. As the temperature rises, the oxidation reaction rate accelerates; however, excessively high temperatures may lead to saturation of the oil recovery effect [47,48]. Therefore, it is essential to optimize the oil displacement conditions based on reservoir temperature to balance oxidation efficiency and operational safety.

Oxygen concentration plays a decisive role in the LTO process of crude oil. Higher oxygen concentrations can intensify the oxidation reaction but may also result in pore blockage and safety risks. In oxygen-reduced air flooding, controlling the oxygen concentration enables the achievement of optimal LTO efficiency while ensuring the safety of the gas injection process (Figure 6) [18]. An appropriate oxygen concentration can also delay gas breakthrough, optimize gas flow pathways, enhance sweep efficiency, and improve oil recovery performance [49].

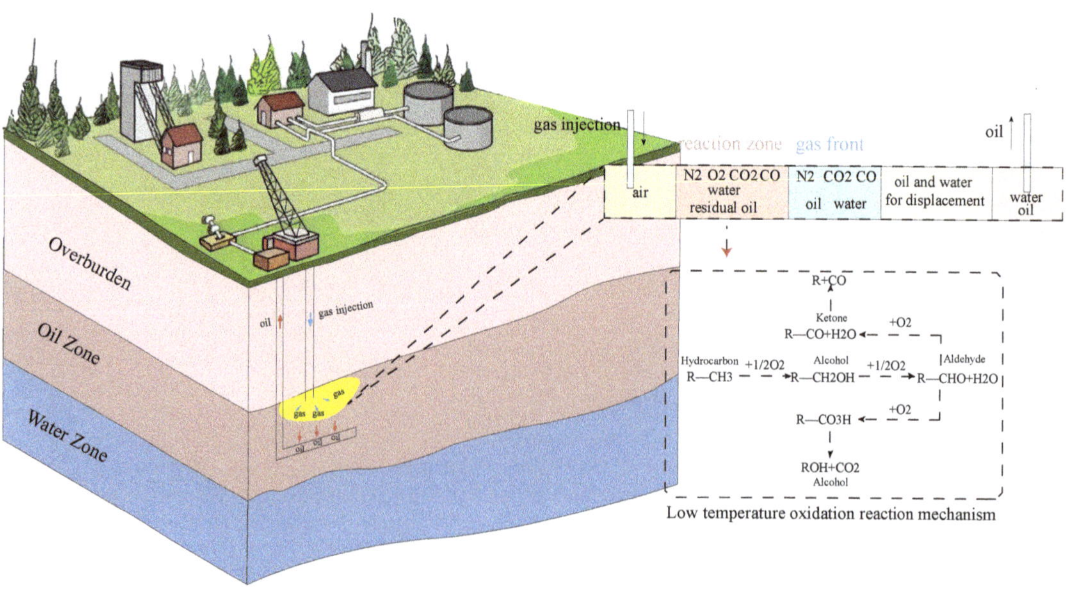

Figure 6. Mechanism of LTO [18].

The products of LTO reactions include heat, flue gases such as CO_2, CO, and CH_4, as well as partial oxidation derivatives such as alcohols, ketones, and organic acids [17,46,48]. These products not only help reduce crude oil viscosity and oil–water interfacial tension but also enhance oil recovery efficiency through gas miscibility and oil expansion. Furthermore, the generated flue gases contribute to maintaining reservoir pressure and delaying gas breakthrough [50]. However, heavy components formed during the oxidation process may deposit locally, impairing the reservoir's permeability and flow capacity, which necessitates effective control measures [51].

Common methods for studying the LTO process include thermogravimetric analysis (TGA), differential scanning calorimetry (DSC), gas chromatography (GC), and Fourier-transform ion cyclotron resonance mass spectrometry (FT-ICR-MS). These techniques are instrumental in revealing heat release, compositional changes, and reaction pathways during the oxidation process [52,53]. In porous media, static and dynamic oxidation experiments combined with kinetic models can be employed to investigate reaction behaviors under varying temperatures, pressures, and oxygen concentrations. These studies facilitate the optimization of injection parameters for oxygen-reduced air flooding.

The reaction pathways of LTO are primarily determined by the reservoir temperature, pressure, and oxygen concentration [20,36]. At temperatures below 120 °C, the reaction products are predominantly oxygenated derivatives. As the temperature increases, fuel deposition and high-temperature combustion gradually become dominant processes (Figure 7) [54,55]. Studies have shown that appropriate oxidation conditions not only improve crude oil mobility but also optimize gas flow pathways through self-correction effects, further expanding the sweep area and enhancing the oil recovery efficiency [56].

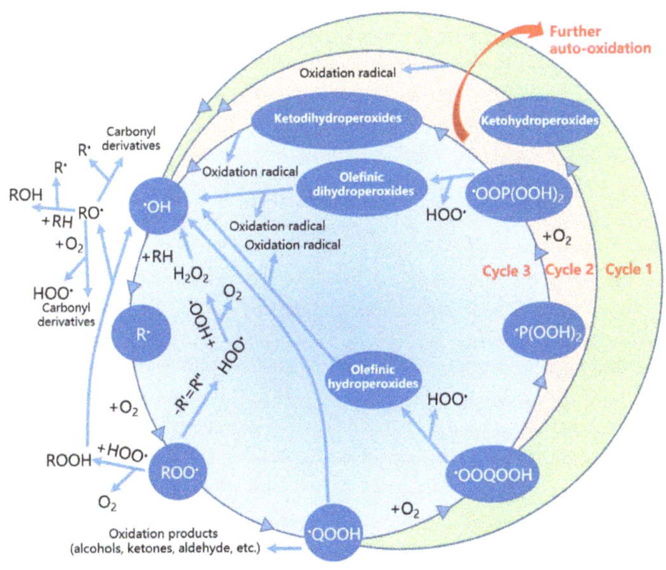

Figure 7. Reaction pathways of crude oil during LTO [55].

3.2. Injection Rate

By carefully designing the injection rate, it is possible to stabilize the oil–gas interface, optimize the displacement efficiency, and meet the requirements of various reservoir conditions. Numerical simulations and experimental studies provide theoretical and practical support for optimizing injection rates, with a focus on balancing the injection production rates, sweep efficiency, and economic benefits.

Proper control of the gas injection rate is crucial for maintaining the stability of the oil–gas interface. Lower injection rates can help establish a stable oil–gas interface, slow the downward movement of the interface, and prevent the occurrence of viscous fingering [57,58]. However, excessively high injection rates can compromise interface stability, accelerating the movement of the gas front and causing gas channeling, which reduces the oil recovery efficiency. This issue is particularly pronounced under miscible displacement conditions (Figure 8) [10,59,60].

Determining the optimal range of injection rates is equally important, as oil recovery typically exhibits a trend of initially increasing and then decreasing with the injection rate [61]. Studies have shown that appropriately reducing the injection rate can decrease residual oil saturation, increase the gas swept volume, and enhance oil recovery [62]. However, excessively low injection rates may lead to insufficient viscous force and capillary trapping, prolonging the breakthrough time and impacting economic efficiency. Conversely, excessively high injection rates can exacerbate gas breakthrough, resulting in a decline in oil recovery (Figure 9) [10].

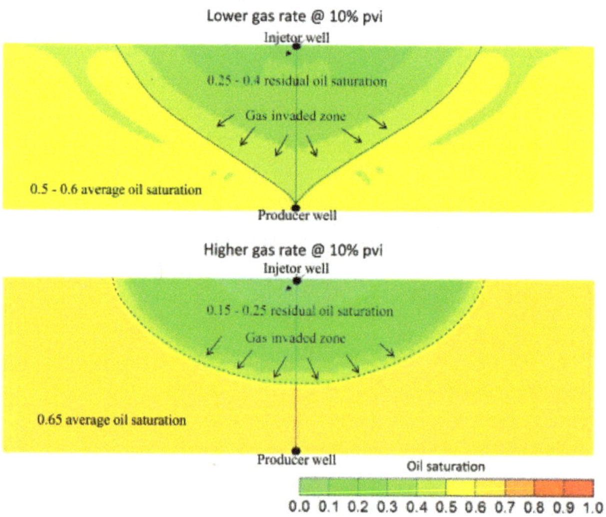

Figure 8. Comparison of the gas injection rate on oil saturation before the breakthrough time [60].

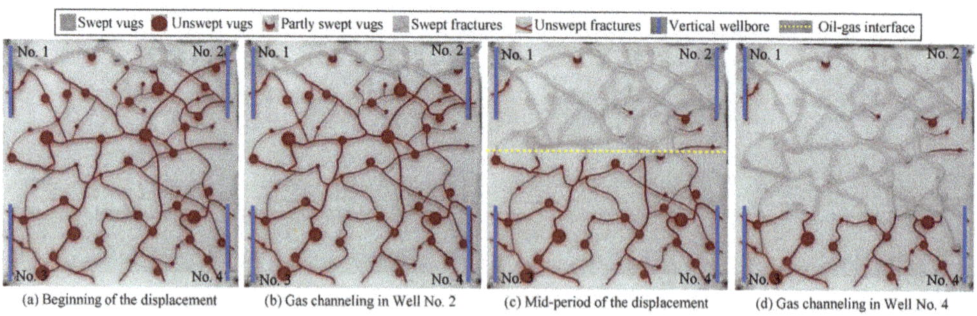

Figure 9. Stable nitrogen flooding process diagram [10].

The design of injection rates must consider reservoir characteristics and the specific stage of displacement. In low-permeability reservoirs, lower injection rates can prevent rapid gas breakthrough and uneven flow. In contrast, in high-permeability reservoirs or light oil reservoirs, moderately higher injection rates can enhance the displacement efficiency, provided they do not compromise interface stability due to excessive rates [63–65]. In addition, factors such as the reservoir temperature, pressure, and oxygen concentration significantly influence the optimal range of injection rates. These parameters must be carefully balanced to achieve effective displacement and maintain reservoir stability.

Through numerical simulation and experimental analysis, the optimization of injection rate parameters can be effectively achieved. Studies have demonstrated that lower injection rates not only enhance oil recovery but also expand the gas swept volume and maintain reservoir pressure more effectively [66]. In experiments conducted in the Qinghai Oilfield, oxygen-reduced air demonstrated significant LTO effects at a moderate injection rate, leading to improved ultimate oil recovery. Similarly, studies in the Honghe Oilfield revealed that optimized injection rates and oxygen concentrations enhanced the oil displacement efficiency. Furthermore, numerical simulations indicate that the impact of the injection rate on the oil recovery efficiency is closely tied to the balance among gravity, viscous forces, and capillary forces. Experimental calibration of the model parameters is essential to achieve optimal displacement performance (Figures 10 and 11) [67,68].

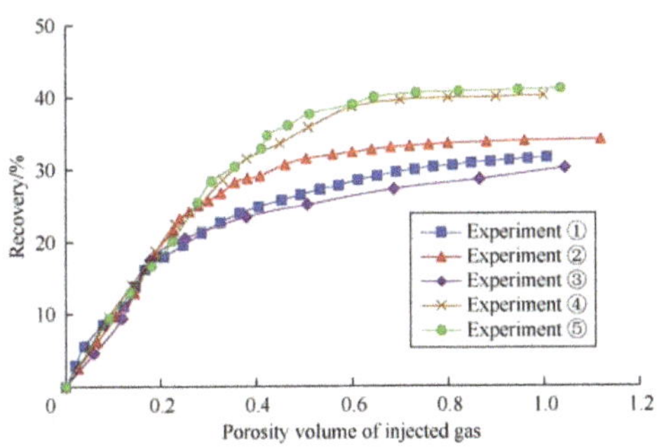

Figure 10. Relationship between oil recovery factor and injected gas volume at different injection rates [67].

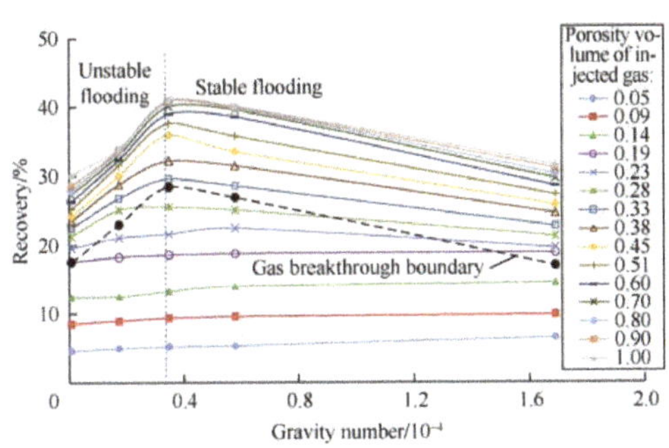

Figure 11. Relationship between gravity number and oil recovery factor in oxygen-reduced air gravity drainage [68].

3.3. Reservoir Inclination

Reservoir inclination is a critical factor in OAGD. High-inclination reservoirs enhance gravity segregation, resulting in a more stable gas cap for oil displacement, reduced gas channeling, and significantly improved oil recovery rates. In practical applications, it is essential to prioritize the evaluation of reservoir inclination and associated geological conditions. Optimizing well placement and gas injection parameters is key to leveraging the advantages of reservoir inclination effectively [61].

In high-inclination reservoirs, gas naturally migrates upward due to its lower density, forming a gas cap. This gas cap utilizes the density contrast between gas and crude oil to drive the oil–gas interface downward, effectively displacing residual oil located at the top and near faulted areas, commonly referred to as "attic oil" [69–71]. As the reservoir dip angle increases, the gravitational differentiation effect becomes more pronounced, significantly enhancing the stability of the oil–gas interface, reducing viscous fingering, and expanding the swept volume, thereby improving recovery efficiency. In reservoirs with smaller dip angles, gas tends to break through the oil zone, leading to gas channeling and reduced displacement efficiency. In contrast, larger dip angles allow gas to accumulate more effectively at the reservoir top, delaying the onset of gas channeling.

Experimental and numerical simulations indicate that the reservoir dip angle has a direct impact on displacement efficiency. As the dip angle increases, the gravitational force component becomes more pronounced, significantly EOR efficiency. The data in Table 6 show that as the dip angle increases from 0° to 80°, the ultimate recovery factor improves from 31.87% to 55.64%, representing a significant increase of 23.77% [72]. In addition, the gas displacement front in high-dip reservoirs is more stable, facilitating the accumulation of residual oil at the front to form an oil bank. This enhances the displacement efficiency, extends the displacement duration, and expands the sweep area [73].

Table 6. Displacement experiment results under different reservoir inclination angles [72].

No.	Inclination Angle (°)	Hydrocarbon Pore Volume	Displacement Efficiency Before Gas Breakthrough (%)	Final Displacement Efficiency (%)	Efficiency Improvement (%)
1	0	0.18	19.08	31.87	0
2	30	0.21	31.95	41.85	9.98
3	45	0.24	39.95	46.80	14.93
4	60	0.31	42.21	50.52	18.65
5	80	0.35	44.79	55.64	23.77

In high-dip reservoirs, a greater reservoir dip angle corresponds to a higher gravitational stability number (e.g., Non-Dimensional Gravity-Assisted Gravity Index (NGAGI)), which promotes the formation of a stable oil–gas interface [62]. Studies indicate that when the reservoir dip angle reaches 13.8°, the gravitational stability number exceeds 1, signifying that the gas injection process can achieve stable gravity-driven oil recovery. Enhanced gravitational stability models further demonstrate that an increased dip angle not only improves the efficiency of gas migration to the reservoir top but also reduces the residual oil saturation at the top, thereby enhancing displacement stability and recovery efficiency [74].

In the practical application of oxygen-reduced air injection, the advantages of high-dip reservoirs have been validated through experiments and numerical simulations [45,46]. For example, in the Weizhou Oilfield experiment, arranging injection wells at the top of the reservoir and production wells at the bottom enabled gravity-stabilized gas flooding, which ultimately enhanced the recovery factor [62]. Meanwhile, studies have also shown that reservoir dip angle and injection rate are the most critical factors influencing oil recovery efficiency. Parameters such as the production rate, vertical-to-horizontal permeability ratio, and crude oil viscosity significantly impact efficiency, whereas capillary pressure and reservoir heterogeneity have relatively minor effects.

3.4. Reservoir Types

The applicability of oxygen-reduced-air-assisted gravity drainage (OAGD) largely depends on the geological and fluid properties of the reservoir. Studies indicate that reservoirs with moderate to steep dip angles (5–36°) are most suitable for OAGD [61]. These reservoirs significantly enhance gravity segregation, stabilizing the oil–gas interface during gas injection. For instance, the Weeks Island Reservoir (26° dip angle) and Bay St. Elaine Reservoir (36° dip angle) demonstrate strong gas accumulation capabilities, delaying gas breakthrough and improving recovery efficiency. This highlights the critical role of dip angles in gas distribution and displacement stability, where overly low angles may weaken gravity segregation, and excessively high angles could lead to uneven gas flow.

Reservoir thickness is another crucial factor influencing the sweep efficiency and stability of the injection front. OAGD technology performs exceptionally well in reservoirs with thicknesses ranging from 15 to 290 m. For example, the Wizard Lake Reservoir [75] (198 m thick) showcases the advantages of thick reservoirs for displacement, while the Handil Main Reservoir [76] (15–25 m thick) demonstrates that medium-thickness reservoirs can also achieve significant recovery improvements with optimized injection production strategies. However, thinner reservoirs may face limitations in sweep efficiency, requiring adjustments in their injection rates and gas distribution paths to maximize effectiveness.

Permeability plays a decisive role in determining the efficiency of gas flow and oil displacement. OAGD technology has been successfully applied across a wide permeability range from 10 to 3400 mD. High-permeability reservoirs like the Hawkins Dexter Reservoir (3400 mD) exhibit excellent gas flow performance, while low-permeability reservoirs such as the Donghe 1 Reservoir [77] (60 mD) have achieved notable recovery improvements through optimized injection parameters. Nevertheless, gas flow resistance in low-permeability reservoirs necessitates careful regulation of the injection pressure and oxygen concentration, while high-permeability reservoirs require measures to prevent gas channeling.

The viscosity and density of crude oil directly impact the efficiency of OAGD. Research shows that low- to medium-viscosity oils (viscosity below 50 mPa·s) are better suited for gravity-driven segregation processes, while lower-density oils (API gravity above 25°) enhance gas displacement and viscosity reduction effects. These fluid characteristics make OAGD particularly advantageous in low-permeability and complex reservoirs. For high-viscosity oils, additional measures such as thermal recovery or chemical modification may be needed to enhance applicability.

Reservoir lithology is another critical factor for the successful application of OAGD. Both sandstone and carbonate reservoirs exhibit good adaptability. Sandstone reservoirs (e.g., Weeks Island and Donghe 1) typically offer higher gas permeability, whereas carbonate reservoirs (e.g., Westpem Nisku) excel under steep dip angles due to their fracture development, providing superior gravity-driven displacement. However, fractures may also lead to premature gas breakthrough, requiring optimized injection production designs for mitigation.

In summary, OAGD is well suited for reservoirs with moderate to steep dip angles, significant thickness, medium to high permeability, and low- to medium-viscosity and -density oils. By aligning reservoir characteristics with injection production strategies, such as adjusting the injection rates, optimizing the oxygen concentration, and controlling the injection pressure, the efficiency and adaptability of OAGD can be further enhanced. This provides a robust theoretical foundation and practical guidance for its application in complex reservoir conditions.

4. Conclusions and Outlook

4.1. Challenges

1. While extensively studied, LTO in OAGD faces critical challenges. Current research predominantly examines single-factor effects, such as temperature, oxygen concentration, and pressure, under idealized laboratory conditions. This approach limits the understanding of dynamic coupling mechanisms in complex reservoirs, leading to discrepancies between experimental results and field applications. Key issues include the insufficient exploration of the combined effects of reservoir heterogeneity, rock wettability, and pressure variations on LTO reactions and gas transport. Furthermore, the impact of oxidation byproducts, such as heavy component deposition causing pore blockage and permeability reduction, is poorly understood. Long-term gas injection studies remain underdeveloped, with inadequate insights into reservoir temperature, oxygen concentration, and reaction rate dynamics.

2. The injection rate significantly influences oil–gas interface stability, sweep efficiency, and recovery. However, challenges persist in understanding multi-factorial effects, including reservoir heterogeneity, temperature, pressure, and the oxygen concentration. Most studies focus on initial stages, overlooking temporal variations in injection rates and their impact on reservoir pressure distribution and fluid flow. Additionally, field validation of injection production strategies is limited, particularly in complex reservoir conditions.

3. High dip angles improve gravity segregation, stabilize gas caps, and enhance recovery efficiency. However, existing research focuses on homogeneous reservoirs, neglecting the effects of fractures, connectivity, and permeability contrasts in heterogeneous systems. The interplay between the dip angle and injection rate is poorly understood, and long-term oil–gas interface stability remains inadequately analyzed.

4.2. Recommendations and Future Perspectives

1. Future studies should develop integrated coupling models that account for reservoir heterogeneity, thermodynamic, and kinetic factors to uncover the synergistic effects of gas transport and LTO reactions. Comprehensive evaluations of oxidation byproduct distribution and deposition mechanisms are necessary to mitigate pore blockage and maintain reservoir permeability. Prolonged injection experiments should focus on optimizing the oxygen concentration, pressure, and temperature for enhanced recovery stability.

2. Research should prioritize multi-factor coupling mechanisms using experimental and numerical simulation approaches to assess dynamic injection rate effects in complex reservoirs. Long-term monitoring and analysis are essential to establish robust, field-applicable injection strategies tailored to diverse reservoir conditions.

3. Investigations should emphasize the influence of heterogeneity, including fractures and permeability contrasts, on gravity segregation and gas migration in steeply dipping reservoirs. The coupled effects of dip angle and injection rate should be systematically studied to optimize injection production strategies. Enhanced monitoring of oil–gas interface stability during extended injection processes is critical for providing precise theoretical guidance and practical solutions.

4.3. Summary and Conclusions

OAGD presents significant potential for enhancing oil recovery, particularly in complex reservoir conditions. LTO remains a pivotal process, with oxygen concentration control being critical to balancing reaction intensity and safety. Injection rate optimization can stabilize oil–gas interfaces and improve the recovery efficiency, while the reservoir dip angle enhances gravity segregation and the swept volume. Addressing challenges related to multi-factor coupling, byproduct impacts, and long-term dynamics will enable OAGD to achieve its full potential. Comprehensive research integrating experimental, simulation, and field data is vital for advancing the technology and providing actionable insights for practical applications.

Funding: This research was supported by the National Key Research and Development Program of China (2023YFF0614100), the Major Project of EOR of PetroChina (2023ZZ04-01; 2023ZZ04-03), and the Major Project of CCUS of PetroChina (2021ZZ01-03).

Conflicts of Interest: Authors Hongwei Yu and Ming Gao were employed by the China National Petroleum Corporation. Peifeng Yan was employed by the PetroChina Huabei Oilfield Company. The remaining authors declare that the research was conducted in the absence of any commercial or financial relationships that could be construed as a potential conflict of interest.

Abbreviations

OAGD: oxygen-reduced-air-assisted gravity drainage; EOR: enhanced oil recovery; LTO: low-temperature oxidation; GAGD: gas-assisted gravity drainage; NGAGI: Non-Dimensional Gravity-Assisted Gravity Index; PSA: pressure swing adsorption; TGA: thermogravimetric analysis; DSC: scanning calorimetry; GC: gas chromatography.

References

1. Dakuang, H. On concepts, strategies and techniques to the secondary development of China's high water-cut oilfields. *Pet. Explor. Dev.* **2010**, *37*, 583–591. [CrossRef]
2. Han, D. Precisely predicting abundant remaining oil and improving the secondary recovery of mature oilfields. *Acta Pet. Sin.* **2007**, *28*, 73–78.
3. Martirosyan, A.V.; Ilyushin, Y.V.; Afanaseva, O.; Kukharova, T.V.; Asadulagi, M.-A.M.; Khloponina, V. Development of an Oil Field's Conceptual Model. *Int. J. Eng.* **2025**, *38*, 381–388. [CrossRef]
4. Habib, S.H.; Yunus, R.; Zakaria, R.; Awang Biak, D.R.; Mohamed Jan, B.H.; Amir, Z. Chemical enhanced oil recovery: Synergetic mechanism of alkali, surfactant and polymer with overview of methyl ester sulfonate as a green alternative for EOR surfactant. *Fuel* **2024**, *363*, 130957. [CrossRef]
5. Zhao, M.; Liu, K.; Meng, X.; Ma, Z.; Dai, C. Review on principles, influence and applications of nanomaterials in enhancing oil recovery. *Fuel* **2024**, *371*, 131985. [CrossRef]
6. Wang, F.; Liao, G.; Su, C.; Wang, F.; Ma, J.; Yang, Y. Carbon emission reduction accounting method for a CCUS-EOR project. *Pet. Explor. Dev.* **2023**, *50*, 989–1000. [CrossRef]
7. Liu, Z.-X.; Liang, Y.; Wang, Q.; Guo, Y.-J.; Gao, M.; Wang, Z.-B.; Liu, W.-L. Status and progress of worldwide EOR field applications. *J. Pet. Sci. Eng.* **2020**, *193*, 107449. [CrossRef]
8. Long, A.; Qi, Q.; Chen, X.; Li, Y.; Lu, S.; Zhang, P.; Li, X. Full-Diameter Physical Simulation of Oxygen-Reducing Air Assisted Gravity Drainage:A Case Study of Gasikule E13 Reservoir in Qinghai Oilfield. *Xinjiang Pet. Geol.* **2021**, *42*, 565–571.
9. Ping, G.U.O.; Zhi-wang, Y.; Guang-zhi, L. Status and enlightenment of international gas injection EOR technology. *Nat. Gas Ind.* **2009**, *29*, 92–96.
10. Wang, J.; Ji, Z.; Liu, H.; Huang, Y.; Wang, Y.; Pu, Y. Experiments on nitrogen assisted gravity drainage in fractured-vuggy reservoirs. *Pet. Explor. Dev.* **2019**, *46*, 355–366. [CrossRef]
11. Hu, Y.; Hao, M.; Chen, G.; Sun, R.; Li, S. Technologies and practice of CO_2 flooding and sequestration in China. *Pet. Explor. Dev.* **2019**, *46*, 716–727. [CrossRef]
12. Al-Mudhafar, W.J. Statistical Reservoir Characterization, Simulation, and Optimization Of Field Scale-Gas Assisted Gravity Drainage (GAGD) Process with Uncertainty Assessments. Ph.D. Dissertation, Louisiana State University, Baton Rouge, LA, USA, 2016. [CrossRef]
13. Ji, Y.; Li, B.; Han, Z.; Wang, J.; Li, Z.; Li, B. Study on Flow Characteristics of Flue Gas and Steam Co-Injection for Heavy Oil Recovery. *Processes* **2023**, *11*, 1406. [CrossRef]
14. Chen, C.; Tang, X.; Qin, M.; Zhou, R.; Ding, Z.; Lian, G.; Qi, H.; Chen, X.; Liu, Z.; Li, Y. Experimental Evaluation of Shale Oil Development Effectiveness by Air Injection. *Energies* **2022**, *15*, 9513. [CrossRef]

15. Fassihi, M.R.; Moore, R.G.; Mehta, S.A.; Ursenbach, M.G. Safety Considerations for Air Injection into Light Oil Reservoirs. In Proceedings of the SPE Improved Oil Recovery Symposium, Tulsa, OK, USA, 12–16 April 2014.
16. He, W.; He, H.; Zheng, H.; Liu, P. Experimental Study on Low-Temperature-Oxidation Parameters and Simulations of Exothermic Process during Air Injection in Light Oil Reservoirs. *SPE J.* **2024**, *29*, 4232–4247. [CrossRef]
17. Ren, S.R.; Greaves, M.; Rathbone, R.R. Air Injection LTO Process: An IOR Technique for Light-Oil Reservoirs. *SPE J.* **2002**, *7*, 90–99. [CrossRef]
18. Luan, J.; Dong, P.; Zheng, J. The Experimental Studies on the EOR Laws of Air Injection Displacement in the Low Permeable Oil Reservoirs. *Preprints* **2020**, 2020020012.
19. Gutiérrez, D.; Taylor, A.R.; Kumar, V.K.; Ursenbach, M.G.; Moore, R.G.; Mehta, S.A. Recovery Factors in High-Pressure Air Injection Projects Revisited. *SPE Reserv. Eval. Eng.* **2008**, *11*, 1097–1106. [CrossRef]
20. Kang, W.-L.; Zhou, B.-B.; Issakhov, M.; Gabdullin, M. Advances in enhanced oil recovery technologies for low permeability reservoirs. *Pet. Sci.* **2022**, *19*, 1622–1640. [CrossRef]
21. Huang, L.; Wang, Y.; Yang, Z.; Wang, Q.; Pei, S.; Zhang, L.; Ren, S. Flammability and Explosion Characteristics of Methane in Oxygen-Reduced Air and Its Application in Air Injection IOR Process. *Energy Fuels* **2019**, *33*, 11850–11860. [CrossRef]
22. Chen, S. Comparison of nitrogen producing processes of pressure shift adsorption (PSA) and cryogenic separation. *Mod. Chem. Ind.* **2013**, *33*, 76–78.
23. Feng, S.; Lu, J.; Liu, W.; Jiang, J.; Liu, S. Separation performance of hollow fiber membrane for on-board inerting gas generating system. *J. Aerosp. Power* **2012**, *27*, 1332–1339.
24. Shu, Q.; Guo, Y.; Sun, Z.; Cong, G.; Xu, Y. Research and application of percolation mechanism in extra-low permeability oil reservoir. *Pet. Geol. Recovery Effic.* **2016**, *23*, 58–64.
25. Feilong, G.U.; Lijun, Z.; Dong, C. Research and application of PSA technology. *Chem. Ind. Eng. Prog.* **2007**, *26*, 1356–1358.
26. Gu, M.; Xian, X. Development of Pressure Swing Adsorption Technique. *Guangzhou Chem.* **2006**, *31*, 60–65.
27. Shi, W.; Tian, C.; Ding, Z.; Han, Z.; Zhang, D. Review on Simulation, Optimization and Control of Pressure Swing Adsorption. *J. Chem. Eng. Chin. Univ.* **2018**, *32*, 8–15.
28. Parrish, D.R.; Pollock, C.B.; Craig, F.F. Evaluation of cofcaw as a Tertiary recovery method, sloss field, nebraska. *J. Pet. Technol.* **1974**, *26*, 676–686. [CrossRef]
29. Gillham, T.H.; Cerveny, B.W.; Fornea, M.A.; Bassiouni, Z. Low Cost IOR: An Update on the W. Hackberry Air Injection Project. In Proceedings of the SPE/DOE Improved Oil Recovery Symposium, Tulsa, OK, USA, 19–22 April 1998.
30. Kumar, V.K.; Fassihi, M.R.; Yannimaras, D.V. Case History and Appraisal of the Medicine Pole Hills Unit Air-Injection Project. *SPE Reserv. Eng.* **1995**, *10*, 198–202. [CrossRef]
31. Gutiérrez, D.; Miller, R.J.; Taylor, A.R.; Thies, B.P.; Kumar, V.K. Buffalo Field High-Pressure Air Injection Projects: Technical Performance and Operational Challenges. In Proceedings of the SPE Symposium on Improved Oil Recovery, Tulsa, OK, USA, 19–23 April 2008; Volume 28.
32. JPT Staff. Horse creek air-injection project: An overview. *J. Pet. Technol.* **1998**, *50*, 78–80. [CrossRef]
33. Kulkarni, M.M. Multiphase Mechanisms and Fluid Dynamics in Gas Injection Enhanced Oil Recovery Processes. Ph.D. Thesis, Louisiana State University, Baton Rouge, LA, USA, 2005.
34. Clara, C.; Durandeau, M.; Quenault, G.; Nguyen, T.-H. Laboratory Studies for Light-Oil Air Injection Projects: Potential Application in Handil Field. *SPE Reserv. Eval. Eng.* **2000**, *3*, 239–248. [CrossRef]
35. Badics, B.; Vető, I. Source rocks and petroleum systems in the Hungarian part of the Pannonian Basin: The potential for shale gas and shale oil plays. *Mar. Pet. Geol.* **2012**, *31*, 53–69. [CrossRef]
36. Limón-Hernández, T.; De-la-Fuente, G.; Garza-Ponce, G.; Monroy-Hernández, M. Overview of Cantarell field development offshore Mexico. *World Oil* **1999**, *220*, 45–50.
37. Chaoguang, L.U.O.; Gaofu, W.; Zuoshan, Z.; Guijun, S.U.N.; Shaomin, R.A.N. Pilot Study on Foam Assisted Miscible Gas Displacement in Block Lun 16. *J. Oil Gas Technol.* **2006**, *28*, 132–135.
38. Xiao, Z.; Yang, S.; Ma, X.; Han, J.; Wang, M. Adaptability Evaluation of Nitrogen Foam Flooding After Water-Flooding in Low-Permeability Light Oil Reservoirs. *Xinjiang Pet. Geol.* **2020**, *41*, 729–734.
39. Jianfen, D.U.; Ping, G.U.O.; Zhonglin, W. Laboratory research of accelerating rate calorimeter test of high-pressure air injection at light oil reservoirs. *J. Southwest Pet. Institute. Nat. Sci. Ed.* **2007**, *29*, 17–21.
40. Yang, G.; Lin, Y.; Liu, J.; Bao, J. Suitability Study of Nitrogen-Foam Flood in Heavy Oil Reservoir-An example of Liaohe oilfield. *Xinjiang Pet. Geol.* **2004**, *25*, 188–190.
41. Zhang, H.; Yu, X.; Ma, L.; Chen, G.; Zheng, X.; Zhan, S. Potentials of the gas flooding for jilin low-permeability oil reservoirs. *Pet. Geol. Oilfield Dev. Daqing* **2014**, *33*, 130–134.

32. Bai, F.; Shen, Y.; Meng, Q.; Song, F.; Weng, Y. Field Trials of Nitrogen Injection for Enhanced Oil Recovery in Yanling Oilfield. *Acta Pet. Sin.* **1998**, *19*, 61.
33. Honggang, M.I.; Xiao, L.E.I. Research on gas injection & gravity assisted stable and displacement mechanism in Weizhou12-1 oilfield. *Oil Drill. Prod. Technol.* **2007**, *29*, 28–31.
34. Wu, Z.; Li, H.; Niu, Z.; Gong, S.; Guo, C.; Chen, H. Research on the Factors Influencing the Residual Resistance Coefficient of Air Foam Flooding. *Oilfield Chem.* **2015**, *32*, 511–514.
35. Liao, G.; Wang, H.; Wang, Z.; Tang, J.; Wang, B.; Pan, J.; Yang, H.; Liu, W.; Song, Q.; Pu, W. Oil oxidation in the whole temperature regions during oil reservoir air injection and development methods. *Pet. Explor. Dev.* **2020**, *47*, 334–340. [CrossRef]
36. Qi, H.; Li, Y.; Chen, X.; Long, A.; Wei, L.; Li, J.; Luo, J.; Sun, X.; Tang, X.; Guan, C. Low-temperature oxidation of light crude oil in oxygen-reduced air flooding. *Pet. Explor. Dev.* **2021**, *48*, 1393–1402. [CrossRef]
37. Xiong, Y.; Sun, L.; Sun, L. Reasonable Velocity of N_2 Injection NOnmiscible Flooding In Tilting Multilayer Reservoir. *J. Southwest. Pet. Inst.* **2002**, *24*, 34–36.
38. Luo, F.Q.; Wang, Q.Y.; Geng, W.S.; Hou, J.; Zou, N.A.; Tan, J. Establishment of Stability Evaluation Criteria for the Oil-Gas Interface during Gas-Cap Gravity Flooding in Oil Reservoirs with Strong Edge-Water Drive. *Acs Omega* **2023**, *8*, 14915–14925. [CrossRef] [PubMed]
39. Tong, K.; Liu, H.; Zhang, Y.; Wang, J.; Ge, L.; Dai, W.; Hong, C.; Meng, Q. Three-dimensional physical modeling of waterflooding in metamorphic fractured reservoirs. *Pet. Explor. Dev.* **2015**, *42*, 538–544. [CrossRef]
40. Jiang, Y.; Zhang, Y.; Liu, S.; Guan, W.; Chen, Y.; Liu, S. Displacement mechanisms of airection in low permeability reservoirs. *Pet. Explor. Dev.* **2010**, *37*, 471–476.
41. Jia, H.; Sheng, J.J. Numerical modeling on air injection in a light oil reservoir: Recovery mechanism and scheme optimization. *Fuel* **2016**, *172*, 70–80. [CrossRef]
42. Wang, L.; Wang, T.; Bai, Z.; Yuan, C.; Song, W.; Chen, Y.; Wang, J. Thermo-oxidative behavior and kinetics analysis of light and heavy oils based on TG, DSC, and FTIR. *Geoenergy Sci. Eng.* **2023**, *223*, 211525. [CrossRef]
43. Kok, M.V.; Gul, K.G. Thermal characteristics and kinetics of crude oils and SARA fractions. *Thermochim. Acta* **2013**, *569*, 66–70. [CrossRef]
44. Chen, X.Y.; Ba, Z.C.; Lu, Z.Y.; Gao, Y.H.; Zhou, Y.; Li, X.R. Evaluation of low-temperature oxidation analysis and the development effect of high-pressure air injection in low-permeability reservoirs. *Front. Earth Sci.* **2024**, *12*, 1416824. [CrossRef]
45. Zhao, S.; Pu, W.; Peng, X.; Zhang, J.; Ren, H. Low-temperature oxidation of heavy crude oil characterized by TG, DSC, GC-MS, and negative ion ESI FT-ICR MS. *Energy* **2021**, *214*, 119004. [CrossRef]
46. Turta, A.T.; Singhal, A.K. Reservoir engineering aspects of oil recovery from low permeability reservoirs by air injection. In Proceedings of the SPE International Oil and Gas Conference and Exhibition in China, Beijing, China, 2–6 November 1998. [CrossRef]
47. Hou, S.; Liu, Y.; Yu, H.; Niu, B.; Ren, S. Kinetics of low temperature oxidation of light oil in air injection process. *J. China Univ. Petroleum. Ed. Natrual Sci.* **2011**, *35*, 169–173.
48. Zheng, W.K.; Li, Z.Y.; Huang, H.; Cao, X.P.; Wu, G.C.; Wang, J.; Kang, H. Evaluation and improvement of CO2 assisted gravity drainage for enhanced oil production in China. *Energy Rep.* **2023**, *10*, 4465–4473. [CrossRef]
49. Hasanzadeh, M.; Madani, M. Deterministic tools to predict gas assisted gravity drainage recovery factor. *Energy Geosci.* **2024**, *5*, 100267. [CrossRef]
50. Bautista, E.V.; Barillas, J.L.M.; Dutra, T.V.; da Mata, W. Capillary, viscous and gravity forces in gas-assisted gravity drainage. *J. Pet. Sci. Eng.* **2014**, *122*, 754–760. [CrossRef]
51. Feng, K.; Mu, L.; Yan, Y.; Wu, Z.; Li, Q.; Zhang, Y. Numerical simulation of the enhanced oil recovery by gas-assisted gravity drainage (GAGD). *Pet. Geol. Oilfield Dev. Daqing* **2020**, *39*, 77–86.
52. Ren, S.; Liu, Y.; Zhang, L.; Cui, G.; Gong, Z.; Wang, Y.; Han, B. Gravity assisted gas injection:assessment model and experimental study. *J. China Univ. Petroleum. Ed. Natrual Sci.* **2018**, *42*, 59–66.
53. Zhou, W.; Zhang, J.; Tang, Y.; Chai, X.; Zhou, Y. Application of Top Gas Injection-Assisted Gravity Drainage in Bottom Water Reservoirs. *J. Southwest Pet. Univ. Sci. Technol. Ed.* **2017**, *39*, 92–100.
54. Qasem, F.H.; Nashawi, I.S. Simulation and performance prediction of partially naturally fractured reservoirs under solution gas drive primary recovery and gas injection processes. *J. Pet. Explor. Prod. Technol.* **2024**, *14*, 1259–1282. [CrossRef]
55. Mohammed, N.; Abbas, A.J.; Enyi, G.C.; Edem, D.E.; Suleiman, S.M. Enhanced gas recovery by nitrogen injection: The effects of injection velocity during natural gas displacement in consolidated rocks. *J. Nat. Gas Sci. Eng.* **2020**, *83*, 103513. [CrossRef]
56. Xiao, Z.; Qi, H.; Zhang, Y.; Li, Y.; Yao, S.; Liu, T. Characteristics and Influencing Factors of Natural Gas Gravity Drainage in Sanjianfang Formation Reservoir of Pubei Oilfield. *Xinjiang Pet. Geol.* **2023**, *44*, 334–340.

67. Chen, X.; Li, Y.; Liao, G.; Zhang, C.; Xu, S.; Qi, H.; Tang, X. Experimental investigation on stable displacement mechanism and oi recovery enhancement of oxygen-reduced air assisted gravity drainage. *Pet. Explor. Dev.* **2020**, *47*, 780–788. [CrossRef]
68. Rostami, B.; Pourafshary, P.; Fathollahi, A.; Yassin, M.R.; Hassani, K.; Khosravi, M.; Mohammadifard, M.; Dangkooban, A. A new approach to characterize the performance of heavy oil recovery due to various gas injection. *Int. J. Multiph. Flow* **2018**, *99* 273–283. [CrossRef]
69. Zhang, L.; Li, H.; Wu, H.; Wang, Y.; Liu, H. Fluid and Development Characteristics of Near Critical Oil and Gas Reservoirs with High Dip Angle. *Spec. Oil Gas Reserv.* **2017**, *24*, 100–104.
70. Jiang, Y.; Yang, J.; Chen, D. Research and application of gravity drainage in steam flooding:an example from the high-dip angle area of Qi-40 block in Liaohe oilfield. *Oil Gas Geol.* **2012**, *33*, 938–943.
71. He, H.; Liu, W.; Chen, Y.; Liu, H.; Liu, H.; Luo, G. Synergistic mechanism of well pattern adjustment and heterogeneous phase combined flooding on enhancing oil recovery in mature fault-block reservoirs. *J. Pet. Explor. Prod. Technol.* **2022**, *12*, 3387–3398 [CrossRef]
72. Al-Mudhafar, W.J.; Rao, D.N. Effect of Gas Injection Pressure on the Performance of CO_2-Assisted Gravity Drainage Process in Heterogeneous Clastic Reservoirs. In Proceedings of the SPE Asia Pacific Oil and Gas Conference and Exhibition, Brisbane Australia, 23–25 October 2018.
73. Han, H.; Chen, X.; Ji, Z.; Li, J.; Lv, W.; Zhang, Q.; Gao, M.; Kang, H. Experimental characterization of oil/gas interface self-adjustment in CO2-assisted gravity drainage for reverse rhythm reservoir. *Energies* **2022**, *15*, 5860. [CrossRef]
74. Lei, M.; Qu, Y.; Wan, C.; Yang, B.; He, D.; Ma, G.; Gong, X. Analysis of Influencing Factors on the Development Effect of CO_2 Assisted Gravity Flooding. *Spec. Oil Gas Reserv.* **2022**, *29*, 133–140.
75. Douglas, J.L.; Weiss, M. Wizard lake—reservoir quality as a key to successful miscible displacement. *J. Can. Pet. Technol.* **1991**, *30* 86–94. [CrossRef]
76. Gunawan, S.; Caié, D. Handil field: Three years of lean-gas injection into waterflooded reservoirs. *Spe Reserv. Eval. Eng.* **2001**, *4* 107–113. [CrossRef]
77. Tao, X.; Li, M.; Jia, J.; Ma, D.; Zhang, L. Geochemical Characteristics of Natural Gases and Type Study of Donghe Sandstone Reservoir,Hadexun Oilfield in Tarim Basin. *Nat. Gas Geosci.* **2014**, *25*, 70.

energies MDPI

Article

Prediction of Key Development Indicators for Offshore Oilfields Based on Artificial Intelligence

Ke Li [1,2], Kai Wang [1,2], Chenyang Tang [1,2], Yue Pan [1,2], Yufei He [1,2], Shaobin Cai [1,2], Suidong Chen [3] and Yuhui Zhou [4,5,*]

1 Development Research Institute, China National Offshore Oil Corporation Research Institute, Beijing 100028, China; like2@cnooc.com.cn (K.L.); wangkai43@cnooc.com.cn (K.W.); tangchy4@cnooc.com.cn (C.T.); panyue@cnooc.com.cn (Y.P.)
2 State Key Laboratory of Offshore Oil Exploitation, Beijing 100028, China
3 Hubei Key Laboratory of Oil and Gas Exploration and Development Theory and Technology, China University of Geosciences, Wuhan 430100, China
4 School of Petroleum Engineering, Yangtze University, Wuhan 430100, China
5 Western Research Institute, Yangtze University, Karamay 834000, China
* Correspondence: zhyhtree@yangtzeu.edu.cn

Abstract: As terrestrial oilfields continue to be explored, the difficulty of exploring new oilfields is constantly increasing. The ocean, which contains abundant oil and gas resources, has become a new field for oil and gas resource development. It is estimated that the total amount of oil resources contained in ocean areas accounts for 33% of the global total, while the corresponding natural gas resources account for 32% of the world's resources. Current prediction methods, tailored to land oilfields, struggle with offshore differences, hindering accurate forecasts. With oilfield advancements, a vast amount of rapidly generated, complex, and valuable data has piled up. This paper uses AI and GRN-VSN NN to predict offshore oilfield indicators, focusing on model-based formula fitting. It selects highly correlated input indicators for AI-driven prediction of key development metrics. Afterwards, the Shapley additive explanations (SHAP) method was introduced to explain the artificial intelligence model and achieve a reasonable explanation of the measurement's results. In terms of crude-oil extraction degree, the performance levels of the Long Short-Term Memory (LSTM) neural network, BP neural network, and ResNet-50 neural network are compared. LSTM excels in crude-oil extraction prediction due to its monotonicity, enabling continuous time-series forecasting. Artificial intelligence algorithms have good prediction effects on key development indicators of offshore oilfields, and the prediction accuracy exceeds 92%. The SHAP algorithm offers a rationale for AI model parameters, quantifying input indicators' contributions to outputs.

Keywords: offshore oilfield; development indicator prediction; artificial intelligence; SHAP

1. Introduction

Oilfield development indicators are data produced along with the production of oilfields to record oilfield production conditions. After years of research by scholars, some indicator data that are closely related to oilfield development effects have been identified, and through the changing rules of these indicator data and possible future change trends, they provide guidance for oilfield development [1–3]. These data are usually divided into static indicators and dynamic indicators based on whether they change with production time. Such static indicators mainly include geological attributes such as structural type, sedimentary type, lithology, interlayer type, porosity, permeability, and saturation. They will change slightly over time, but this change is minimal, and they can be regarded as unchanging static indicators. On the contrary, indicators such as oil production speed, liquid production speed, water consumption rate, and production gas–oil ratio continue to change throughout the entire oilfield production process. Such indicators need to be collected at fixed time intervals in order to record oilfield production changes, such as in oilfield production daily reports, oilfield production monthly reports, and oilfield production annual reports. In the actual production process, in addition to static indicators and dynamic indicators, special circumstances such as engineering conditions and manual adjustments need to be taken into consideration. In order to eliminate the influence of these factors, the study adopts part of the "Oilfield Development Level Classification" SY-T-6219:2023 [4]. Based on the indicators in this standard, the offshore oilfield development management indicators of the China National Offshore Oil Corporation are introduced, and quantitative characterization of these factors is carried out. The technical indicators include energy maintenance level, water-flooding reserve control degree, water-flooding reserve utilization degree, etc. (Table 1).

Table 1. Classification of oilfield development indicators.

Dynamic Indicators	Static Indicator	Management Indicators
Gas–oil ratio	Oilfield classification	Oil well measures are efficient
Water consumption rate	Sedimentary phase type	Water injection well-injection rate
Production measures to increase oil volume	Reserve abundance	Injection qualification rate of sub-injection well section
Moisture content	Effective thickness	Comprehensive hourly rate of oil and water wells
Adjustments in the number of wells	Porosity	Energy level
Annual oil production	Penetration	Dynamic detection plan completion rate
Oil production rate	Saturation	Waterflooding reserves control degree
	Ground crude-oil density	Water-drive reserve utilization
	Viscosity	
	Reservoir type	
	Drive type	
	Medium-depth reservoir	

Traditional methods are suitable for scenarios with relatively small amounts of data and relatively simple geological conditions. For example, in the exploration stage or the early stage of new oilfield development, traditional methods can quickly provide preliminary prediction results. However, traditional methods have limited ability to deal with complex geological conditions and large amounts of data, and the prediction accuracy of traditional methods may gradually decline as field development deepens and geological conditions change. Artificial intelligence methods are suitable for scenarios with sufficient data volume or complex geological conditions, or those requiring high-precision prediction. In the middle and late stages of oilfield development or scenarios requiring refined management, artificial intelligence methods can provide prediction results which are more accurate.

In response to the above problems, this paper takes the prediction method of key development indicators of oilfields as the research object and carries out research on the prediction of key development indicators of oilfields based on artificial intelligence. By constructing neural network models such as BP and ResNet-50, this paper predicts the degree of oilfield recovery. At the same time, this paper also interprets the model based on the artificial intelligence model interpretation tool SHAP and fits the empirical formulas of each key development indicator for the target oilfield.

2. Key Development Indicator Prediction Method

2.1. Key Development Indicator Prediction Based on Traditional Methods

The prediction of oilfield development indicators is based on the historical data of the oilfield. By studying the historical development indicators of the oilfield, analyzing and clarifying their changing rules, and combining the existing data to predict the changing trends of future development indicators, timely production adjustments are made based on the prediction results in order to obtain better production and development effects. The traditional prediction methods of oilfield development indicators can be divided into four categories based on their theoretical bases:

2.1.1. Classical Formula Prediction

Based on the classic formula prediction method, a variety of prediction curves have been derived, including decline curves, water-drive curves, injection–production relationship curves, oil and gas ratio curves, etc. Among them, water-drive characteristic curves are classified into four types: A, B, C, and D. The applicability of each water-flooding curve type is related to the viscosity of the crude oil. Specifically, the Type D water-flooding curve is more suitable when the crude-oil viscosity is lower, the Type A (or possibly another type, depending on the specific classification) water-flooding curve is more suitable when the crude-oil viscosity is medium, and the Type B water-flooding curve is more suitable when the crude-oil viscosity is higher. It is more suitable for type A and type C water-flooding characteristic curves [5]. Different curves can only predict specific oilfield development indicators. For example, the oil–gas ratio curve method is used to predict the production oil–gas ratio, while the water-drive curve is employed to predict indicators such as water cut and liquid production. However, the application of these characteristic curves needs to be under specific conditions. Only in this way can better accuracy be achieved. For instance, the water-drive characteristic curve is suitable for the medium-to-high water cut development stage of the oilfield, and the prediction accuracy will be affected in other stages [6]. Zhu Mingxia et al. (2022) [7] used the oil–water two-phase seepage theory to derive a new type of water-drive curve, and greatly improved the prediction accuracy for water content and geological reserves, reducing the average relative error levels to 2.1% and 5.2%. Zhu Lang (2022) [8] constructed a set of water-flooding characteristic curves for activated water flooding of heavy oil in offshore oilfields and used this characteristic curve to predict the increase in oil values. Deng Jingfu (2023) [9] used multiple nonlinear regression to fit and predict the production, water cut, and decline of the Bohai S oilfield. Zhang Jianda (2024) [10] combined the phase permeability curve and the water-flooding curve to predict oil production and water content. Ma Chao et al. (2022) [11] used water-drive characteristic curves to predict the recovery factor of an offshore oilfield.

2.1.2. Prediction Using the Hydrodynamic Formula Method

The hydrodynamic formula method is based on fluid-mechanics formulas. This part of the prediction method encompasses the seepage mechanics prediction model, the equivalent seepage resistance model, the piston flow method, and the non-piston flow method. This prediction method is primarily used in the early stage of oilfield development. Although it has a solid theoretical basis, when it is combined with practice, the prediction effects will greatly differ due to variations in actual conditions. Zong Huifeng (2007) [12] used the hydrodynamic formula to optimize the development-based effects of water-driven oilfields and effectively improve the recovery rate. Guo Wenmin (2016) [13] combined the hydrodynamic control improvement measures and other oilfield characteristics to construct a hydrodynamic method for injection and production regulation in the ultra-high-water-content period to improve the water-drive control intensity. Gao Min (2021) [14] optimized the development method for the high-water content period by using fault-block reservoirs based on hydrodynamics (Table 2).

Table 2. Classic formula.

Formula Name	Functional Relationship
Type A water-drive curve	$logWp = a + b \times Np$
Type B water-drive curve	$logLp = a + b \times Np$
Type C water-drive curve	$Lp/Np = a + b \times Lp$
Type D water-drive curve	$Lp/Np = a + b \times Np$
Hyperbolic decline curve	$q(t) = q'\left(1 + nDt\right)^{\frac{1}{n}}$
Injection–production relationship curve	$lg(W_I - F) = C + DN_p$

2.1.3. The Material Balance Equation Method of Prediction

The material balance equation method considers the oilfield development process as a container, in which oil, gas, and water are the substances. This method postulates that these three substances always adhere to the material balance equation throughout the entire development process [15]. In the prediction process, the material balance equation is divided into the material balance equation for closed elastic drive reservoirs and the material balance equation for unclosed elastic drive reservoirs. The principle of using the material balance equation for prediction is simple, but its effectiveness in refined predictions is not very ideal. Wang Di et al. (2021) [16] constructed the material balance equation of a buried-hill condensate gas reservoir using the material balance principle and estimated the dynamic reserves of the corresponding work area. Gu Hao et al. (2022) [17] modified the material balance equation of the ultra-deep reservoir, estimated the dynamic geological reserves of the work area, and predicted the change of dynamic geological reserves after the reservoir pressure drop increased (Table 3).

Table 3. Material balance equation formula.

Formula Name	Functional Relationship	Describe
Material balance of closed elastic flooding reservoir	$N_pB_o = \Delta V_W + \Delta V_P + \Delta_O$	Elastic cumulative oil production = expansion volume of crude oil + expansion volume of bound water + shrinkage volume of rock pores
	$N_pB_o + W_pB_W = CB_{oi}N\Delta p + W_e$	Cumulative oil production of the reservoir + cumulative water production = total elastic expansion of the reservoir + edge water intrusion

2.1.4. Reservoir Numerical Simulation Prediction

Reservoir numerical simulation employs computers to solve mathematical models of oilfields, simulating the flows of oil and water within underground reservoirs. Through model selection, sensitivity testing, data input, and history matching, oilfield development indicators such as water cut and production can be dynamically predicted [18]. This method can simulate the oil and water flow in various heterogeneous reservoirs and is suitable for development planning and adjustment.

2.2. *Development Indicator Prediction Method Combined with Artificial Intelligence*

With the continuous development of artificial intelligence, oilfield development indicator prediction methods combined with artificial intelligence have been continuously proposed in recent years. Han Rong et al. (2000) [19] proposed a method using a BP neural network to quickly predict the production of a single well in an oilfield. The predicted results from the prediction model for the liquid production, oil production and gas production of the oil well show that the BP neural network prediction model can improve the prediction accuracy of the liquid production, oil production and gas production of the oil well; Ren Baosheng (2008) [20] proposed an insensitive support vector machine and introduced it into the prediction model of oilfield development dynamic indicators, which effectively solved the overfitting problem caused by using too limited an amount of sample data in the traditional method and effectively improved the generalization ability of the model; Ma Linmao et al. (2015) [21] used a genetic algorithm to optimize the BP neural network and applied it to the production prediction of the high water-content period in the BED test area of Daqing Oilfield; Zhao Ling et al. (2018) [22] proposed a process support vector regression machine algorithm (PSVR) for process parameter optimization using the turbine algorithm, predicted the liquid production and water content, and obtained good prediction results; Zhang Yuhang (2016) [23] proposed an improved particle swarm discrete process neural network model through comparative research, predicted the oil production and liquid production of the oilfield, and obtained better prediction results; Chen Chenglong (2022) [24] used the BP neural network, improved by a genetic algorithm, to predict the water content, cumulative oil production, and recovery rates of production wells in the eastern transition zone of North Zone 1 of the Sazhong Development Zone of Daqing Oilfield, and potential wells were identified based on this prediction; Zhong Yihua et al. (2020) [25] used deep learning convolutional neural networks and recurrent neural networks to mine the reservoir-characteristic patterns and development-dynamic change laws of the oilfield development system, and, using the ELMo-based residual multi-head selection joint extraction model of deep learning entities and relationships, proposed a method, knowledge base, and model library for mining the best prediction model based on reservoir type and development stage; Li Tiening (2016) [26] used the Elman network, optimized by an improved genetic algorithm, to predict the water content of a single well, and used the double hidden layer process neural network combined with particle swarm algorithm to predict the oilfield production; Ha and Nguyen et al. (2002) [27] used an MNN neural network to predict monthly production; in Hu et al. (2019) [28], the GRU neural network, improved by principal component analysis, was used to predict oil production, and compared with the BP neural network, as improved by principal component analysis, with the results showing that the PCA-GRU neural network achieved higher accuracy. Dang Chen (2023) [29] used LSTM, the LSTM algorithm as improved by genetic algorithm, and a particle swarm optimization algorithm to develop a warning model for four oilfield development indicators. Zhu Bilei (2024) [30] constructed a CNN-Bi LSTM oil production and water content prediction model. Qu Qing (2024) [31] used the C bi GRU-Attention model to predict oil production and water production. Sun Dongming (2021) [32] predicted oilfield development indicators based on a radial basis process neural network. Fan Sen (2023) [33] predicted the injection volume of stratified water injection based on CNN-LSTM (Table 4).

Table 4. Development indicator prediction methods combined with artificial intelligence.

Researchers	Predictive Indicators	Method
Han Rong et al. (2000) [19]	Oil well liquid production, oil production, gas production	BP neural network
Ma Linmao et al. (2015) [21]	Yield prediction during high water content period	GA-BP
Zhang Yuhang (2016) [23]	Oil production, liquid production	Improved discrete process neural network model using particle swarm
Li Tiening (2016) [26]	Moisture content, oilfield production	Elman network optimized by improved genetic algorithm, Double hidden layer process neuron network combined with particle swarm algorithm

Table 4. *Cont.*

Researchers	Predictive Indicators	Method
Zhao Ling et al. (2018) [22]	Liquid production, moisture content	Turbine algorithm optimization, Process support vector regression machine algorithm (PSVR)
Chen Chenglong (2022) [24]	Water content, cumulative oil production, recovery factor	GA-BP
Ha et al. (2002) [27]	Monthly production	MNN neural network
Hu et al. (2019) [28]	Oil production	GRU neural network improved by principal component analysis

3. Feature Correlation Analysis and Key Development Indicator Prediction Algorithm

3.1. Feature Correlation Analysis

In the actual prediction processes for key development indicators, a greater number of development indicators participating in the prediction does not mean higher accuracy. The participation of low-correlation or non-correlation development indicators will reduce the prediction accuracy values for key development indicators. Therefore, before the key development indicator prediction work is carried out, it is also necessary to conduct a feature correlation analysis between the selected indicator system and the key development indicators. Currently, commonly used correlation analysis methods include conventional calculation methods such as the correlation coefficient method and grey correlation analysis method. There are also feature correlation analysis algorithms based on deep learning.

3.1.1. Correlation Coefficient Method

The correlation coefficient method usually evaluates the correlation between features by constructing a linear functional relationship between two features. However, in the actual application process, it will be found that the correlation between features is not always a simple linear relationship, but also includes many other types of correlations, such as exponential correlation and polynomial correlation. Therefore, there are many ways to calculate the correlation coefficient. Commonly used correlation coefficients include the Pearson correlation coefficient, Spearman rank correlation coefficient, and Kendall rank correlation coefficient, etc.

Pearson's correlation coefficient is calculated in such a way that, firstly, the mean values of features X, Y need to be calculated, and then the covariance $Cov(X, Y)$ between the features X, Y as well as the standard deviation σ_x and σ_y for each of X, Y are computed, and the correlation coefficients are obtained:

$$E(X) = \frac{\sum_{I=1}^{n} x_i}{n}, E(Y) = \frac{\sum_{I=1}^{n} y_i}{n} \tag{1}$$

$$Cov(X, Y) = \frac{\sum_{i=1}^{n} (X_i - E(X))(Y_i - E(Y))}{n} \tag{2}$$

$$\sigma_x = \sqrt{\frac{\sum_{I=1}^{n} \left(X_i - E(X) \right)^2}{n}}, \sigma_y = \sqrt{\frac{\sum_{I=1}^{n} \left(Y_i - E(Y) \right)^2}{n}} \tag{3}$$

$$Pearson = \frac{Cov(X, Y)}{\sigma_x \sigma_y} \tag{4}$$

The calculation of the Spearman's correlation coefficient needs to be evaluated first by ranking the data separately, i.e., ranking the size of each data element after arranging them in an order from smallest to largest, and calculating the correlation coefficient through the grade difference of the corresponding positions of the two groups of data:

$$d_i = X_i - Y_i \tag{5}$$

$$Spearman = 1 - \frac{6 \times \sum_{I=1}^{n} d_i^2}{n(n^2 - 1)} \tag{6}$$

Kendall's rank correlation coefficient evaluates the correlation by measuring the consistency of the ranking or ordering of the variables, similar to Spearman's rank correlation coefficient. However, the key difference lies in how they assess the relationship between data pairs. Spearman's rank correlation coefficient considers the relative distance error between the ranks of corresponding data pairs, meaning it focuses on the differences in the positions of the data pairs in their respective rankings. In contrast, Kendall's rank correlation coefficient assesses the concordance of the changes in the rankings of the two sets of variables, specifically by counting the number of concordant and discordant pairs. A concordant pair refers to a situation in which the ranks of the corresponding elements in two sets of variables agree in their ordering (i.e., both increase or both decrease), whereas a discordant pair refers to a situation in which their ordering disagrees. Kendall's tau coefficient is calculated based on the difference between the number of concordant pairs and the number of discordant pairs, divided by the total number of possible pairs. This approach makes Kendall's coefficient less sensitive to outliers and more robust in certain scenarios compared to Spearman's coefficient. For two random variables X, Y take the corresponding data pairs (x_i, y_i) and (x_j, y_j), where $i < j$. If $x_i < x_j$ and $y_i > y_j$ or $x_i > x_j$ and $y_i > y_j$, it is determined that there is a consistency between this group of variables; otherwise there is no consistency. As for the special case of $x_i = x_j$ and $y_i = y_j$, it is considered inconsistent and not contradictory. For Kendall's correlation coefficient the formula is as follows:

$$Kendall = \frac{n_c - n_d}{n(n-1)/2} \tag{7}$$

where n_c is the number of data pairs with consistency, n_d is the number of data pairs without consistency, and n is the total number of data pairs.

3.1.2. Grey Correlation Analysis

Grey relational analysis (GRA) is a method in grey system theory [34] which can be used to determine the correlations between various indicators. Unlike the correlation coefficient method, which analyzes data pairwise, the grey relational analysis method treats all indicators as a whole, and the analysis of correlation is based on the whole system [35–38]. Grey relational analysis requires, first, clarifying the parent sequence and the subsequence. In this study, the parent sequence is the key development indicator, and the subsequences are the selected development indicators within the system. After clarifying the parent sequence and the subsequences, each sequence needs to be de-dimensionalized using normalization and averaging methods, and then the correlation coefficients are calculated.

Firstly, it is necessary to perform the absolute difference calculation for the subsequence and the parent sequence after preprocessing, respectively, i.e., $|x_i - x_0|, i \in [1, n]$, x_0 is the value of the parent sequence. Interpolation of all data needs to be filtered to determine the global minimum and global maximum:

$$a = min_i min_j |x_0(k) - x_i(k)| \tag{8}$$

$$b = max_i max_j |x_0(k) - x_i(k)| \tag{9}$$

The grey correlation coefficient is calculated:

$$\zeta_i(k) = \frac{a + \rho \cdot \mathrm{b}}{|x_0(k) - x_i(k)| + \rho \cdot \mathrm{b}} \tag{10}$$

where ρ is the discrimination coefficient, taking a value between [0, 1], and usually taking the value of 0.5. The mean value of $\zeta(k)$ obtained after the calculation is calculated subsequently, and the calculated value is the grey correlation between the current sequence and the parent sequence. The closer the grey correlation is to 1, the stronger the correlation between the two variables.

3.1.3. Artificial Intelligence Correlation Analysis

The correlation between features can also be analyzed using an artificial neural network model. Typically, a specific neural network is employed to assess the feature contribution and weight between input indicators and output indicators, and this then serves as the basis for evaluating feature correlation. In this article, we utilize a method proposed by Oxford University and Google, one based on Gated Residual Networks (GRN) and Variable Selection Networks (VSN), to perform feature correlation analysis.

The process of using neural networks to conduct feature correlation analysis is similar to the process of predicting key development indicators. First, the dataset needs to be preprocessed. The processed data then enter the Variable Selection Network (VSN), where variables are selected, and multiple groups of training subsets are created. These subsets are then put into the Gated Residual Network (GRN) for training, and finally, the importance of each feature is judged based on the accuracy of all subsets during verification (Figure 1).

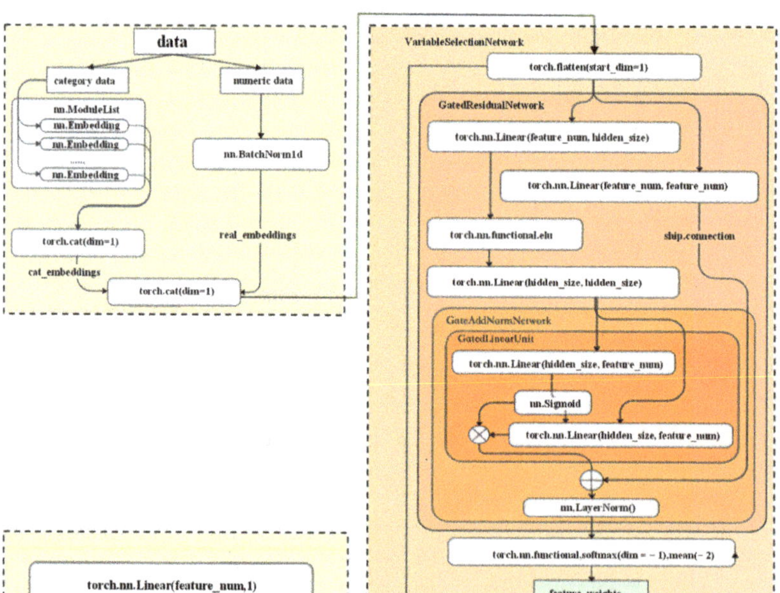

Figure 1. GRN-VSN algorithm flow.

The GRN controls the residual module by introducing the concept of a gate, which can effectively prevent the problems of gradient descent stagnation and gradient explosion, ensuring the accuracy of the constructed function. Additionally, it plays a precise role in constructing complex functions. Typically, the gate is composed of a Sigmoid function.

3.2. Key Development Indicator Prediction Algorithm

3.2.1. Residual Network (Res Net)

Res Net was proposed by Microsoft Research in 2015. The algorithm was initially used for classification and object detection. Compared with the traditional convolutional neural network (CNN), it introduces a residual structure module and batch normalization to address the problem of model degradation. The residual is the difference between the observed value and the predicted value. The residual neural network primarily utilizes CNN to extract data features. Compared with the traditional fully connected neural network, the convolutional neural network introduces the concepts of a convolutional layer and a pooling layer, which can accelerate the convergence speed of the network.

In the early days, it was widely believed that the more convolutional and pooling layers a CNN had, the better the model would perform. However, in actual research, it was discovered that an increase in the number of layers did not necessarily improve the accuracy, and could instead degrade the overall performance of the model. This phenomenon is known as model degradation. The reason for this is that as the number of layers increases, the gradient can gradually diminish during back-propagation, rendering the model unable to effectively adjust its weights. In response to this phenomenon, Res Net introduced short-circuit connections to solve the problem of model degradation, as shown in Figure 2, below.

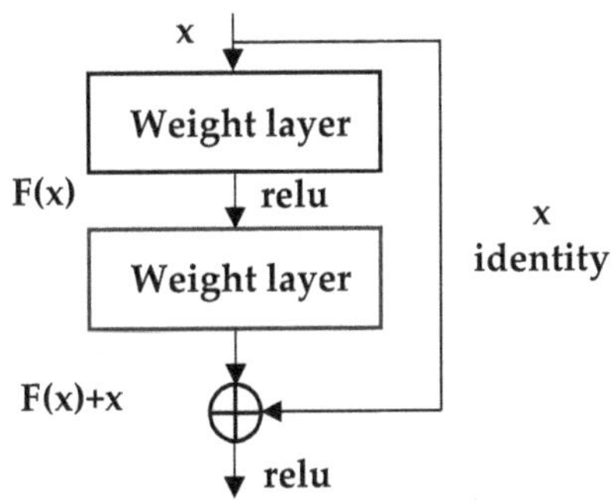

Figure 2. Res Net residual module.

In the two hidden layers depicted in the figure, Res Net adds a shortcut connection before the activation function in the second hidden layer, causing the input value of the activation function to change from the original F(x) to F(x) + x. This ensures that the network can continue to learn effectively, even when F(x) approaches zero. This setting helps to avoid the problem of network degradation. Secondly, Res Net also employs batch normalization to replace the original global normalization algorithm, normalizing the same batch of data input into the network for each training iteration. This can mitigate the issues of gradient vanishing or gradient explosion. The introduction of shortcut connections prevents Res Net from degrading in performance when stacking more layers, allowing Res Net to have ultra-deep architectures which are unmatched by other networks.

Residual networks are able to construct a very deep network structure by stacking multiple residual blocks to learn richer feature representations. These feature representations can better reflect the complex relationship between oil recovery and various influencing factors. Residual networks can adjust the network structure and parameters according to the needs of specific tasks in order to adapt to different datasets and prediction goals. This flexibility makes residual networks more advantageous in complex tasks such as oil recovery rate prediction.

3.2.2. Long Short-Term Memory (LSTM)

LSTM is based on the recurrent neural network (RNN) and adds memory units to each hidden layer neural unit to achieve controllable memory of information in a time series. It is suitable for processing and predicting important events with relatively long intervals and delays in time series [39,40].

The LSTM neural network features a repeating chain structure. The hidden neurons in LSTM differ from those in the single neural network layer of the RNN chain structure, as LSTM incorporates four distinct neural network layers. The relationship between these four neural network layers is intricate. The specific internal structure is illustrated in Figure 3. The red part represents the neural network layers, the yellow part represents operational symbols, the plus sign module indicates vector addition, and the arrow lines depict the transfer of vector information.

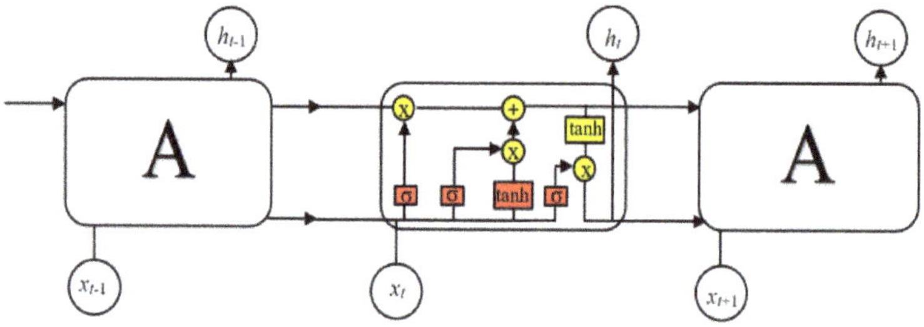

Figure 3. LSTM repeated chain structure diagram.

Oil recovery prediction involves a large amount of time-series data, such as oil well production, water injection, formation pressure, etc. These data vary over time and there may be dependencies between the data at different time points. An LSTM network, as a special recurrent neural network, is particularly good at dealing with this kind of time-series data, and is able to capture the long-term dependencies in the data, thus predicting the future crude-oil recovery rate more accurately. There are complex nonlinear relationships between oil recovery rates and a variety of geological, engineering-based, and economic factors, and LSTM networks, through their internal memory units and gating mechanisms, can learn and model these nonlinear relationships in order to more accurately reflect the actual situation.

3.2.3. Back-Propagation Neural Network (Back-Propagation, BP)

The BP neural network (back-propagation) [41], also known as the back-propagation neural network, is named for its use of the gradient descent method to modify the weights and biases of each node based on the error computed after each training iteration, until an optimal result is obtained. The BP neural network has a simple structure and consists of three main components: the input layer, the hidden layer(s), and the output layer. It was first proposed in 1986 and has since achieved good results in many fields (Figure 4).

BP neural networks are known for their ability to handle complex nonlinear relationships by means of powerful nonlinear mapping. In crude-oil recovery prediction, there are often complex nonlinear relationships among geological, engineering-based, and economic factors, and BP neural networks can effectively capture these relationships in order to predict crude-oil recovery more accurately. The BP neural network has the abilities of self-learning and self-adaptation, and it can automatically adjust the network parameters by learning the laws in the training samples, so it can quickly adapt and give accurate prediction results when facing a new oilfield or new production conditions.

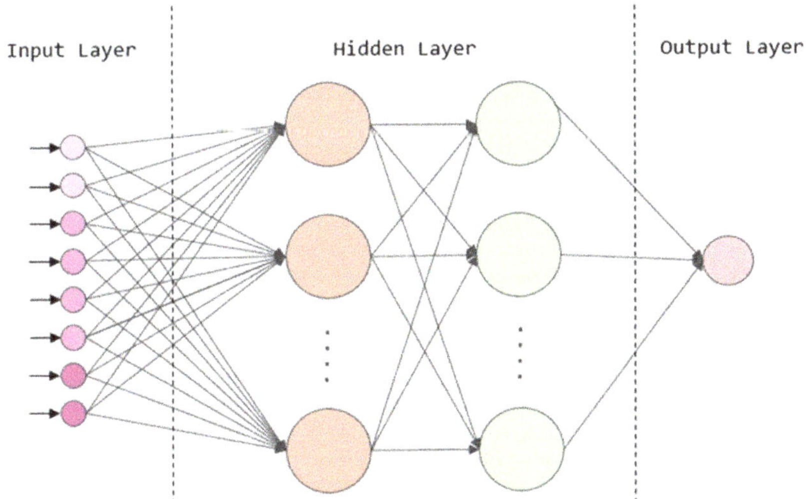

Figure 4. BP neural network architecture.

In order to prevent the BP neural network model from falling into a local optimal solution instead of the global optimal solution, and to enhance the accuracy and robustness of the particle swarm model, the genetic algorithm (GA) and optimization (PSO) algorithm were employed to optimize the BP neural network (Figure 5). As a result, the GA-BP and PSO-BP models were developed and subsequently applied to the prediction of key development indicators (Figure 6).

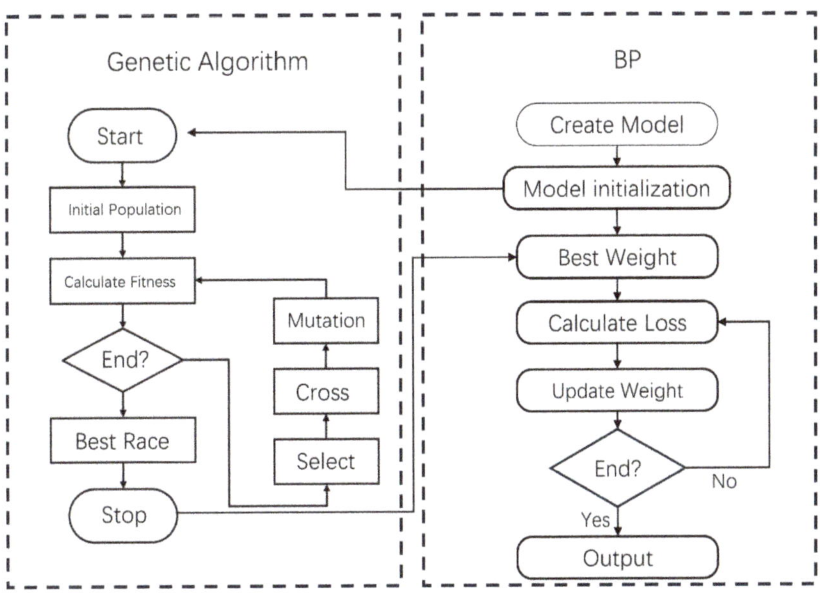

Figure 5. Genetic algorithm optimization neural network process.

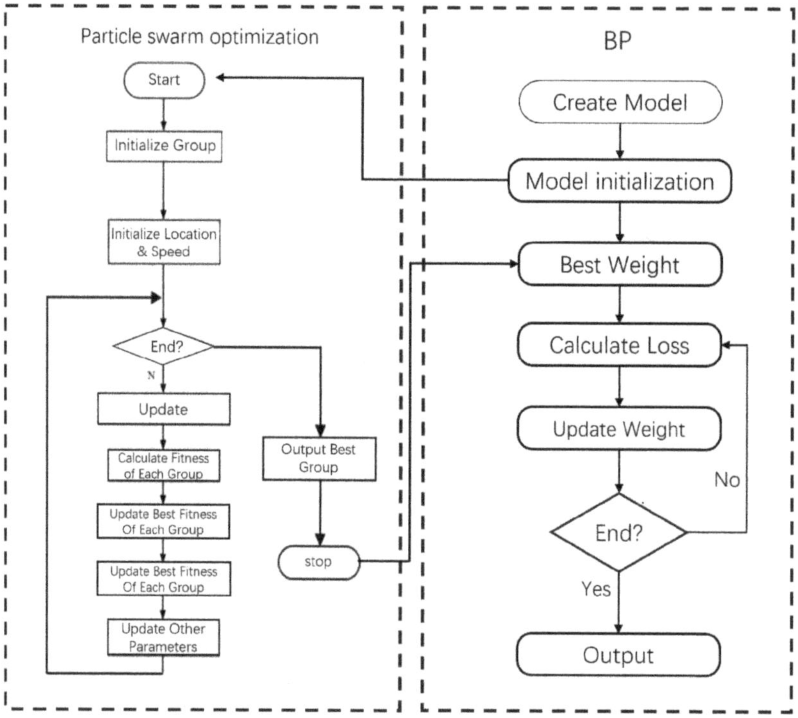

Figure 6. Particle swarm optimization algorithm process.

4. Forecast of Key Development Indicators

The degree of crude-oil recovery serves as one of the crucial indicators for assessing the effectiveness of oilfield development. Predicting this value can provide insights into both the current development level of the oilfield and its subsequent development potential.

4.1. Data Preprocessing

The selected dataset needs to be preprocessed first; the specific process includes data interpolation, data cleaning, and discretization.

4.1.1. Data Interpolation

The purpose of data interpolation is to eliminate the possibility of accidental missing values in the data. There is no need to interpolate the missing data before the first valid data point or after the last valid data point. A scientific and reliable missing-data interpolation method not only does not affect the overall accuracy of the data, but also significantly improves the learning accuracy of the deep learning model. Currently, the commonly used data interpolation methods include linear interpolation, least squares interpolation, and inverse distance interpolation.

In this paper, inverse distance interpolation is employed to interpolate missing data. This method utilizes the inverse of the distance between a known point and an unknown point as a factor that influences the weights assigned to the known points. The data value at the unknown point is then determined by computing a weighted sum of the data values at the known points, where the weights are determined as described below:

$$\omega_i = \frac{\frac{1}{d_i}}{\sum_{i=1}^{n} \frac{1}{d_i}} \tag{11}$$

$$y = \sum_{i=1}^{n} \omega_i x_i \quad (12)$$

In the above equation, d_i is the distance from the known point to the unknown point and x_i is the actual value of the known point.

4.1.2. Data Cleaning

In the deep learning process, the input model receives data consisting of one or more input indicators. The deep learning model then studies the relationship between these data and the indicator data to be predicted. The data cleaning process primarily focuses on identifying and handling invalid data and outliers. Invalid data typically refers to data with too many missing values within a single piece of data, or which cannot effectively reflect the correlation between the input indicators and the indicators to be predicted due to the limited amount of valid data. When such data is input into the model for learning, it can negatively impact the model's accuracy. Outliers, on the other hand, are values that significantly deviate from the normal pattern of change in the data, and their inclusion in the model can also hinder its learning ability.

4.1.3. Discretization

The purpose of de-scaling (or normalization) is to eliminate the magnitude differences between different data points, as the order-of-magnitude differences between the data can directly affect the weight allocation in deep learning, thereby seriously impacting the accuracy and robustness of the deep model. Data de-scaling methods typically involve normalization or homogenization.

The purpose of data normalization is to uniformly scale each data point to the range [0, 1], thereby unifying the influence size of each indicator at the initial stage to a common level. This helps the algorithm converge to the optimal solution more quickly. At the same time, it is necessary to record the maximum and minimum values of different indicators for a subsequent anti-normalization operation. This process, known as anti-normalization, aims to restore the prediction results to their original scale, enabling better comparison with real data and thereby assessing the model's prediction accuracy.

$$X' = \frac{X - X_{min}}{X_{max} - X_{min}} \quad (13)$$

4.2. Calculation of Indicator Correlation

The selected indicators and crude-oil recovery degree data are analyzed using two correlation detection algorithms: grey correlation analysis and the GRN-VSN neural network. The correlations calculated by the two methods are combined to determine the indicators used for oil recovery prediction. The optimization is performed, and the indicators with low correlation are discarded, with the following results (Figure 7).

After performing grey correlation analysis, the correlations between dynamic indicators, static indicators, and management indicators, as well as the degree of crude-oil production, were obtained. Among the dynamic indicators, A7 and A5 are the two most-correlated indicators, namely, oil production speed and the number of adjusted wells, with correlation degrees of 0.8197 and 0.8212, respectively. However, the correlations of other indicators, excluding water content, are also strong. Among the static indicators, B9, B12, B3, and B4 constitute the top 30% of the correlated indicators, and correspond to reserve abundance, formation type, viscosity, and effective thickness. Among these, viscosity has the highest correlation, at 0.791. For the management indicators, the two indicators with the highest correlation are C6 and C3, namely, the completion rate of the dynamic detection plan and the injection qualification rate of the sub-injection well section, with correlation degrees of 0.812 and 0.804, respectively (Figure 8).

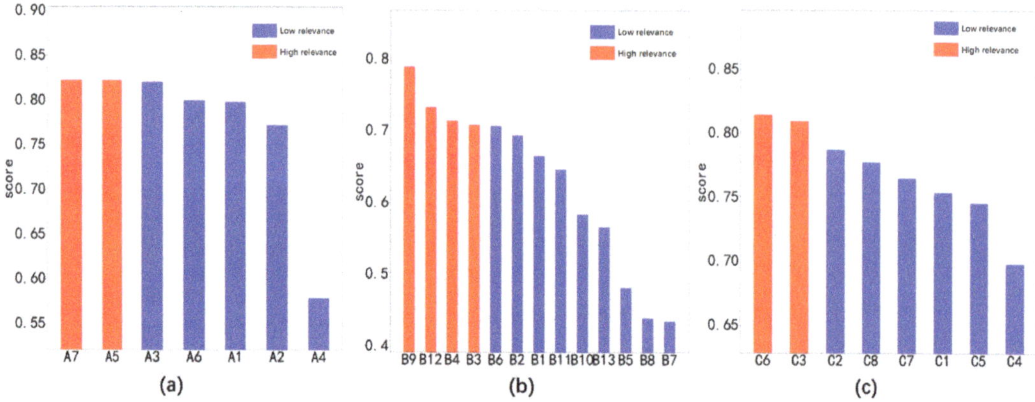

Figure 7. Grey correlation algorithm correlation calculation results of crude-oil recovery degree; (**a**): dynamic indicators, (**b**): static indicators, and (**c**): management indicators.

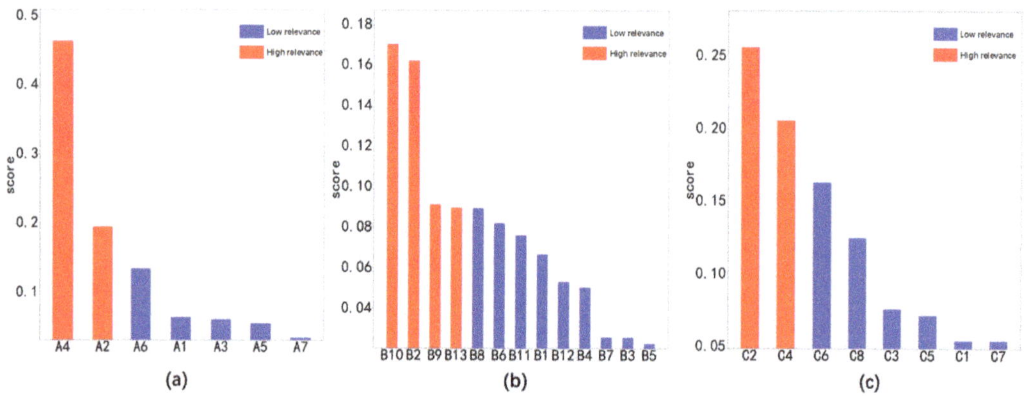

Figure 8. GRN-VSN neural network crude-oil recovery degree correlation calculation results; (**a**): dynamic indicators, (**b**): static indicators, and (**c**): management indicators.

The correlation results obtained through GRN-VSN neural network analysis differ significantly from those derived using the grey correlation algorithm. Through GRN-VSN neural network analysis, it is found that the dynamic indicators with the highest correlation are water content and water consumption rate. The static indicators with the strongest correlation are sedimentary facies type, viscosity, reservoir type, and reservoir depth. The management indicators with the greatest correlation are the water injection well-injection rate and the oil and water well comprehensive hourly rate.

Since the two algorithms have very different analytical results of indicator correlations, the two algorithms are combined, and the one or more indicators with the greatest correlation are used to predict the degree of crude-oil production. The results are as follows (Figure 9):

In the dynamic indicators, there are seven kinds of indicators: gas–oil ratio (A1), water consumption rate (A2), production measures to increase the amount of oil (A3), water content (A4), the number of wells to be adjusted (A5), annual oil production (A6), and recovery speed (A7). From among these, the water content (A4) and annual oil production (A6) were selected as input indicators.

There are thirteen static indicators, including field classification (B1), sedimentary phase type (B2), viscosity (B3), effective thickness (B4), porosity (B5), permeability (B6), saturation (B7), surface crude-oil density (B8), reserve abundance (B9), surface crude-oil density (B10), drive type (B11), stratigraphic layer (B12), medium-depth reservoir (B13), and effective thickness (B14). From among these, viscosity (B3), reserve abundance (B9), mid-depth reservoir (B13), and effective thickness (B14) are selected as input indicators.

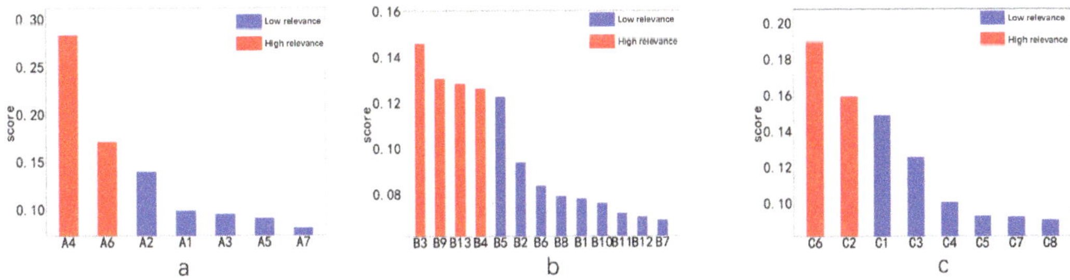

Figure 9. Correlations between post-mean indicators and crude-oil production degree; (**a**): dynamic indicators, (**b**): static indicators, (**c**): management indicators.

There are eight kinds of management indices, including the effective rate of well measures (C1), the injection rate of water injection wells (C2), the qualified rate of injection in the layer section of water injection wells (C3), the integrated time rate of oil and water wells (C4), the energy retention level (C5), the rate of completion of the dynamic testing program (C6), the degree of control of water-driven reserves (C7), and the degree of water-driven reserves utilization (C8); from among these, the injection rate of water injection wells (C2) and the completion rate of dynamic testing program (C6) are selected as the input indices. The completion rate of the dynamic testing program (C6) was also selected as an input indicator.

Based on the average results, the final indicators for the degree of crude-oil recovery are selected, as shown in Table 5:

Table 5. Selection of prediction indicators for crude-oil recovery degree.

Indicator Type	Select Indicator
Dynamic indicators	Water content (A4), annual oil production (A6)
Static indicators	Viscosity (B3), reserve abundance (B9), medium-depth reservoir (B13), effective thickness (B14)
Management indicators	Dynamic detection plan completion rate (C2), water injection well-injection rate (C6)

4.3. Index Prediction Results

The optimized indicator data are divided into a training set and a prediction set, with the training set comprising 70% of the total data volume. Different network models are trained, and the efficiency and accuracy of each network are statistically analyzed (Table 6).

The learning rate is 10×10^{-4}, the batch size is 16, and the optimizer used is Adam.

Table 6. Comparison of algorithm running time and accuracy at different crude-oil recovery levels.

Model Name	Iterations	Run Time	Error (RMSE)
LSTM	500	600 s	0.003
Res Net	2000	550 s	0.015
BP	2000	400 s	0.028
GA-BP	2000	650 s	0.024
PSO-BP	2000	630 s	0.036

The crude-oil recovery rate of the oilfield is predicted, and the results are shown in the figure below (Figure 10).

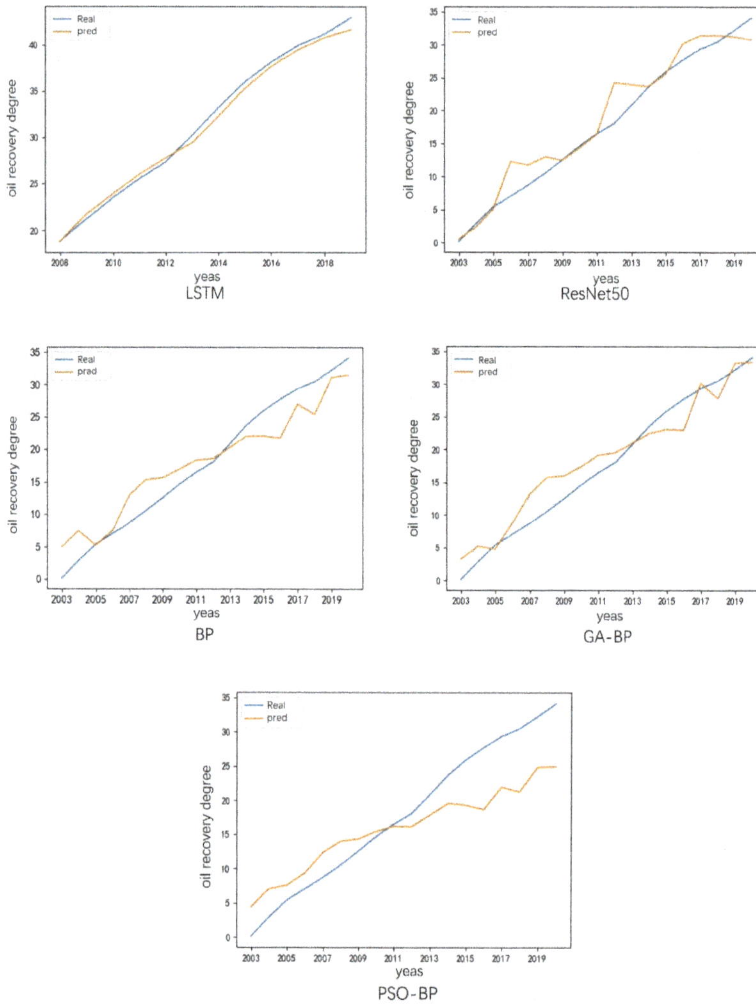

Figure 10. Prediction results of crude oil recovery degree of different models.

For the prediction of crude-oil production levels, the LSTM neural network exhibits significant advantages. This is due to the fact that crude-oil production levels are monotonically increasing. For LSTM, the learning of the index change pattern is relatively straightforward, enabling it to achieve better prediction results. The accuracy levels of ResNet-50 and the BP neural network are similar, but ResNet-50 achieves slightly higher accuracy. By applying genetic algorithm optimization and particle swarm optimization to the BP neural network, the accuracy of the BP neural network optimized using the genetic algorithm has been significantly improved. However, the accuracy of the BP neural network based on the particle swarm optimization algorithm has decreased. This may be related to the particle swarm algorithm's tendency to fall into a local optimal solution.

4.4. Model Interpretability Analysis Based on the SHAP Algorithm

First, each model needs to be explained using the SHAP explanation tool. However, during the actual research process, it was found that some models have a greater impact on the interpretability of features due to their inherent learning mechanisms. The LSTM model, while having the best prediction effect, poses challenges due to its loop structure and gate settings in the training process. The functional relationship corresponding to the model's internal workings is extremely complex, and the relationship between features cannot be well-captured when interpreting the model. ResNet-50, which has the second-highest prediction accuracy, requires convolution and pooling of features in its training process, merging the features into new features and continuing its learning. During the pooling process, since the selection of features is random and nonlinear (e.g., max pooling, min pooling, median pooling), this down sampling method can significantly affect the reconstruction of the original indicator characteristics. This, in turn, impacts the correlation between the original input indicators and the predictive indicators (Figure 11). Therefore, in the actual relationship function fitting process, the BP neural network is selected for function fitting. In the learning process, the BP neural network has clear formula methods for feature conversion and new feature generation, making it more mathematically rigorous and theoretically supported in the explanation process.

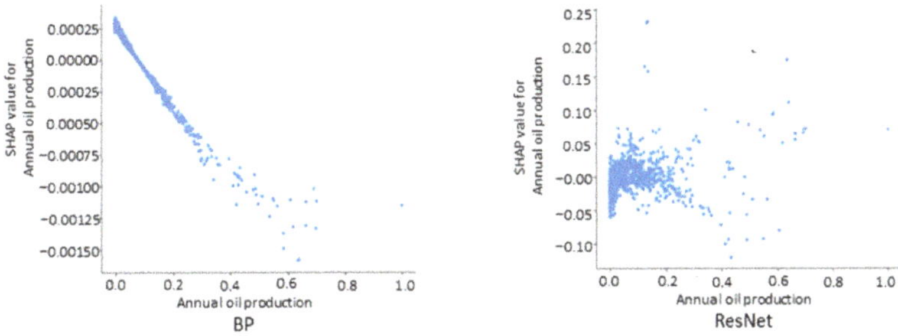

Figure 11. Comparison of the correlation between the Res Net model and BP model as to total annual oil production.

This article uses the GA-BP model, optimized by genetic algorithm, to explain the degree of crude-oil recovery. GA-BP also has high accuracy in prediction accuracy. First, the contribution proportion of each indicator in the overall prediction process is obtained, as shown in the figure below (Figure 12).

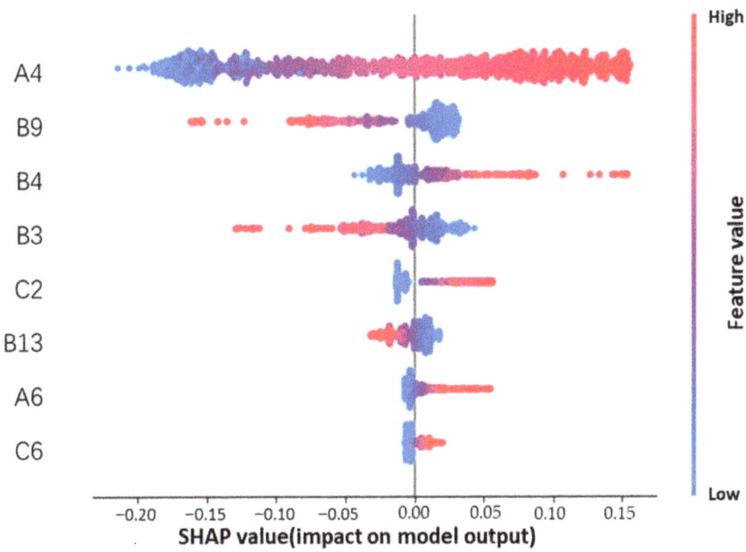

Figure 12. Contribution of various indicators in predicting crude-oil recovery level.

From the figure below (Figure 13), we can clearly see the impacts of different indicators on the prediction results when they change. The specific changes obtained by analyzing the correlation function of each indicator are as follows:

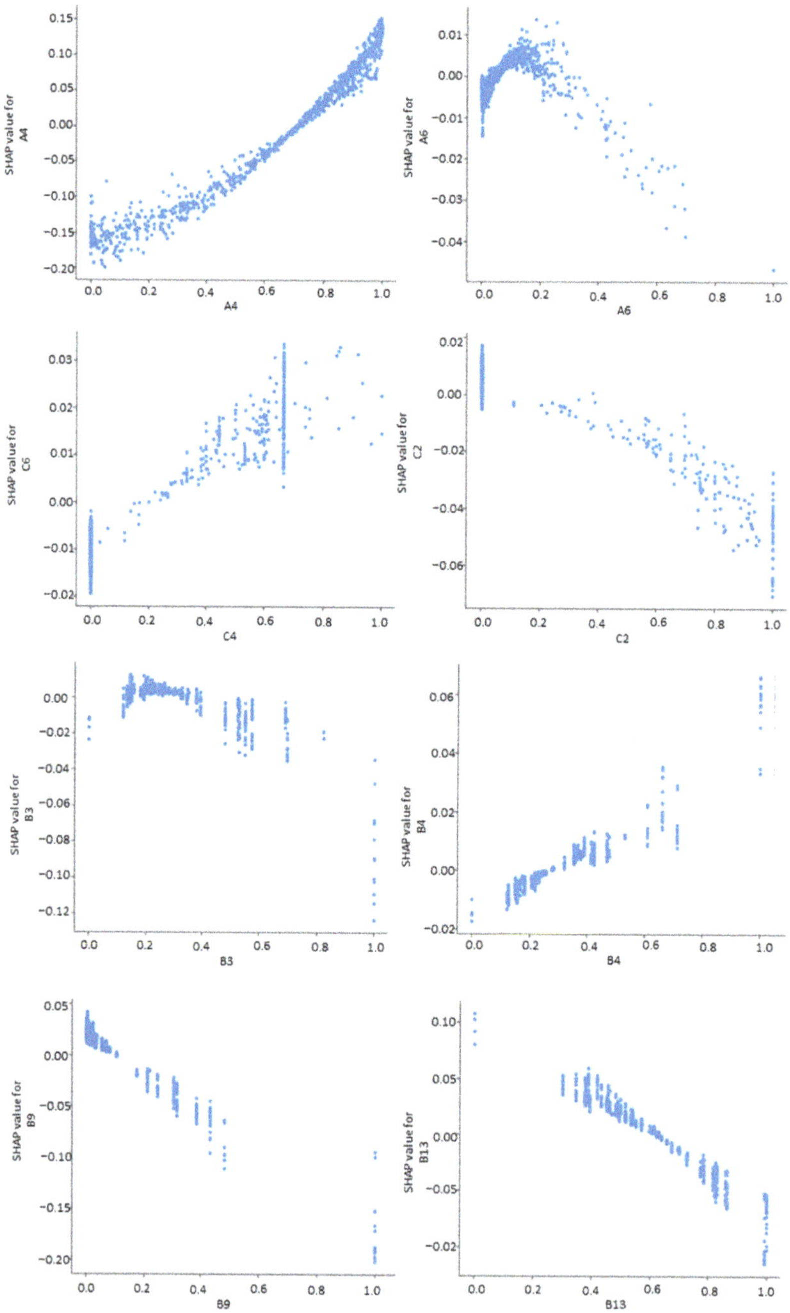

Figure 13. The functional relationship between various indicators and the degree of crude-oil recovery.

Based on the outcomes of the correlation analysis, it is evident that there exists a quadratic relationship between the total annual oil production and the degree of crude-oil recovery. Notably, this relationship is asymmetrical with respect to the Y-axis, suggesting that its functional relationship can be approximately modeled as

$$y = a(x - b)^2 + c \tag{14}$$

The relationship between water cut and crude-oil recovery degree is exponential, and it deviates from the standard exponential function. Therefore, the direct relationship between water cut and crude-oil recovery degree can be roughly fitted as

$$y = e^{ax+b} + b \tag{15}$$

The remaining indicators have a linear relationship with the degree of crude-oil recovery:

$$y = kx + b \tag{16}$$

These relationships are fitted using real oilfield data. Since all data are normalized during the model training process, a rash return to the original dimensions may weaken the functional relationship between indicators. Therefore, this article derives a specific empirical formula by fitting between the normalized indicators. When using this formula, all data must be normalized prior to calculation, and the calculation results are then denormalized back to the original dimensions. The obtained empirical formula and prediction results are as follows:

$$\begin{aligned} \textit{Oil Recovery factor} \\ &= 0.0365 \times e^{1.1745 * A4 + 1.6471} - 0.3449 \times B9 \\ &-0.2728 \times (A6 - 0.5356)^2 - 0.0952 \times B13 + 0.2422 \times B4 \\ &-0.2078 \times B3 + 0.0036 \times C6 + 0.0583 \times C2 - 0.0302 \end{aligned} \tag{17}$$

The correlation coefficient between the results fitted by the empirical formula and the real value can reach 0.884 (Figure 14), and the correlation coefficient can also reach 0.816 when single-well data are used for verification (Figure 15).

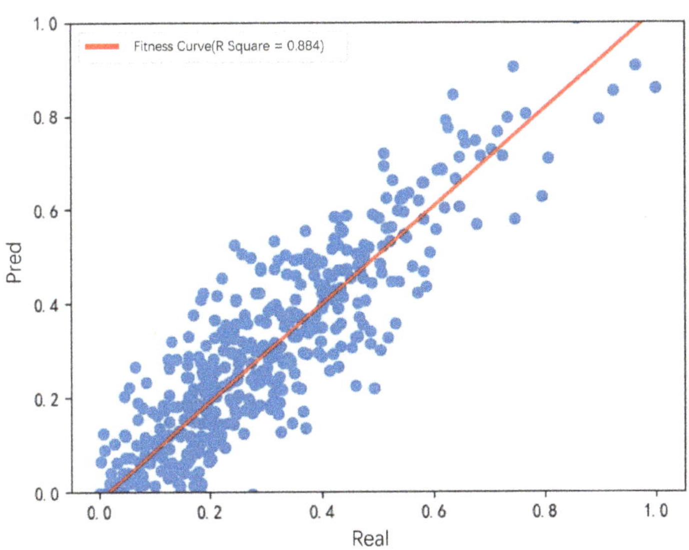

Figure 14. Scatter plot of crude-oil recovery degree calculated by empirical formula and real value.

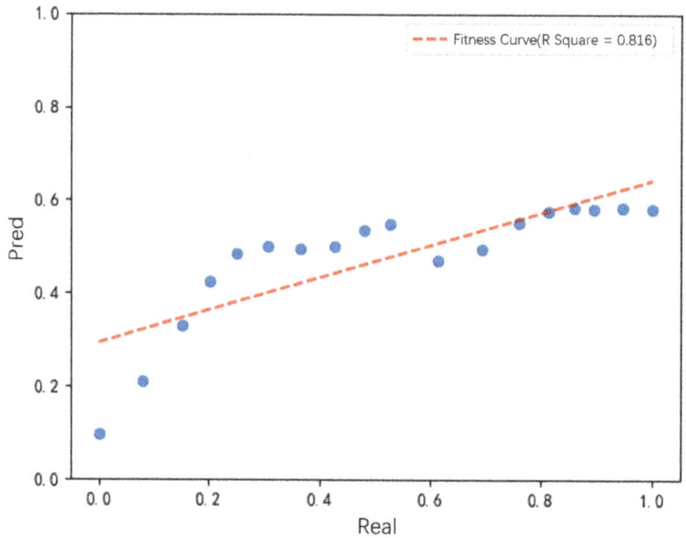

Figure 15. Validation effect of the empirical formula for crude-oil recovery degree.

5. Conclusions

The prediction of oilfield development indicators plays a crucial role in the overall development process of the oilfield. Accurate prediction of these indicators can not only enhance the overall economic benefits that the oilfield can generate, but also enable proactive adjustments to be made to oilfield development methods and means based on the prediction results. This, in turn, can help to prevent in advance the slowing down of overall oilfield production progress and the deterioration of production quality.

In the process of predicting key development indicators, this study first used the grey correlation theory and GRN-VSN algorithm to optimize the selection of input data and reduce the dimensions of input features.

In this study, different artificial intelligence algorithms are used to predict the degree of oilfield recovery, and the predicted results from the different algorithms are compared in order to select the optimal prediction model for key development indicators.

A fitting method for the corresponding empirical formula of the oilfield was constructed using SHAP. SHAP was employed to analyze the correlation between the input index and the output index. Given the high prediction-accuracy of the model, the accuracy of this corresponding relationship can also be ensured, which significantly reduces the difficulty of manual analysis. This method is, furthermore, applicable to other oilfields as well.

Author Contributions: Conceptualization, K.L. and K.W.; methodology, C.T.; software, Y.H.; validation, Y.Z. and S.C. (Suidong Chen); formal analysis, K.L.; investigation, K.W.; resources, K.W. and Y.P.; data curation, C.T. and S.C. (Shaobin Cai); writing—original draft preparation, K.L.; writing—review and editing, C.T.; visualization, K.W.; supervision, K.L.; project administration, K.L.; funding acquisition, Y.Z. All authors have read and agreed to the published version of the manuscript.

Funding: This research was funded by the Tianshan Innovation Team Plan of Xinjiang Uygur Autonomous Region (Grant number: 2023D14011), the National Natural Science Foundation of China (Grant number: 52274030), the Key Research and Development Program Project of Karamay (Grant number: 2024jjldsqld0001) and "Tianchi Talent" Introduction Plan of Xinjiang Uygur Autonomous Region (2022).

Data Availability Statement: The data presented in this study are available on request from the corresponding author.

Conflicts of Interest: Authors Ke Li, Kai Wang, Chenyang Tang, Yue Pan, Yufei He and Shaobin Cai were employed by the company China National Offshore Oil Corporation. The remaining authors declare that the research was conducted in the absence of any commercial or financial relationships that could be construed as a potential conflict of interest.

References

1. Li, B.; Bi, Y.-B.; Pan, H.; Wang, Z.-K.; Zhang, S.-Z. Combination Method for Selecting Comprehensive Oilfield Development Effect Evaluation Targets. *Pet. Sci. Technol. Forum* **2012**, *31*, 38–41, 50.
2. Xu, H. Technical limits of water flooding development index. *Petrochem. Ind. Technol.* **2017**, *24*, 127.
3. Yang, T. Study on Variation of Development Index and Reasonable Production Allocation of Putaohua Oilfield. Master's Thesis, Northeast Petroleum University, Daqing, China, 2021. [CrossRef]
4. *SY/T 6219-2023*; Oilfield development level classification. Petroleum Industry Press: Beijing, China, 2024.
5. Xu, Y. The Suitable Conditions and Application of Water Drive Characteristic Curves. *China Sci. Technol. Inf.* **2009**, *21*, 32–33.
6. Li, Z.; Sun, L.; Deng, H.; Zhang, J.; Li, Y. The research of suitable conditions to water drive characteristic curve. *Comput. Tech. Geophys. Geochem. Explor.* **2012**, *34*, 143–146.
7. Zhu, M.; Shi, L.; Xue, Y.; Wen, J.; Li, S.; Liu, M.; Xin, C. Study and application of new water drive characteristic curve. *Unconv. Oil Gas* **2022**, *9*, 65–70.
8. Zhu, L. New Chemical Flooding Tracking Evaluation and Water Flooding Characteristic Curve Study in Offshore Oil Fiel. Master's Thesis, China University of Petroleum, Beijing, China, 2022. [CrossRef]
9. Deng, J.; Wu, X.; Zhu, Z.; Zhang, L.; Gao, Y. Study on the prediction method of horizontal well development index in the S Oilfield of Bohai Sea. *Complex Hydrocarb. Reserv.* **2023**, *16*, 211–214.
10. Zhang, J. Application Research of New Reservoir Engineering Method in the Late Stage of Ultra High Water Cut in Xingbei Development Area. Master's Thesis, Northeast Petroleum University, Daqing, China, 2024. [CrossRef]
11. Ma, C.; Chen, K.; Zhang, H. Application research on the evaluation system for the development effect of a certain offshore oilfield. *West-China Explor. Eng.* **2022**, *34*, 83–86.
12. Zong, H. Research and application of methods for improving oil recovery in high water cut oil reservoirs. *Inn. Mong. Petrochem. Ind.* **2007**, 198–201. [CrossRef]
13. Guo, W. Hydrodynamics Intensity Method of Well Patterns and Rate Adjustment in Ultra—High Water Cut State. Ph.D. Thesis, China University of Geosciences (Beijing), Beijing, China, 2016. [CrossRef]
14. Gao, M. Study on Optimization of Hydrodynamic Development Mode at High Water-Cut Stage in Fault Block Reservoirs. Master's Thesis, China University of Petroleum (East China), Dongying, China, 2019. [CrossRef]
15. Wu, N.; Shi, S.; Zheng, S.; Zhao, H.; Wang, H. Formation pressure calculation of tight sandstone gas reservoir based on material balance inversion method. *Coal Geol. Explor.* **2022**, *50*, 115–121.
16. Wang, D.; Jiang, Y.; Huang, L.; Wu, Y. Research on oil ring determination and dynamic reserve calculation method of buried hill condensate gas reservoir. *Petrochem. Appl.* **2021**, *40*, 26–30+34. [CrossRef]
17. Gu, H.; Zheng, S.; Zhang, D.; Yang, Y. Modification and application of material balance equation for ultra-deep reservoirs. *Acta Pet. Sin.* **2022**, *43*, 1623–1631.
18. Ma, Q.; Yang, Z.D.; Zheng, P.Y.; Yu, C.; Guo, Q. Design and application of water injection development adjustment scheme for complex faultblock reservoir. *Pet. Plan. Des.* **2016**, *27*, 19–23.
19. Han, R.; Qi, D.; Wu, Z.; Yan, G. Application of BP neural network in predicting production changes of Shinan 31 oilfield. *Inn. Mong. Petrochem.* **2010**, *36*, 170–172.
20. Ren, B.; Zhao, M.; Liu, Z.; Wang, J. Support vector machine prediction of oilfield development dynamic indicators. *Pet. Plan. Des.* **2008**, 12–15+48. [CrossRef]
21. Ma, L.; Li, D.; Guo, H.; Li, W. Application of BP neural network optimized by genetic algorithm in crude oil production prediction: A case studyof BED test area in Daqing Oilfield. *Math. Pract. Theory* **2015**, *45*, 117–128.
22. Zhao, L.; Li, X.; Xu, S.; Xia, H. Oilfield developmentindex prediction model based on process support vector regression machine. *Math. Pract. Theory* **2018**, *48*, 83–88.
23. Zhang, Y. Oilfield Development Index Prediction and Status Assessment Method Based on Reservoir Modeling Results. Master's Thesis, Northeast Petroleum University, Daqing, China, 2016.
24. Chen, C. Research on Oilfield Development Data Analysis and Prediction Based on Artificial Neural Network. Master's Thesis, Northeast Petroleum University, Daqing, China, 2022. [CrossRef]
25. Zhong, Y.; Wang, S.; Luo, L.; Yang, J.; Yue, Y. Using deep learning to mine knowledge of oilfield development index prediction model. *J. Southwest Pet. Univ.* **2020**, *42*, 63–74.
26. Li, T. Research on Oilfield Development Data Analysis and Prediction Model Based on Dynamic Neural Network. Master's Thesis, Northeast Petroleum University, Daqing, China, 2016.
27. Nguyen, H.H.; Chan, C.W.; Wilson, M. Prediction of oil well production using multi-neural networks. In Proceedings of the IEEE CCECE2002. Canadian Conference on Electrical and Computer Engineering. Conference Proceedings (Cat. No.02CH37373), Winnipeg, MB, Canada, 12–15 May 2002; pp. 798–802.

28. Hu, H.; Feng, J.; Guan, X. A Method of Oil Well Production Prediction Based on PCA-GRU. In Proceedings of the 2019 IEEE 10th International Conference on Software Engineering and Service Science (ICSESS), Beijing, China, 18–20 October 2019; pp. 710–713
29. Dang, C. Research on Theapplication of Deep Learning in Oilfield Development Indicator Early Warning. Master's Thesis, Xi'an Shiyou University, Xi'an, China, 2024. [CrossRef]
30. Zhu, B. Research on Injection and Production Control Model and Algorithm of Water Drive Reservoir Based on Computational Intelligence. Master's Thesis, Northeast Petroleum University, Daqing, China, 2024. [CrossRef]
31. Qu, Q. Design and Development of Intelligent Analysis System for Dynamic Indicators of Reservoir Polymer Flooding Development. Master's Thesis, Northeast Petroleum University, Daqing, China, 2024. [CrossRef]
32. Sun, D. Research Onmultidisciplinary Data Analysis Methods and Applications for Oilfield Development Evaluation. Master's Thesis, Shandong University of Science and Technology, Qingdao, China, 2021. [CrossRef]
33. Fan, S. Layered Water Injection Prediction and Downhole Injection System Design Based on CNN-LSTM. Master's Thesis, Harbin University of Science and Technology, Harbin, China, 2024. [CrossRef]
34. Liu, S.; Forrest, J.Y.-L. *Grey System Theory and Its Application*, 5th ed.; Springer: Berlin/Heidelberg, Germany, 2010.
35. Chen, H.; Zhang, Y. Application of Grey Correlation Analysis Method in Bayan Chagan Reservoir Evaluation. *Pet. Geol. Eng.* **2023**, *37*, 45–51.
36. Chen, N.; Wang, H.; Guo, P.; Li, Y.; Zhang, B. Research on evaluation of factors affecting oil well production based on grey correlation analysis. *Petrochem. Appl.* **2023**, *42*, 87–91+113.
37. Wang, C.; Du, H.; Sun, X.; Dai, C.; Yang, J.; Chen, R. Comprehensive evaluation method of shale oil sweet spot based on grey correlation analysis: A case study of Bonan Sag in Bohai Bay Basin. *Pet. Drill. Technol.* **2023**, *51*, 130–138.
38. Liang, Y.; Li, N.; Liu, L.; Han, J.; He, P.; Ai, X. Evaluation of tight gas field gathering and transportation technology based on multi-level grey correlation analysis method. *Nat. Gas Explor. Dev.* **2024**, *47*, 104–111.
39. Hou, C. Oil production prediction method for new wells in oil fields based on longshort-term memory neural network. *Oil Gas Geol. Recovery Effic.* **2019**, *26*, 105–110.
40. Wang, H.; Lin, X.; Jiang, L.; Liu, Z. Oilfield production prediction based on clustering and long short-term memory neural network. *Pet. Sci. Bull.* **2024**, *9*, 62–72.
41. Huang, Q. Research on the Improvement and Application of BP Algorithm. Master's Thesis, Southwest Jiaotong University, Chengdu, China, 2010. [CrossRef]

energies

Article

Asynchronous Injection–Production Method in the High Water Cut Stage of Tight Oil Reservoirs

Jianwen Chen [1], Dingning Cai [2,*], Tao Zhang [1], Linjun Yu [1], Dalin Zhou [1] and Shiqing Cheng [2,*]

[1] No. 12 Oil Production Plant, Changqing Oilfield Company, China National Petroleum Corporation (CNPC), Qingyang 745400, China
[2] College of Petroleum Engineering, China University of Petroleum (Beijing), Beijing 102249, China
* Correspondence: dn.cai@outlook.com (D.C.); chengsq973@163.com (S.C.)

Abstract: Asynchronous injection–production cycle (AIPC) in a horizontal–vertical well pattern is an efficient strategy for enhancing water injection in tight reservoirs. However, current studies lack consideration of waterflood-induced fractures (WIFs) caused by long-term water injection. This paper takes block Z in the Ordos Basin, China, as the research object and first clarifies the formation conditions of WIFs considering the horizontal principal stress and flow line. Then, the pressure-sensitive permeability equations for the induce-fracture region between wells are derived. Finally, the WIFs characteristics in a horizontal–vertical well network with different injection modes are discussed by numerical simulation. The results show that WIFs preferentially form where flow aligns with the maximum principal stress, influencing permeability distribution. Controlling the injection rate of vertical wells on the maximum principal stress and flow line and cyclically adjusting the production rate of horizontal wells can regulate the appropriate propagation of WIFs and expand the swept areas. The parallel injection mode (PIM) and the half-production injection mode are superior to the full-production injection mode. This study can provide theoretical support for the effective development of tight oil reservoirs.

Keywords: asynchronous injection–production cycle (AIPC); waterflood-induced fracture; horizontal–vertical well network; numerical simulation

1. Introduction

Tight reservoirs have extremely low porosity and permeability, requiring techniques such as hydraulic fracturing for economic extraction [1–3]. In China, tight oil reservoirs, exemplified by the Ordos Basin, typically have permeability lower than 1×10^{-3} μm^2 [4], and can even be as low as 0.01×10^{-3} μm^2. These reservoirs are characterized by small pore throats, intricate pore structures, and insufficient natural energy and usually have weakly laminated interfaces and natural fractures [5–7]. These reservoirs do not have the ability to undergo depletion development, and their exploitation mainly relies on artificial stimulation techniques such as hydraulic fracturing. In the early stages of tight reservoir development, natural fractures are not the main transport channels for fluids, and fracture features are rarely identified in the data [8,9]. However, as development progressed, production performance data revealed characteristics indicative of fracture influence [10–12]. The main reason for the aforementioned phenomenon is that the buildup of formation pressure due to poor water absorbing capacity of low-permeability reservoirs leads to the generation and continuous extension of waterflood-induced fractures (WIFs) [13]. Unlike natural or artificial fractures, WIFs constantly change in shape and properties over time, and are significantly influenced by the development conditions such as water injection pressure, injection/production ratio, and so on [14,15].

Given the constraints posed by conventional water injection methods, there is an immediate imperative to explore effective water injection techniques that are appropriate for fractured formations [16]. These methods should consider the dynamic propagation of waterflood-induced fractures (WIFs) and mitigate issues such as one-way water sighting and flooding of highly water-cut wells [17,18]. The current state of fractured tight reservoir development is analyzed from two aspects: well pattern types and water injection strategies. Common well patterns for tight reservoirs include five-spot and seven-spot configurations, with the inverted five-spot pattern showing better performance [19]. The enhanced recovery observed in the mixed well pattern with vertical water-inject wells and horizontal oil-produce wells [20] is attributable to the greater contact area of the horizontal well [21,22]. The five-spot well pattern in the mixed well type has a higher water drive ripple coefficient [23], indicating better water drive efficiency. However, the above studies focus on the injection–production relationship and do not consider the phenomenon where water injection exacerbates reservoir heterogeneity, thus affecting development effectiveness. Zhang et al. [24] incorporated fracture parameters and stress field orientation, demonstrating the dynamic characteristics of water-induced fractures in a numerical simulation of diamond-shaped vertical well patterns. The simulated WIF propagation is constrained due to focusing only on the principal stress direction. The omission of fluid flow direction consideration restricts its applicability in optimizing various well patterns.

To address both energy supplementation and water control requirements, unstable water injection methods, such as huff-and-puff [25] and asynchronous injection [26], are applied to tight reservoirs. He [27] carried out asynchronous injection and recovery tests for multi-stage and multi-direction water-seeing horizontal wells. The method of "no production during injection and no injection during production" was adopted, and numerical simulation was combined with several rounds of field tests to optimize different injection and extraction parameters. The primary oil recovery mechanisms in AIPC technology are pressure perturbation and flow line deviation [28]. Oil saturation around the injection wells decreases more rapidly compared to cyclic water injection [29]. AIPC has proven effective in practical applications [30,31], including gas injection and other scenarios [32,33]. Besides, some studies provided improved AIPC to simulate inter-fracture injection–production in horizontal fractured wells [34,35]. However, it does not account for the potential natural fracture opening induced by pressure disturbances, nor do the fractures evolve dynamically throughout the simulation.

In this paper, we focus on a horizontal–vertical well pattern in the Z block that is in a high water-cut stage. To determine whether an appropriate asynchronous water injection method can constrain water-induced fractures and achieve higher oil recovery rates, we have developed a model and optimized the parameters of AIPC. Firstly, the production data and injection indication curves are used to identify whether induced fractures have developed in the reservoir. Then, we calculated the pressure threshold of WIF, which was updated during the model simulation. Finally, the asynchronous injection and extraction parameters of the horizontal well–straight well network are optimized by numerical simulation. AIPC is applied to the tight oil field with a horizontal–vertical well network to control the proper opening of waterflood-induced fractures, which improves the sweep efficiency and provides guidance for further tapping of remaining oil.

2. Methodology

The two mechanisms of water-induced fracture formation are the activation of NFs and the formation of new fractures. This study fully considers activation pressure, fracture pressure, and flow line relationships, providing a formula for induced fracture formation. Three assumptions are made for the method in this study:

1. Fracture propagation behavior excludes the influence of chemical, biological, and other external factors.
2. The two mechanisms can occur independently or simultaneously.
3. In each simulation round, the principal stress direction and magnitude remain constant with fixed activation pressures and fracture pressures. Parameters can be adjusted for subsequent simulations if needed.

Figure 1 shows the flow of the process.

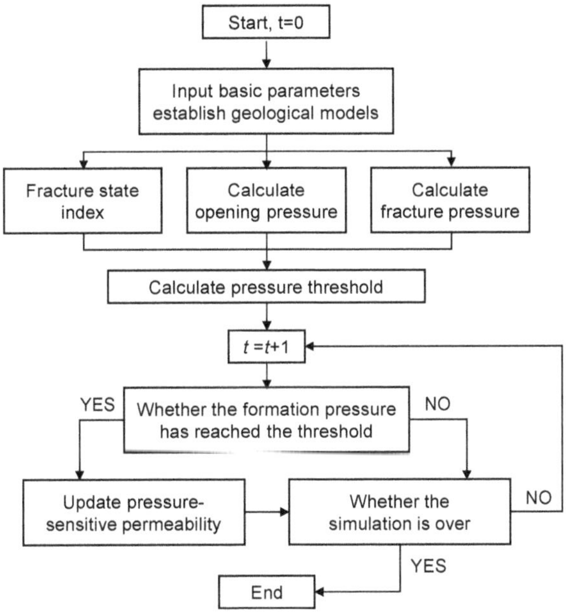

Figure 1. Flow of the process.

2.1. *Waterflood-Induced Fracture*

WIFs are dominant, highly permeable seepage channels centered on water wells that form in tight reservoirs after long-term water injection. The extension orientation of the WIF generally coincides with the direction of the main seepage fracture or the direction of the maximum horizontal principal stress. Consequently, the rapid flow of injected water along these fractures is prone to swiftly inundating wells positioned along the fracture direction. However, the sweep area on the two sides of the main streamline is small, resulting in low oil displacement efficiency [36].

WIFs are formed dynamically during water injection. In the tight oil reservoir with a relatively steep gradient of drive pressure, pre-existing natural fractures in the formation open when the water injection pressure exceeds the formation pressure [37]. This leads to a rapid increase in water content and liquid production. These newly opened natural fractures enhance the formation's permeability. When the injection pressure has not reached formation fracture pressure, the production wells maintain stable production, and the water content and fluid production remain relatively stable. With the increase in water injection time, formation pressure gradually increases until it reaches the formation fracture pressure. At this point, fractures in the formation propagate and connect, forming a high-permeability channel, and the water content rises rapidly [38]. The formation of an interconnected high-permeability zone stabilizes the water content rate at a high value, leading to a significant reduction in the effects of sweep area.

The mechanism of WIFs is usually considered to be the activation of natural fractures and the formation of new fractures. The pressure conditions for formation are water injection pressure higher than natural fracture opening pressure ($p_w > p_i$), water injection pressure higher than closure pressure of hydraulic fractures (HFs) ($p_w > p_c$), or water injection pressure higher than formation fracture pressure ($p_w > p_f$) [39]. When the maximum horizontal principal stress and the direction of the fracture are consistent, the fracture closing pressure and the natural fracture opening pressure are approximately the same.

2.1.1. Single Fracture

(1) Natural fracture activation. When the water injection pressure exceeds the opening pressure of HFs, the HFs will open and propagate and eventually form large-scale WIFs [40,41]. The open pressure of a natural fracture is as follows:

$$p_{\mathrm{i}} = \frac{\nu}{1-\nu} p_{\mathrm{o}} sin\theta + p_{\mathrm{o}} \cos\theta - p_{\mathrm{p}} + \sigma_{\mathrm{H}} sin\beta + \sigma_{\mathrm{h}} \cos\beta \quad (1)$$

where p_{i} is the open pressure of naturally fractures, ν is the Poisson ratio, p_{o} is overburden pressure, p_{p} is pore pressure, σ_{H} and σ_{h} are the maximum and minimum horizontal stress, respectively, θ is fracture dip angle, and β is the angle of maximum horizontal stress and fracture trend.

According to Equation (1), the opening sequence of cracks at different inclination angles is different. As the angle between the direction of flow and the maximum horizontal principal stress decreases, the opening pressure of natural cracks decreases. Directional enhancement of water absorption occurs around the well after the natural fracture opens, and the injected water advances along the natural fracture in the direction of the maximum principal stress.

(2) New fracture formation: When the pressure is higher than the crack pressure, a new crack is formed. The fracture pressure of the reservoir is calculated by the Hubbert-Willis empirical equation [42]:

$$p_{\mathrm{f}} = 3\sigma_{\mathrm{h}} - \sigma_{\mathrm{H}} + \sigma_{\mathrm{f}} - p_{\mathrm{o}} \quad (2)$$

where p_{f} is the fracture pressure and σ_{f} is tensile strength.

When neither natural nor artificial fractures are developed in the formation around an injection well and water is continuously injected with an injection pressure close to or above the formation rupture pressure, an induced seam along the maximum horizontal principal stress will form from around the injection well.

2.1.2. Fracture Area

The formation conditions of a single fracture can be determined by Equations (1) and (2), while the water injection-induced fracture is a high-permeability area where multiple fractures communicate with each other. In this paper, different fracture formation mechanisms are combined to represent the calculation formula of regionally induced fracture pressure threshold and equivalent permeability in a unified form.

We define the fracture state index for area (A_{f}). When there are no natural fractures and no hydraulic fractures, the A_{f} is 0. The closer A_{f} is to 1, the higher the development degree of natural fractures or hydraulic fractures. The pressure threshold function is as follows:

$$p_{\mathrm{s}} = (A_{\mathrm{f}} p_{\mathrm{i}} + (1 - A_{\mathrm{f}}) p_{\mathrm{f}}) e^{1-\cos\gamma} \quad (3)$$

where γ is the angle between the injection–production connection line (main flow line) and the maximum horizontal principal stress. In formations with the same fracture development status, WIF is more likely to form between injection and production wells.

WIFs are a highly permeable fracture zone connecting oil and water wells.

$$K_{\mathrm{f}}(p) = \begin{cases} K_{\mathrm{i}} & p < p_{\mathrm{s}} \\ c_1 K_{\mathrm{i}} e^{c2(p-p_{\mathrm{s}})} & p \geq p_{\mathrm{s}} \end{cases} \quad (4)$$

where K_{f} is the permeability of the WIF-influenced area, K_{i} is the initial permeability, p is pressure, and c_1 and c_2 are the coefficients that adjust the pressure-sensitive relationship.

The process of WIFs is influenced by both geological and engineering factors, and its direction is related to the maximum horizontal stress and the direction of injection streamlines. The smaller the angle between the injection and extraction well connecting line and the direction of the maximum horizontal principal stress, the lower the threshold of injection pressure for multi-directional WIF opening.

2.2. Asynchronous Cyclic Waterflooding for the Horizontal–Vertical Well Pattern

Horizontal well–straight well asynchronous injection and recovery development methods can change the direction of injected water seepage, play the role of seepage suction and differential pressure unblocking in low-permeability reservoirs, improve the efficiency of oil repulsion, and reduce the water content rate [43].

Cyclic water injection offers an enhanced displacement effect by alternating between periods of high-intensity injection and shut-in well percolation [44,45]. Figure 2a depicts a mixed five-point well unit with four straight wells and a horizontal well. Conventional continuous injection creates areas between injection wells where residual oil is difficult to recover, such as the yellow portion of Figure 2a. Figure 2b represents 1/4 of the unit. The background figure illustrates the water-driven ripple area during conventional continuous injection. The red solid line indicates the displacement front for each of the three cycles, and the blue dashed line marks the change in displacement front during the shut-in periods in the current cycles. Compared to conventional continuous water injection, cyclic water injection can delay the onset of water breakthrough and increase the waterflood-swept volume. This water injection method avoids the occurrence of a single dominant channel and ultimately improves the displacement efficiency of the reservoir.

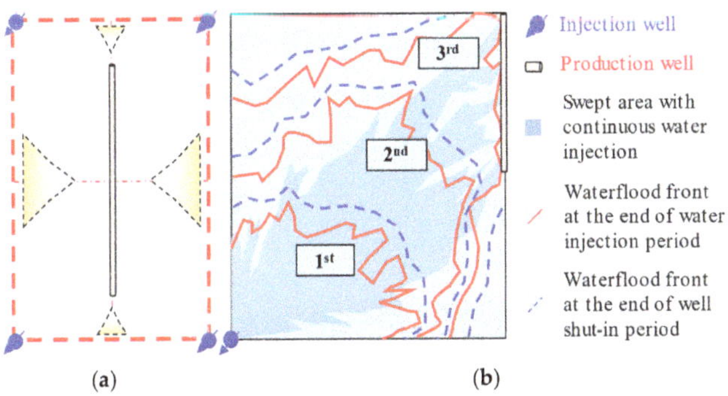

Figure 2. Schematic diagram of the displacement front in cyclic water injection. (**a**) A five-point well unit in which straight wells inject water and horizontal wells produce oil; (**b**) waterflood front of the 1/4 well network unit over 3 cycles.

The oil enhancement mechanism of asynchronous cycle water injection in a 5-point horizontal–vertical well pattern is similar to the cycle water injection described. In each cycle, by alternately turning on and off the wells in each cycle, the dead area between injection wells is reduced, which would be caused by continuous injection. Asynchronous cyclic waterflooding can be divided into the following four stages.

Stage 1—Water injection: The two straight wells near the toe end of the horizontal wells are opened and the two straight wells near the heel end of the horizontal wells are shut down. The horizontal production wells are shut down. Injected water enters the low-permeability reservoir along the microfractures. Because the external pressure of the injected water in the crack is much greater than the external extrusion force of the matrix pores, the seepage on the injection side is strengthened, the remaining oil in the matrix pores is displaced, and the flow distance of the injected water is expanded.

Stage 2—Oil production: All four injection wells are closed and production commences in half of the horizontal sections. The unidirectional valves associated with the hydraulic fractures in the horizontal sections proximal to the toe end are closed while the unidirectional valves in the hydraulic fractures of the horizontal sections near the heel end are opened. This stage enhances seepage on one side of the horizontal well, so that residual oil can flow from the matrix pore into the horizontal wellbore through the microfractures.

Stage 3—Water injection: Unlike Stage 1, the strategy for water injection wells is completely reversed. The two straight wells near the toe end of the horizontal wells are shut down and the two straight wells near the heel end of the horizontal wells are opened. The horizontal well is completely closed. This stage allows the residual oil to flow on the other side of the horizontal well. The alternate injection of water injection wells can avoid the possible flow line disturbance between injection wells and reduce the dead oil area.

Stage 4—Oil production: All four injection wells are closed and production commences in half of the horizontal sections. The open horizontal section and closed horizontal section of the unidirectional valve are opposite to stage 2. At this stage, the seepage is strengthened from the toe-end side to the low-permeability direction to drive out the residual oil in the pore space of the matrix in the low-permeability direction.

Repeating the above four-stage working system, the asynchronous injection and production method changes the direction of seepage of injected water, inhibits the flow of high-permeability hydraulic fracture, promotes the flow in the low-permeability direction, and drives out the crude oil endowed in the channels with lower pore permeability. At the stage of well closure, the injected water invades the matrix system along the small pore throats under imbibition effect and drives out the remaining oil along the large pore space, which improves the oil-driving efficiency. Periodic alteration of the seepage field in the reservoir causes an unstable pressure drop in the formation, which provides power to the crude oil attached to the surface of the rock particles and promotes the flow of crude oil through the multi-scale pore structures of the formation, and finally flows into the wellbore through the fractures to be extracted.

3. Results and Discussion

3.1. Reservoir Model

Block Z of the Heshui area is located in the southwestern part of the Ordos Basin, NW China, and the main reservoir is Chang6 member of the Triassic Yanchang Formation (Figure 3a). The study area develops natural fractures with a NE 70° direction. The maximum principal stress is NEE-SWW [46]. The main cause of WIFs in this area is that the water injection pressure is higher than the fracture opening pressure for a long time. Referring to the Chang 6 rupture pressure gradient of 0.022~0.023 MPa/m in other oil fields, the fracture pressure in this area is calculated to be at 35.2 MPa. The theoretical fracture pressure calculated from Equation (2) is 37 MPa. By injecting water for a long time at close to the fracture pressure, the induced fracture of a small area may be formed around the compression fracture.

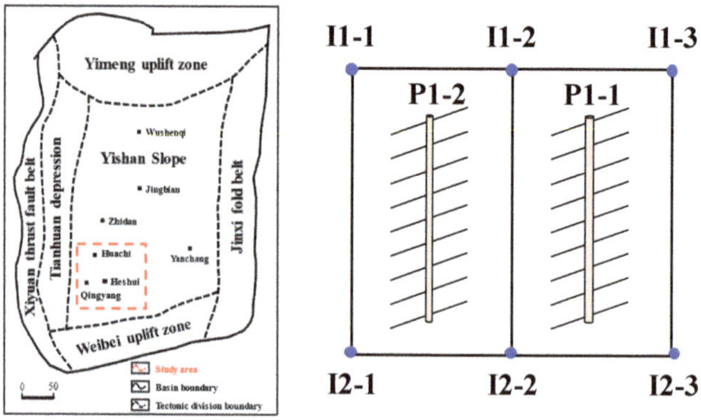

(a) (b)

Figure 3. Structural location of the study area. (**a**) Location map of Heshui oilfield; (**b**) a mixed 5-point well pattern of vertical and horizontal wells.

In this paper, we focus on the typical well pattern type in block Z, characterized by a mixed 5-point well pattern of vertical and horizontal wells. Two units within the study area, currently in the high water cut stage (Figure 3b), were selected and modeled with tnavigator. This simulation platform provides advanced tools for geoscience, reservoir, and production engineering. The basic information is as follows: the horizontal wells have a length of 700 m, the dimension of the established model is 2250 m × 1950 m × 53 m, permeability ranges from 0.01 × 10^{-3} μm^2 to 0.48 × 10^{-3} μm^2, porosity ranges from 0.18% to 12.42%, the total number of grids is 63,180, of which 39,663 are effective. The simulation parameters are detailed in Table 1.

Table 1. Simulation parameters.

Parameter	Unit	Value
Maximum principal stress, σ_h	MPa	28
Minimum principal stress, σ_H	MPa	36
Tensile strength σ_f	MPa	5
Overburden pressure, p_o	MPa	21
Poisson ratio, ν	-	0.25
Pore pressure, p_p	MPa	14.5
Angle of maximum horizontal stress and fracture, β	°	[0, 5]
Fracture state index for area, A_f	-	[0, 1]
Initial permeability, K_i	10^{-3} μm^2	0.36

3.2. Result Analysis

This study focused on well patterns in the high water cut stage. Referring to historical data, we initially simulated a 10-year period with an injection rate of 3 m^3/d and a production rate of 6 m^3/d. At the end of the simulation, the bottom hole pressure of the water injection well is 30.8~32.7 MPa, and the water saturation in wells P1-1 and P1-2 reached 91% and 96%, respectively. The bottom-hole pressure of the injection well and the water saturation of the production well exhibited deviations of less than 5% from the actual data, confirming the method's validity.

The development of waterflood-induced fractures was then assessed. Figure 4a presents the field map of threshold pressure. The figures indicate that lower pressure levels are needed to form WIFs along the flow lines of the injection and production wells. Notably, the pressure required in the region between the two fractured horizontal wells is slightly higher than in the heel and toe sections of the horizontal wells. This is attributed to the fluid's preferential flow along the hydraulic fracture, leading to limited fluid washout in this area. Consequently, the threshold pressure is higher here than in other fractured areas but lower than in less fractured areas. Using Equation (4), we obtained the distribution of high-permeability channels at this stage (Figure 4b). These high-permeability strips exhibit directionality consistent with the maximum principal stress and the orientation of hydraulic fractures.

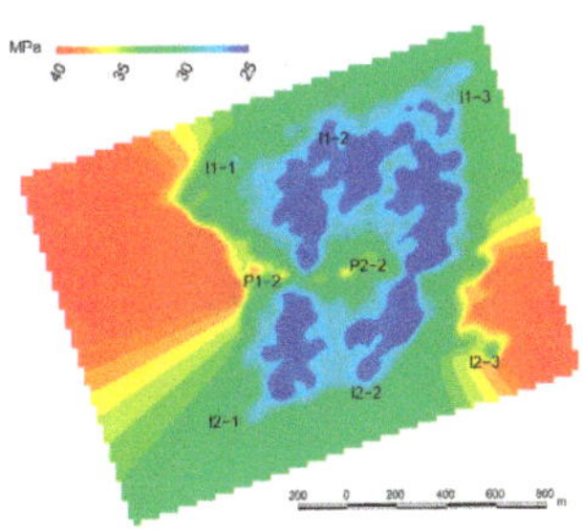

(a)

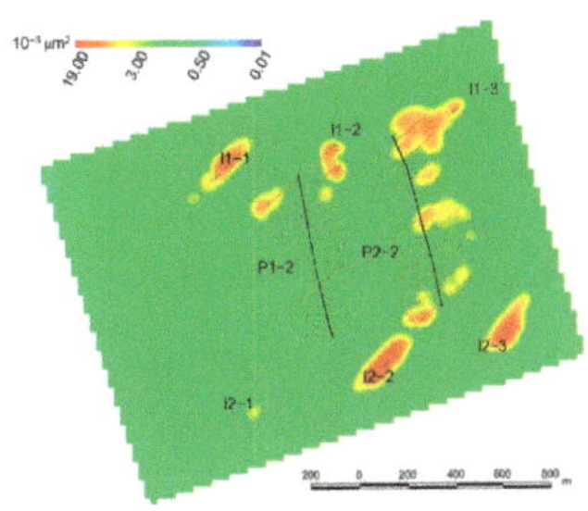

(b)

Figure 4. Pressure threshold and distribution of WIFs after a decade of water injection. (**a**) The pressure threshold field plot calculated according to Equation (3); (**b**) distribution of high-permeability channels.

3.2.1. Asynchronous Mode of the Full-Production-Injection Well

Different asynchronous modes affect the flow field in various ways. The asynchronous modes of the five-point horizontal–vertical well pattern can be classified into parallel injection mode (PIM) and diagonal injection mode (DIM). In the PIM, water invades from one side of the horizontal well; in the DIM, water invades from both corner sides of the horizontal well. Each cycle consists of four stages: injection–production–injection–production, with different wells working during the two injection stages. Based on these asynchronous modes, we categorize the various classes of injection wells into Group 1 and Group 2, as shown in Table 2.

Table 2. Injection wells groups in different modes.

Mode	Group 1	Group 2
PIM	I1-1, I1-2, I1-3	I2-1, I2-2, I2-3
DIM	I1-1, I2-2, I1-3	I2-1, I1-2, I2-3

The injection duration (D_i) was fixed, while only the production duration (D_p) was varied to explore the effect of different ratios of injection–production duration (r_{ipd}) on the performance of enhanced oil recovery. The cumulative oil production was simulated for six r_{ipd} values (0.25, 0.5, 0.75, 1, 1.25, 1.5). Meanwhile, the injection/production ratio was maintained at 1:1, and the total injection volume remained constant. In these scenarios, we simulated 20 cycles. The cumulative oil production for the different cases is shown in Figure 5. With the increase in the injection time interval, the cumulative oil production first increased and then decreased.

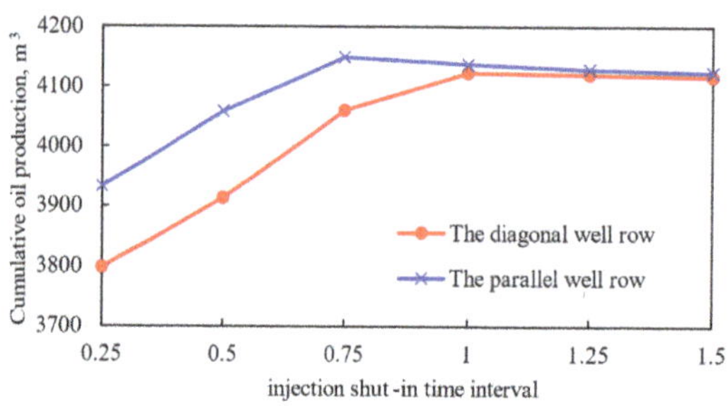

Figure 5. Cumulative oil production of the full-production-injection well.

AIPC objectively alters the flow direction of injected water, effectively preventing it from channeling along fractures towards the production well. For both methods, a too high r_{ipd} is detrimental to recovery. Due to the reservoir's high density, pressure and fluid propagation are slow. Long injection cycles tend to maintain high pressure at the bottom of the injection well leading to the formation of induced fractures. In addition, short production cycles with high recovery rates hinder fluid supply and exacerbate reservoir heterogeneity. After multiple cycles, the WIFs and hydraulic fractures facilitate flow communication, making it difficult to utilize the residual oil in the matrix.

The cumulative oil production increases and then decreases as the injection interval increases. PIM has the best program effect at r_{ipd} = 0.75, and DIM has the best program effect at r_{ipd} = 1. Overall, PIM is more effective than DIM. Due to the maximum horizontal principal stress direction being NE 72°, diagonal injection leads to a significant pressure increase in the direction of the maximum principal stress, resulting in the formation of seepage channels. Although this enhances the water absorption capacity of the injection wells, the swept area remains small, leading to low cumulative oil production.

The pressure disturbances caused by asynchronous injection–production promote the penetration of injected water into deeper pore throats within the matrix. Figure 6 illustrates the bottomhole pressures in the production wells, showing that P1-2 is significantly more effective at recharging than P1-1. Analyzing the well network and fracturing patterns, P1-2 is situated lower compared to P1-1, and the fractures and injection wells (I2-1 and I1-2) are closer together. This proximity gives P1-2 an advantage in pressure retention. However, it also increases the likelihood of water breakthrough. The DIM exacerbates this phenomenon. Figure 7 shows a field map of the high permeability region at five cycle intervals with PIM. The high permeability region around well P1-2 is much more connected to the hydraulic fracture.

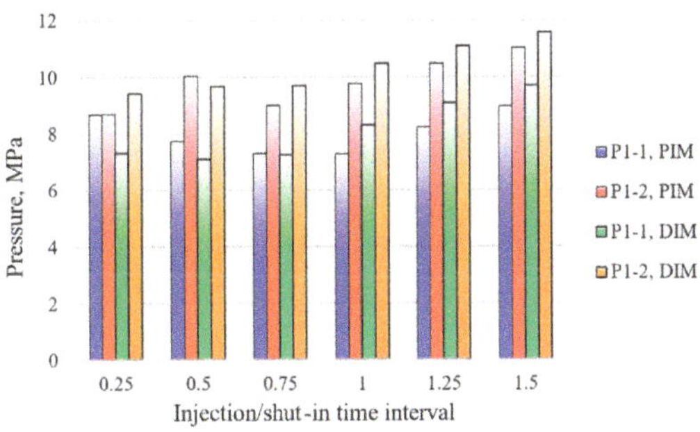

Figure 6. Bottom pressure of the full-production wells.

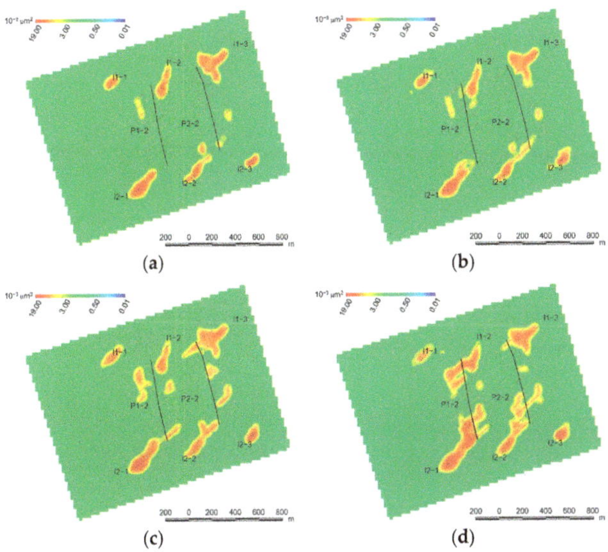

Figure 7. High permeability region at five cycle intervals (PIM, r_{ipd} = 0.75). (**a**) 5th cycle; (**b**) 10th cycle; (**c**) 15th cycle; (**d**) 25th cycle.

In this section, it is shown that the best production performance can be achieved by using the PIM, considering the principal stress direction and natural fracture distribution in the region. The optimal asynchronous mode and r_{ipd} are parallel injecting mode and 0.75, respectively.

3.2.2. Asynchronous Mode of the Half-Production-Injection Well

Horizontal wells with part of the well section open allow for a greater flow line overlap. This section simulates the asynchronous mode of the half-production-injection well. Each cycle consists of four stages of injection–production–injection–production, where the grouping of wells working in the two injection stages is the same as in the previous series of simulations. The two production stages open only half of the horizontal wells near the heel or near the toe, respectively.

Figure 8 shows the cumulative oil production at the end of 20 cycles for different cases. Comparing the simulation results of the asynchronous mode of the production-injection well, the cumulative oil production of the half-production-injection well is higher under each r_{ipd}. Since the horizontal well section opened in the production stage is the well section in the opposite direction of the previous injection stage, the swept area of the injected fluid is larger compared to the asynchronous mode of production-injection wells. The oil between the injection wells is driven closer to the production wells, ultimately increasing cumulative oil production.

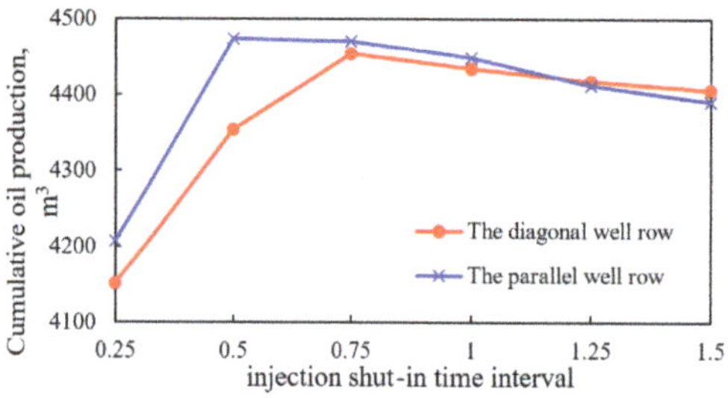

Figure 8. Cumulative oil production of the half-production-injection well.

The present scheme also demonstrates superior performance in terms of pressure maintenance. As illustrated in Figure 9, there is an increase in bottomhole pressure of approximately 2 MPa per well, indicating more effective pressure maintenance. The pressure at P1-1 is essentially the same for both injection modes. However, the DIM simulation results in a slightly higher bottomhole pressure for P1-2 compared to PIM, aligning with previous simulation findings. Notably, the optimal ripd for DIM is 0.5, whereas for PIM it is 0.75. The AIPC for half-production-injection well shortens the length of individual cycles, achieving a higher recovery degree within a shorter simulation duration.

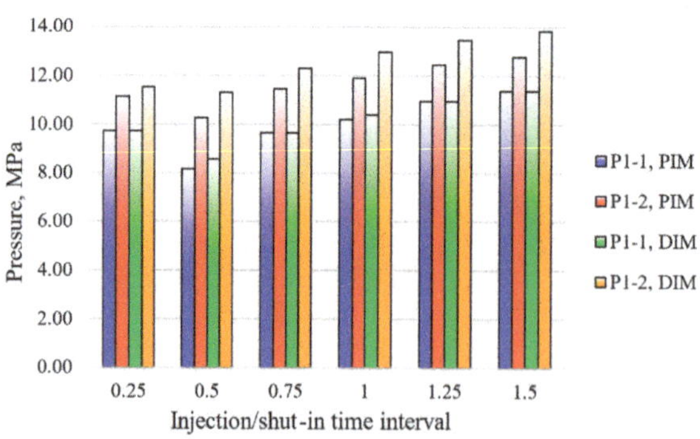

Figure 9. Bottom pressure of half-production wells.

Figure 10 shows a field map of the high permeability region at five cycle intervals with PIM. In contrast to Figure 7, the morphology of the high permeability zones with the half-production–injection well method between the fractured joints of the production wells is more diffuse and does not communicate directly with the injection wells. This avoids directional seepage of injected water and controls the water content of the production wells.

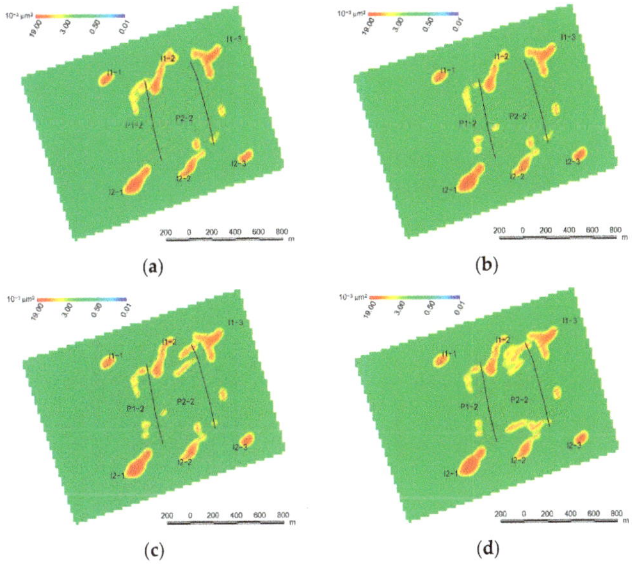

Figure 10. High permeability region at five cycle intervals (PIM, r_{ipd} = 0.5). (**a**) 5th cycle; (**b**) 10th cycle; (**c**) 15th cycle; (**d**) 25th cycle.

4. Conclusions

In this paper, we consider the WIFs caused by long-term water injection, take the high water cut, typical horizontal–vertical well pattern unit of Block Z as an example and propose to control the proper openings of WIFs by the asynchronous injection–production cycle method. This method provides an effective adjustment method for the development of fractured tight reservoirs at the late stage of development, improves the sweeping efficiency, and guides the further exploitation of the remaining oil.

1. The use of a 5-point horizontal–vertical well pattern enhances recovery through an asynchronous injection–production method. The main recovery mechanism involves periodically altering the pressure distribution to mobilize residual oil in the matrix through pressure perturbation and changes in flow lines, thereby expanding the swept area of the injected water.
2. WIFs are more likely to form where the angle between the flow line and the maximum horizontal principal stress is small. In the full-production injection mode, high permeability regions are concentrated between the injection wells and the fractures. In the half-production injection mode, high permeability regions are more dispersed and appear between the fractures.
3. Cumulative oil production initially increases and then decreases with the rise in the ratios of injection–production duration (r_{ipd}). The optimal r_{ipd} ratios were 0.75 and 0.5 for the parallel injection mode in the full-production injection case and half-production injection case, respectively, and 1 and 0.75 for the diagonal injection mode, respectively.
4. The parallel mode (r_{ipd} = 0.5) of the half-production injection well fully utilized the imbibition mechanism and maximized oil recovery. It limited the expansion of WIFs and prevented water breakthroughs.
5. This method is suitable for reservoirs where WIFs have already been identified; otherwise, the lack of reference pressure-sensitive permeability may reduce the accuracy of predictions.

Author Contributions: Conceptualization, J.C. and S.C.; data curation, T.Z. and L.Y.; methodology, J.C.; software, D.C. and D.Z.; validation, T.Z. and L.Y.; visualization, D.Z.; writing—original draft, J.C., D.Z. and D.C.; writing—review and editing, S.C. All authors have read and agreed to the published version of the manuscript.

Funding: This research received no external funding.

Data Availability Statement: The original contributions presented in the study are included in the article, further inquiries can be directed to the corresponding authors.

Conflicts of Interest: Authors Jianwen Chen, Tao Zhang, Linjun Yu and Dalin Zhou were employed by the company Changqing Oilfield Company. The remaining authors declare that the research was conducted in the absence of any commercial or financial relationships that could be construed as a potential conflict of interest.

Abbreviations

The following abbreviations are used in this manuscript:

AIPC	asynchronous injection–production cycle
DIM	diagonal injection mode
HF	hydraulic fractures
PIM	parallel injection mode
WIF	waterflood-induced fracture

Nomenclature

The following variables are used in this manuscript:

p_i	open pressure of naturally fractures
p_p	pore pressure
p_o	overburden pressure
p_f	the fracture pressure
p	pressure
ν	Poisson ratio
σ_H	maximum horizontal stress
σ_h	minimum horizontal stress
σ_f	tensile strength
θ	fracture dip angle
β	the angle of maximum horizontal stress and fracture trend
γ	the angle between the injection–production connection line (main flow line) and the maximum horizontal principal stress
K_f	the permeability of WIF-influenced area
K_i	the initial permeability
c_1 and c_2	coefficient that adjusts the pressure-sensitive relationship

References

1. Zou, C. Tight-Sandstone Oil and Gas. In *Unconventional Petroleum Geology*; Elsevier: Amsterdam, The Netherlands, 2013; pp. 61–109.
2. Li, N.; Zhu, S.; Li, Y.; Zhao, J.; Long, B.; Chen, F.; Wang, E.; Feng, W.; Hu, Y.; Wang, S.; et al. Fracturing-flooding technology for low permeability reservoirs: A review. *Petroleum* **2023**, *10*, 202–215. [CrossRef]
3. Cao, L.; Xiu, J.; Cheng, H.; Wang, H.; Xie, S.; Zhao, H.; Sheng, G. A New Methodology for the Multilayer Tight Oil Reservoir Water Injection Efficiency Evaluation and Real-Time Optimization. *Geofluids* **2020**, *2020*, 8854545. [CrossRef]
4. Teng, Z.; Zhu, P.; Wang, X.; Qi, R.; He, F.; Ma, T. Characteristics and model of meandering-river-delta reservoir architecture of Member Chang-6 in Fuxian area, Triassic Ordos Basin. *Unconv. Resour.* **2024**, *4*, 100084. [CrossRef]
5. Ma, G.; Li, T.; Wang, Y.; Chen, Y. The Equivalent Discrete Fracture Networks Based on the Correlation Index in Highly Fractured Rock Masses. *Eng. Geol.* **2019**, *260*, 105228. [CrossRef]
6. Zeng, L.; Song, Y.; Liu, G.; Tan, X.; Xu, X.; Yao, Y.; Mao, Z. Natural Fractures in Ultra-Deep Reservoirs of China: A Review. *J. Struct. Geol.* **2023**, *175*, 104954. [CrossRef]
7. Bai, W.; Cheng, S.; Wang, Y.; Cai, D.; Guo, X.; Guo, Q. A Transient Production Prediction Method for Tight Condensate Gas Wells with Multiphase Flow. *Pet. Explor. Dev.* **2024**, *51*, 172–179. [CrossRef]

. Cheng, M.; Xue, W.; Guo, Z.; Hou, M.; Wang, C. Development of Large-Scale Tight Gas Sandstone Reservoirs and Recommendations for Stable Production—The Example of the Sulige Gas Field in the Ordos Basin. *Sustainability* **2023**, *15*, 9933. [CrossRef]

. Liao, J.; Hong, L.; Li, Z.; Tan, K.; Zhao, L.; Yang, J.; Yu, P. Characterization of Ultra-Low Permeability Tight Sandstone Reservoir Properties and Criteria for Hydrocarbon Accumulation in Chang 6 Member, Huaqing Area, Ordos Basin. *Front. Earth Sci.* **2022**, *10*, 1013776. [CrossRef]

0. Azeemuddin, M.; Ghori, S.G.; Saner, S.; Khan, M.N. Injection-Induced Hydraulic Fracturing in a Naturally Fractured Carbonate Reservoir: A Case Study from Saudi Arabia. In Proceedings of the SPE International Conference and Exhibition on Formation Damage Control, Lafayette, LA, USA, 20–21 February 2002.

1. Wang, Y.; Cheng, S.; Zhang, K.; Xu, J.; Qin, J.; He, Y.; Luo, L.; Yu, H. Case Studies: Pressure-Transient Analysis for Water Injector with the Influence of Waterflood-Induced Fractures in Tight Reservoir. In Proceedings of the SPE Improved Oil Recovery Conference, Tulsa, OK, USA, 14–18 April 2018.

2. Wang, Y.; Song, X. Comprehensive Characterization Integrating Static and Dynamic Data for Dynamic Fractures in Ultra-Low Permeability Reservoirs: A Case Study of the Chang 6 Reservoir of the Triassic Yanchang Formation in the Ordos Basin, China. *Minerals* **2022**, *12*, 1277. [CrossRef]

3. Hagoort, J.; Weatherill, B.D.; Settari, A. Modeling the Propagation of Waterflood-Induced Hydraulic Fractures. *Soc. Pet. Eng. J.* **1980**, *20*, 293–303. [CrossRef]

4. Di, S.; Cheng, S.; Bai, W.; Wei, C.; Wang, Y.; Qin, J. Dynamic Fracture Propagation Mechanism and Application in Tight Oil Reservoir. *Chin. J. Theor. Appl. Mech.* **2021**, *53*, 2141–2155.

5. Deng, H.; Steefel, C.; Molins, S.; DePaolo, D. Fracture Evolution in Multimineral Systems: The Role of Mineral Composition, Flow Rate, and Fracture Aperture Heterogeneity. *ACS Earth Space Chem.* **2018**, *2*, 112–124. [CrossRef]

6. Ren, G.; Ma, X.; Zhang, S.; Zou, Y.; Duan, G.; Xiong, Q. Optimization of Water Injection Strategy before Re-Stimulation Considering Fractures Propagation. *Processes* **2022**, *10*, 1538. [CrossRef]

7. Xue, L.; Liu, P.; Zhang, Y. Status and Prospect of Improved Oil Recovery Technology of High Water Cut Reservoirs. *Water* **2023**, *15*, 1342. [CrossRef]

8. Liu, C.; Li, H.; Luo, H.; Lu, Y.; Li, Y.; Su, C.; Chen, S. Investigation and Application of Perforation Optimization Method on Shale Gas Horizontal Well with Numerical Simulation of Multicluster Fracturing under Dense-Segment Pattern. *ACS Omega* **2023**, *9*, 675–691. [CrossRef]

9. Liao, J.; Zhang, Z.; Tang, H.; Yang, J.; Zhang, X.; Wang, B. Simulation of Fracturing and Well Pattern Optimization of Fractured Tight Sandstone Reservoirs. *Front. Earth Sci.* **2022**, *10*, 873617. [CrossRef]

20. Ge, J.; Wang, D.; Qu, D. Combination Production of Vertical and Horizontal Wells. In Proceedings of the SPE Asia Pacific Oil and Gas Conference and Exhibition, Kuala Lumpur, Malaysia, 20–22 March 1995.

21. He, X.; Zhang, P.; He, G.; Gao, Y.; Liu, M.; Zhang, Y.; Fang, D.; Li, Y. Evaluation of Sweet Spots and Horizontal-Well-Design Technology for Shale Gas in the basin-margin Transition Zone of southeastern Chongqing, SW China. *Energy Geosci.* **2020**, *1*, 134–146. [CrossRef]

22. Zhang, D.; Liu, Y.; Luo, H.; Cao, S.; Cao, J.; Li, X. Staged fracturing of horizontal wells in continental tight sandstone oil reservoirs: A case study of Yanchang Formation in Western Ordos Basin, China. *Front. Earth Sci.* **2021**, *9*, 760976. [CrossRef]

23. Wu, B.; Yao, J.; Lü, A. Research on Sweep Efficiency in Horizontal-Vertical Combined Well Pattern. *Acta Pet. Sin.* **2006**, *27*, 85–88.

24. Zhang, F.; Wu, S.; Liu, H. A Waterflood Induced Fracture Simulation Model for Naturally Fractured Low-Permeability Reservoirs. In Proceedings of the SPE Asia Pacific Oil & Gas Conference and Exhibition, Perth, Australia, 25–27 October 2016.

25. Sanchez-Rivera, D.; Mohanty, K.; Balhoff, M. Reservoir simulation and optimization of Huff-and-Puff operations in the Bakken Shale. *Fuel* **2015**, *147*, 82–94. [CrossRef]

26. Di, S.; Cheng, S.; Wei, C.; Bai, W.; Shang, R.; Cao, N.; Zhang, Y. Study on the production mode of horizontal well from water huff and puff to segmented injection and production in tight reservoir. *Geofluids* **2021**, *2021*, 7947911. [CrossRef]

27. He, Y. Experimental Study on Asynchronous Injection-Production in Horizontal Well Area of Low Permeability Reservoir. *J. Phys. Conf. Ser.* **2020**, *1549*, 042045. [CrossRef]

28. Kang, S.; Pu, C.; Wang, Y.; Liu, W.; Wang, K.; Huang, F.; Fan, Q.; Gao, X.; Yang, Q. Parameter Optimization of Asynchronous Cyclic Waterflooding for Horizontal–Vertical Well Patterns in Tight Oil Reservoirs. *ACS Omega* **2022**, *7*, 11226–11239. [CrossRef] [PubMed]

29. Li, J.; Wei, J.; Chen, Y.; Wang, A.; Zhou, X. Analysis of asynchronous/inter-fracture injection-production mechanism under the condition of five-spot vertical and horizontal well combination in low permeability oil reservoirs. *Energy* **2024**, *308*, 132923. [CrossRef]

30. Xu, X.; Wu, W.; Jiang, D.; Wu, J. Development Characteristics and Key Parameter Optimization of High-Pressure Water Injection in Low Permeability Reservoirs. In Proceedings of the International Field Exploration and Development Conference, Qingdao, China, 20–22 October 2021; Springer Nature: Singapore, 2021; pp. 3414–3423.

31. Zhao, J.; Fan, J.; He, Y.; Yang, Z.; Gao, W.; Gao, W. Optimization of Horizontal Well Injection-Production Parameters for Ultra-Low Permeable–Tight Oil Production: A Case from Changqing Oilfield, Ordos Basin, NW China. *Pet. Explor. Dev.* **2015**, *42*, 74–82. [CrossRef]

32. Zhang, Y.; Yang, S.; Bi, L.; Gao, X.; Shen, B.; Hu, J.; Luo, Y.; Zhao, Y.; Chen, H.; Li, J. A technical review of CO_2 flooding sweep-characteristics research advance and sweep-extend technology. *Pet. Sci.* 2024, *in press*. [CrossRef]
33. Hassanpouryouzband, A.; Wilkinson, M.; Haszeldine, R.S. Hydrogen energy futures–foraging or farming? *Chem. Soc. Rev.* **2024**, *53*, 2258–2263. [CrossRef]
34. Gao, S.; Yang, Y.; Liao, G.; Xiong, W.; Liu, H.; Shen, R.; Ye, L.; Zhu, W.; An, H. Experimental research on inter-fracture asynchronous injection-production cycle for a horizontal well in a tight oil reservoir. *J. Pet. Sci. Eng.* **2022**, *208*, 109647. [CrossRef]
35. Yang, Y.; Liao, G.; Xiong, W.; Shen, R.; Zhang, J.; Li, Q.; Wang, S.; Zhang, J.; Tan, L.; Shao, G. Physical and numerical simulation of inter-fracture flooding in heterogeneous tight oil reservoirs. *Energy Rep.* **2022**, *8*, 12970–12978. [CrossRef]
36. Wang, Z.; Muhtar, M.; Xu, D.; Fang, J.; Li, J.; Liu, D.; Zhang, Z.; Gao, L. Seepage Behavior of Fractures in Paleogene Sandstone Reservoirs in Nanpu Sag, Bohai Bay Basin, Eastern China. *Front. Earth Sci.* **2021**, *9*, 718733. [CrossRef]
37. Su, Z.; Li, T.; Bai, M.; Zhou, Z. Influence of Mineral Composition on Initiation Pressure of Waterflood-Induced Fractures in Tight Sandstone Reservoir. *ACS Omega* **2024**, *9*, 9269–9285. [CrossRef] [PubMed]
38. Clark, K.K. Transient Pressure Testing of Fractured Water Injection Wells. *J. Pet. Technol.* **1968**, *20*, 639–643. [CrossRef]
39. Lyu, W.; Zeng, L.; Chen, M.; Qiao, D.; Fan, J.; Xia, D. An Approach for Determining the Water Injection Pressure of Low Permeability Reservoirs. *Energy Explor. Exploit.* **2018**, *36*, 1210–1228. [CrossRef]
40. Liu, H.; Pang, J.; Yu, X. A Mathematical Model for Natural Fracture Evolution in Water-Flooding Oil Reservoir. *Energy Procedia* **2012**, *16*, 1348–1356. [CrossRef]
41. Zhao, X.; Zeng, L.; Wang, X.; Wang, F.; Zhang, Y.; Jiao, J.; Weng, J. Differences of Natural Fracture Characteristics and Their Development Significance in Chang 6, Chang 7 and Chang 8 Reservoir, Ningxian-Heshui Area, Ordos Basin. *Sci. Geol. Sin.* **2015**, *50*, 274–285.
42. Hubbert, M.K.; Willis, D.G. Mechanics of Hydraulic Fracturing. *Trans. AIME* **1957**, *210*, 153–168. [CrossRef]
43. Ji, W.; Yu, H.; Liu, X.; Fu, H.; Yuan, B.; Yan, F.; Luo, C.; Jiang, X.; Dai, C. Oil Production Mechanism of Water Injection Huff-n-Puff for Enhancing Oil Recovery in Tight Sandstone Reservoir. *Energy Fuels* **2023**, *37*, 18867–18877. [CrossRef]
44. Yoon, J.S.; Zimmermann, G.; Zang, A. Numerical Investigation on Stress Shadowing in Fluid Injection-Induced Fracture Propagation in Naturally Fractured Geothermal Reservoirs. *Rock Mech. Rock Eng.* **2015**, *48*, 1439–1454. [CrossRef]
45. Liu, L.; Liu, Y.; Yao, J.; Huang, Z. Mechanistic Study of Cyclic Water Injection to Enhance Oil Recovery in Tight Reservoirs with Fracture Deformation Hysteresis. *Fuel* **2020**, *271*, 117677. [CrossRef]
46. Iltaf, K.H.; Yue, D.; Wang, W.; Wan, X.; Li, S.; Wu, S.; Liu, R.; Weijia, Z.; Mehboob, S.; Shah, S.A.; et al. Facies and Petrophysical Modeling of Triassic Chang 6 Tight Sandstone Reservoir, Heshui Oil Field, Ordos Basin, China. *Lithosphere* **2021**, *2021*, 9230422. [CrossRef]

energies

MDPI

Article

Application of Two-Dimensional NMR for Quantitative Analysis of Viscosity in Medium–High-Porosity-and-Permeability Sandstones in North China Oilfields

Wei Zhang [1], Si Li [1], Shaoqing Wang [1], Jianmeng Sun [2], Wenyuan Cai [1], Weigao Yu [1], Hongxia Dai [1] and Wenkai Yang [2,*]

1 PetroChina Logging Co., Ltd., North China Branch, Renqiu 062550, China; hbzhangw@cnpc.com.cn (W.Z.); hblis@cnpc.com.cn (S.L.); wangshaoq@cnpc.com.cn (S.W.); caiwy@cnpc.com.cn (W.C.); yjy_ywg@petrochina.com.cn (W.Y.); daihx@cnpc.com.cn (H.D.)

2 Logging Department, School of Geosciences and Technology, China University of Petroleum (East China), Qingdao 266580, China; sunjm@upc.edu.cn

* Correspondence: s22010082@s.upc.edu.cn

Abstract: The viscosity of crude oil plays a pivotal role in the exploration and development of oil fields. The predominant reliance on laboratory measurements, which are constrained by manual expertise, represents a significant limitation in terms of efficiency. Two-dimensional nuclear magnetic resonance (NMR) logging offers a number of advantages over traditional methods. It is capable of providing faster measurement rates, as well as insights into fluid properties, which can facilitate timely adjustments in oil and gas development strategies. This study focuses on the loose sandstone reservoirs with high porosity and permeability containing heavy oil in the Huabei oilfield. Two-dimensional nuclear magnetic resonance (NMR) measurements and analyses were conducted on saturated rocks with different-viscosity crude oils and varying oil saturation levels, in both natural and artificial rock samples. This study elucidates the distribution patterns of different-viscosity crude oils within the two-dimensional NMR spectra. Furthermore, the T_1 and T_2 peak values of the extracted oil signals were employed to establish a model correlating oil viscosity with NMR parameters. Consequently, a criterion for determining oil viscosity based on two-dimensional NMR was formulated, providing a novel approach for estimating oil viscosity. The application of this technique in the BQ well group of the Huabei oilfield region yielded an average relative error of 15% between the actual oil viscosity and the computed results. Furthermore, the consistency between the oil types and the oil discrimination chart confirms the reliability of the method. The final outcomes meet the precision requirements for practical log interpretation and demonstrate the excellent performance of two-dimensional nuclear magnetic resonance (NMR) logging in calculating oil viscosity. The findings of this study have significant implications for subsequent exploration and development endeavors in the research area's oilfields.

Keywords: crude oil viscosity; rock samples; two-dimensional NMR; transverse relaxation time; longitudinal relaxation time

1. Introduction

The viscosity of crude oil is a function of the internal frictional resistance encountered during its flow. The viscosity of reservoir crude directly influences its ability to flow through subsurface pore media and pipelines [1,2]. An understanding of the viscosity characteristics of reservoir crude is of significant practical importance for devising development schemes, evaluating well productivity, studying flow mechanisms, and facilitating crude oil transportation [3]. Currently, the primary means of measuring viscosity is through laboratory experiments. This method of viscosity analysis places extremely high demands on the experiments, as the variation in viscosity parameters is influenced by numerous human factors [4]. In the early stages of field development, the lack of viscosity data for crude oil under reservoir conditions often prevents the provision of reliable recommendations for field development plans.

Nuclear magnetic resonance (NMR) logging is widely used in petroleum exploration and production to provide accurate porosity and identify fluid types. The T_1 and T_2 peak values are the most important parameters in nuclear magnetic logging, representing the maximum values of longitudinal relaxation time and transverse relaxation time [5–8]. Nuclear magnetic resonance logging uses the resonance phenomenon that occurs between hydrogen nuclei and an applied magnetic field to detect subsurface oil and gas reservoirs. It is the only logging method capable of providing information on crude oil viscosity [9]. When considered alongside traditional experimental approaches, nuclear magnetic resonance (NMR) stands out for its non-destructive nature and immunity to external influences, such as the composition of the rock matrix. Instead, it exclusively detects and characterizes hydrogen-containing compounds present within the sample, offering a distinct advantage in certain analytical contexts [10]. Nuclear magnetic resonance (NMR) logging offers unique advantages in measuring fluid component properties and reflecting pore structure, facilitating a qualitative assessment of reservoir characteristics [11]. It is commonly used in laboratory studies of fluid properties [12]. One-dimensional nuclear magnetic logging measures only the transverse relaxation time T_2 of formation pore fluids and has significant limitations in identifying and quantitatively evaluating oil, gas, and water. When T_2 is constant, two-dimensional nuclear magnetic resonance logging can differentiate oil, gas, and water by using different T_1 values [13].

In recent years, the development of two-dimensional nuclear magnetic resonance logging has become increasingly mature. Foreign scholars typically use two-dimensional nuclear magnetic resonance logging for rock physics experiments and the identification analysis of fluid components. Research has demonstrated that the T_1–T_2 technology of two-dimensional nuclear magnetic resonance is capable of effectively identifying fluid components and solid organic matter in various states within the pore space [14].

Xie used a 2 MHz NMR spectrometer to study the effect of temperature on the NMR relaxation characteristics of crude oil and provided a relationship formula for relaxation time with temperature and crude oil viscosity changes at 2 MHz. However, the viscosity estimation formula obtained under experimental conditions uses a homogeneous magnetic field, which has significant limitations [15].

He et al. performed variable echo interval (TE) NMR measurements on medium-to-high-viscosity saturated oil samples and established a relationship between crude oil viscosity and the transverse relaxation time T_2 peak of the oil sample. This relationship has been used to estimate subsurface crude oil viscosity using NMR logging, with good practical results. However, these studies only analyzed the one-dimensional NMR measurement results of oil sample fluids and did not consider the effect of the actual formation conditions on the measurement results [16].

Kadkhodaie used three steps to generate capillary pressure and relative permeability curves. First, a cluster analysis was used to classify reservoir rocks into six electrofacies (EFs), with the reservoir quality graded from EF1 to EF6. Secondly, the NMR T_2 distribution arrays were converted into synthetic drainage capillary pressure curves and the results were validated by available laboratory-measured mercury injection capillary pressure curves (MICP). Finally, relative permeability curves were generated from the NMR-derived MICP data using the Wyllie and Gardner equations for each individual electrofacies. A comparison of the results with laboratory data demonstrated the effective role of electrofacies control in the generation of highly accurate capillary pressure and relative permeability curves [17].

Li et al., based on laboratory studies of the NMR mechanism in saturated heavy oil cores, analyzed the difference spectra and migration spectra characteristics of P-type NMR data for heavy oil reservoirs. They summarized methods for identifying heavy oil reservoirs using NMR data, distinguished between heavy oil, medium-viscosity oil, and water layer NMR recording responses, and developed a new method for identifying heavy oil reservoirs. However, the detection depth of NMR logging is shallow, so it is necessary to integrate NMR data with conventional data during reservoir evaluation to accurately assess heavy oil reservoirs [18].

NMR technology provides relaxation times and diffusion coefficients that can be used to calculate porosity [19], permeability [20,21], and oil content [22,23], identify reservoir fluids [24,25], and delineate heavy oil reservoirs [26–28]. Shi and Cai proposed a "five-component" pore interval oil–water discrimination model. Building on this, the present paper further explores the method of assessing crude oil viscosity using two-dimensional NMR T_1–T_2 spectra under conditions of medium-to-high-porosity-and-permeability. The experimental requirements are clean lithology, good physical properties, and minimal interference from bound water [29].

Previous studies mainly focused on one-dimensional nuclear magnetic resonance in the laboratory, and only derived the trend of changes in crude oil viscosity with T_1 and T_2, or did not derive a specific formula for calculating crude oil viscosity using two-dimensional nuclear magnetic resonance, which is not very applicable. This article uses a laboratory nuclear magnetic resonance core analyzer with the same parameters to conduct two-dimensional nuclear magnetic resonance T_1–T_2 experimental measurements on crude oil samples, natural rock samples saturated with different viscosities of crude oil, and artificial rock samples. Full-diameter-core NMR measurements were also conducted on full-diameter-core and crude oil samples at the well site using a full-diameter-core NMR analyzer. By analyzing the characteristics of two-dimensional nuclear magnetic resonance T_1–T_2 spectra, the relaxation characteristics of different-viscosity crude oils and their distribution patterns in two-dimensional nuclear magnetic resonance spectra were elucidated. By utilizing the relationship between the extracted crude oil signal peaks T_1 and T_2 and crude oil viscosity, a two-dimensional NMR (T_1–T_2) rapid discrimination graph of crude oil viscosity was provided for the first time, and an innovative model for calculating crude oil viscosity using two-dimensional NMR was established. This provides a new method for estimating crude oil viscosity and reliable data for fluid evaluation and reservoir development.

2. Sample and Methods

2.1. Crude Oil Samples

To simulate the effect of the crude oil viscosity contained in the core porosity on the NMR response, actual oil samples from the North China Oilfield with viscosities of 70, 136, and 26,000 mPa·s were selected. These viscosities were determined by geochemical pyrolysis experiments. We extracted crude oil from rock samples by first crushing and grinding the samples, followed by centrifugal separation to remove water and solid particles from the crude oil. Pyrolysis experiments volatilized and crack the hydrocarbons in the samples at different temperatures. The hydrocarbon content and pyrolysis parameters of each component in the rock were measured by a detector. Based on the pyrolysis parameters and other analytical data, the viscosity of the crude oil was evaluated using an established chart. These samples were mixed with a low-viscosity base oil (1.5 mPa·s) in the laboratory to produce oil samples of different viscosities, namely 25, 61.7, 519, and 26,000 mPa·s, covering light, medium, and heavy oils. This sample preparation allowed a comprehensive evaluation of the effect of different oil viscosities on the NMR response. Table 1 shows the physical and chemical properties of crude oil samples (Table 1).

Table 1. Basic information of crude oil samples.

Viscosity (mPa·s)	Density (g/cm^3)	Boiling Point (°C)	Sulfur Content (% *w*/*w*)	Asphaltene Content (g)
1.5	0.83	200–300	0.1	4
25	0.85	300–400	0.3	5
75	0.88	400–500	0.5	5
519	0.92	500–600	1.0	6
26,000	1.02	600–700	3.0	10

2.2. Natural and Artificial Rock Samples

Five representative natural rock samples from medium-to-high-porosity-and-permeability reservoirs in the North China Oilfield were selected for the experiments. The physical parameters of the rocks are shown in Table 1. The natural rock core samples indicated that all five samples were relatively loose, with porosity measured by gas displacement ranging from 15.5 to 28.3% and an average porosity of 23.24%. Permeability ranged from 23.39 to 500.16×10^{-3} μm^2, with an average permeability of 269.32×10^{-3} μm^2, classifying the samples as medium-to-high-porosity-and-permeability sandstone reservoirs (Table 2).

Table 2. Basic information of natural rock samples.

Sample	Length (mm)	Diameter (mm)	Volume (mL)	Dry Weight (g)	Helium Porosity (%)	Helium Permeability ($\times 10^{-3}$ μm^2)
7A	42.52	25.06	20.98	47.01	15.5	23.39
9B	42.88	25.07	21.17	46.04	17.2	140.09
11A	42.31	25.12	20.97	39.26	27.8	305.90
13A	42.12	25.11	20.85	39.10	28.3	377.04
14B	41.66	25.13	20.66	39.49	27.4	500.16

To simulate the actual conditions of the formation, different pore structures in rock cores were mimicked using glass sand of different grain sizes. Three sets of artificial rock core experiments were designed using 60 mesh, 120 mesh, and 180 mesh cores corresponding to different sandstone grain sizes (Table 3).

Table 3. Classification table of artificial rock samples by mesh size.

Mesh Size	Equivalent Diamete (μm)	Equivalent Pore Diameter (μm)	Corresponding Particle Grade
60 mesh	250	77	Medium Sand
120 mesh	125	39	Fine Sand
180 mesh	80	25	Silt Sand

2.3. Experimental Methods

This experiment involves nuclear magnetic resonance (NMR) testing of natural rock samples saturated with oil, artificial rock samples saturated with oil, and full-diameter rock cores containing oil at well sites. The objective is to perform a multi-dimensional comparative analysis of crude oil samples, natural rock samples, artificial rock samples, and full-diameter rock cores from well sites. The results will clarify the response characteristics of different viscous fluid components in the T_1–T_2 spectra and establish a crude oil viscosity calculation model.

2.3.1. Natural Rock Sample NMR Experiment

First, five groups of natural rock samples are selected for oil extraction. The extracted core samples are washed to remove salts and residual oil from the pore spaces. The cores are then dried in an oven at 70 °C for 24 h to determine their dry weight. Length and diameter measurements are taken, followed by porosity and permeability tests to obtain gas-measured porosity and permeability data.

Experimental oils of different viscosities (1.5, 70, 136, and 581 mPa·s) are prepared for free-state nuclear magnetic resonance (NMR) experiments. NMR one-dimensional T_2 spectra and two-dimensional T_1–T_2 spectra are obtained for each viscosity.

Finally, the rock samples saturated with oils of different viscosities, especially under bound water conditions, are subjected to NMR experiments. This is undertaken to confirm the positions of oils of different viscosities in the two-dimensional NMR T_1–T_2 spectra of natural rock samples and to establish a model for calculating crude oil viscosity.

2.3.2. Artificial Rock Sample NMR Experiment

First, artificial rock samples are prepared with different pore sizes, covering at least the ranges of medium sandstone, fine sandstone, and siltstone. These samples are saturated with oil and subjected to nuclear magnetic resonance (NMR) measurements.

The cylindrical samples are placed in an SVF rock core vacuum saturation apparatus (Xuan Yu Mechanical and Electrical Technology (Shanghai) Co., Ltd., Shanghai, China). Oil is introduced into the device and after 8 h of vacuum pumping to remove air, the device is sealed and pressurized to 1000 psi to ensure saturation with the crude oil. Pressure stability is maintained throughout this process. After one week, the cores are removed, cleaned on the surface, and subjected to NMR measurements of one-dimensional T_2 spectra and two-dimensional T_1–T_2 spectra.

Finally, the response characteristics of oils of different viscosities (viscosities of 1.5, 70, 136, and 445 mPa·s) in different pore spaces are analyzed. This analysis confirms the positions of oils of different viscosities in the two-dimensional T_1–T_2 NMR spectra of the artificial rock samples and establishes a model for calculating crude oil viscosity.

2.3.3. Full-Diameter Rock Core NMR Experiment

To prevent the escape of hydrocarbons, oil-bearing full-diameter-core samples are first wrapped for containment at the well site. These oil-bearing cores are then measured using a full-diameter nuclear magnetic resonance (NMR) logging tool to determine the position of the oil within the two-dimensional NMR T_1–T_2 spectrum of the full-diameter cores.

Finally, the NMR response characteristics of full-diameter cores saturated with crude oils of different viscosities (viscosities of 1.5, 25, 61.7, 136, 519, 26,000 mPa·s) in different pore spaces are analyzed by integrating the well test data. The aim of this analysis is to establish a model for the calculation of crude oil viscosity.

This approach provides detailed insight into how crude oils of different viscosities are distributed and behave within full-diameter cores in different pore spaces, thereby facilitating the development of a viscosity calculation model based on NMR data and well test analysis.

3. Results and Discussion

3.1. NMR Characteristics and Distribution Patterns of Crude Oil Samples with Different Viscosities

Figure 1 shows the T_1 and T_2 distributions (Figure 1). When observing Figure 1, it can be seen that the longitudinal relaxation time (T_1) and the transverse relaxation time (T_2) are approximately the same for the light crude oil. However, the transverse relaxation time (T_2) is significantly shifted back in Figure 1 for the heavy crude due to the increase in viscosity and decrease in visual hydrogen index, resulting in the NMR recorded response signal being indistinguishable from that of the bound fluid. Both the T_1 and T_2 relaxation times in the NMR measurements decrease significantly with increasing crude oil viscosity, but T_2 decreases more rapidly than T_1, suggesting that T_2 is more sensitive to information reflecting crude oil viscosity.

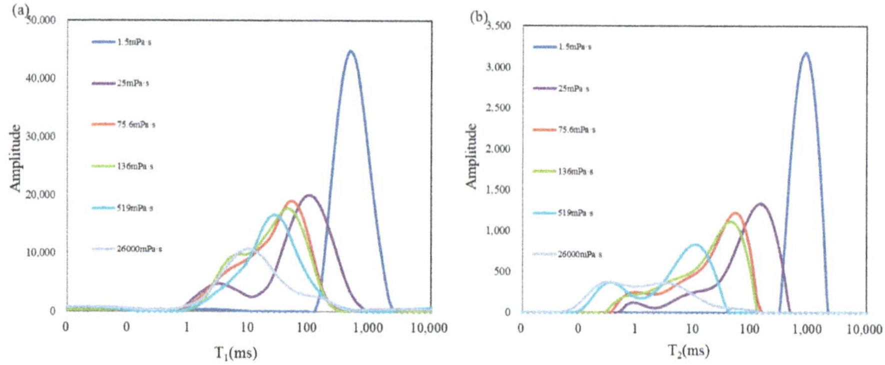

Figure 1. (**a**) T_1 Distribution of crude oils with different viscosities and (**b**) T_2 distribution of crude oils with different viscosities.

Figure 2 shows the two-dimensional nuclear magnetic maps (T_1–T_2) of crude oil samples measured using a laboratory NMR scanner (Figure 2). The white circles in the figure represent the peaks of T_1 and T_2. The three dashed lines from top to bottom represent T_1/T_2 values of 100, 10, and 1, respectively. These three lines can be used to observe the changes in T_1/T_2 near the dashed lines as the viscosity of the crude oil increases. From the six 2D NMR plots of different viscosities in Figure 2, it can be seen that the relaxation times T_1 and T_2 of the crude oil samples in the nuclear magnetic resonance (NMR) measurements are influenced by the viscosity and composition of the crude oil. Both T_1 and T_2 decrease with increasing crude oil viscosity, and the T_1/T_2 ratio increases from about 1 to about 10. In addition, T_2 shows a clear double- or multiple-peak feature in Figure 2, particularly in the differentiation of asphaltenes, heavy hydrocarbon fractions, and light fractions in the T_2 dimension. This suggests that higher concentrations of heavy hydrocarbon fractions correspond to larger T_1/T_2 ratios.

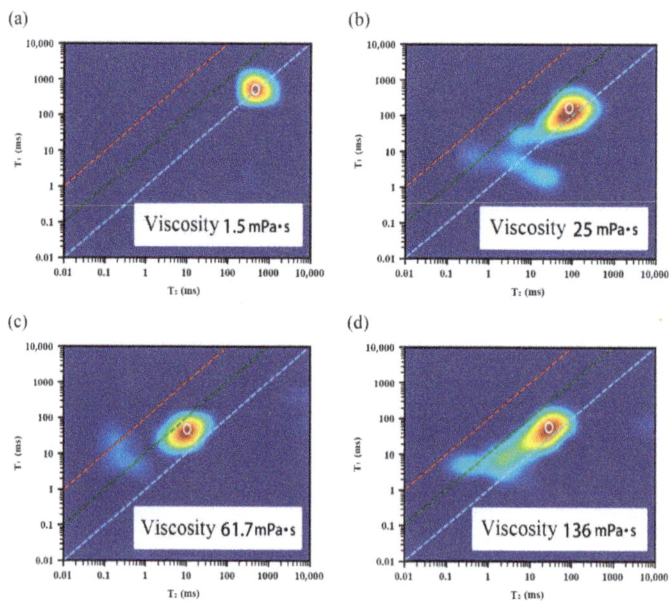

Figure 2. *Cont.*

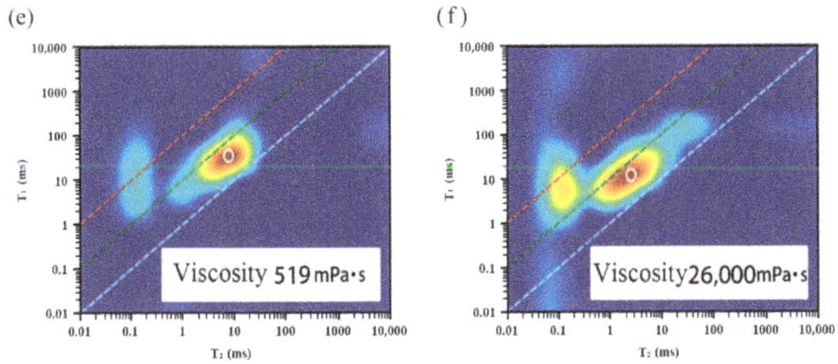

Figure 2. Two-dimensional NMR T_1–T_2 maps of crude oils with viscosities of (**a**) 1.5 mPa·s; (**b**) 25 mPa·s; (**c**) 61.7 mPa·s; (**d**) 136 mPa·s; (**e**) 519 mPa·s; (**f**) 26,000 mPa·s.

3.2. NMR Characteristics and Distribution Patterns of Natural and Artificial Rock Samples with Different Viscosities

Figure 3 illustrates the variation of the T_1 and T_2 NMR relaxation times in the natural cores saturated with oil of different viscosities (Figure 3). Figure 3 shows natural cores containing capillary-bound water, where the signals from the oil gradually overlap with those from the capillary-bound water as the viscosity of the oil increases.

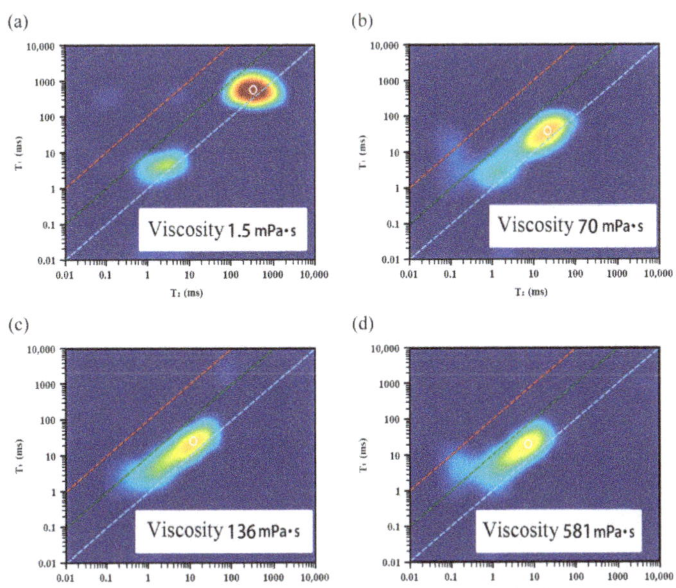

Figure 3. Characteristics of two-dimensional NMR T_1–T_2 maps for crude oil samples with viscosities of (**a**) 1.5 mPa·s; (**b**) 70 mPa·s; (**c**) 136 mPa·s; (**d**) 581 mPa·s in natural cores.

The variation of the T_1 and T_2 NMR relaxation times in the artificial cores saturated with oils of different viscosities is shown in Figure 4, which shows that the position of the oil peaks in the artificial cores is less affected by changes in pore structure. Under the same oil viscosity condition, the oil peak positions are basically consistent at different oil saturations. This consistency suggests that accurate measurement of mobile oil signals in real samples can reliably estimate crude oil viscosity, provided the pore structures are similar. The patterns of the T_1 and T_2 relaxation time changes observed in natural and artificial cores saturated with oil of different viscosities are consistent with those observed for crude oil samples. In particular, the oil peaks move towards lower T_1 and T_2 values with increasing oil viscosity and recombinant content, and the T_1/T_2 ratio tends to increase.

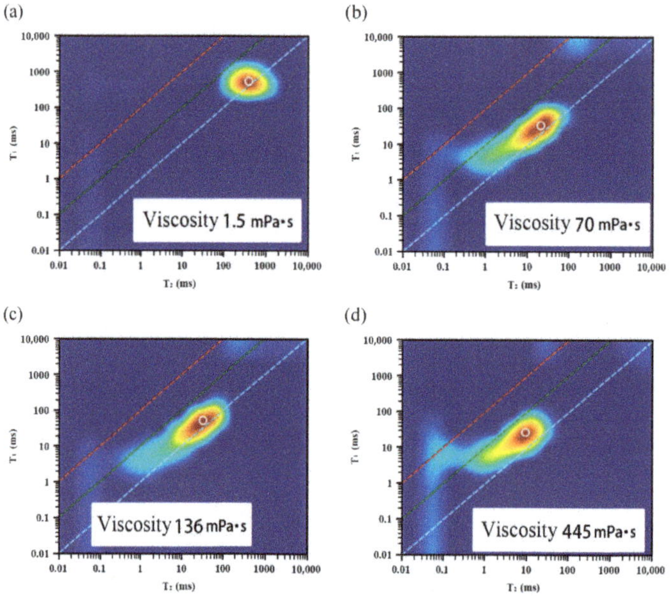

Figure 4. Characteristics of two-dimensional NMR T_1–T_2 maps for crude oil samples with viscosities of (**a**) 1.5 mPa·s; (**b**) 70 mPa·s; (**c**) 136 mPa·s; (**d**) 445 mPa·s in artificial cores.

3.3. *Establishment of Crude Oil Viscosity Calculation Model*

The samples used to establish the crude oil viscosity calculation model are as follows: Oil Sample (Lab)—actual crude oil from the Huabei Oilfield, blended with base oil in the laboratory to achieve different viscosities. Oil Sample (Xinghua)—actual crude oil samples taken from three wells in the Huabei oil field. Actual Sample (Lab)—five natural rock samples saturated with different viscosity crude oils in the laboratory. Lab (Field—Piston)—oil bearing core samples obtained by rotary coring at the well site. Lab (Full-Diameter Core)—oil-bearing full-diameter-core samples obtained by core drilling at the well site. Artificial Samples (60, 120, 180 mesh)—artificial rock samples of various grain sizes (pore sizes) injected with crude oils of various viscosities prepared in the laboratory.

3.3.1. Oil Sample (Laboratory)

In NMR logging, the relaxation times T_1 and T_2 are affected by molecular motion. An increase in viscosity means an increase in intermolecular friction, which affects the rotation and translation of the molecules, thereby affecting T_1 and T_2. This influence is related to the complexity of molecular motion, which usually manifests itself as a non-linear relationship, so we decided to use an exponential relationship to fit it. For the laboratory oil samples, Figure 5 shows the viscosity versus T_1/T_2 cross plot (Figure 5), from which the following formula (1) is derived to relate viscosity and two-dimensional NMR:

$$\mu = 2.5649 * (\frac{T_1}{T_2})^{6.0156} \quad (1)$$

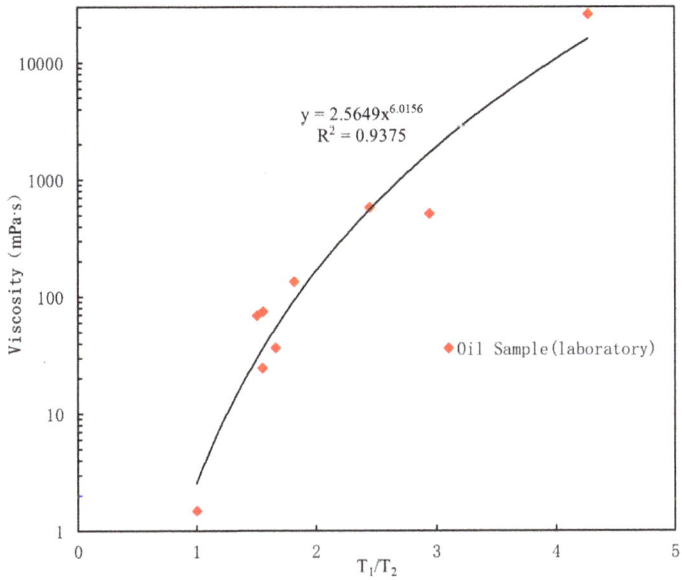

Figure 5. Crude oil viscosity vs. T_1/T_2 cross plot.

3.3.2. Field–Core Plug and Full-Diameter-Core Sample

For the oil-bearing rock cores obtained from the well site, Figure 6 shows the actual field oil-bearing rock viscosity versus T_1/T_2 cross plot (Figure 6), which leads to the following formula (2) for viscosity and two-dimensional NMR:

$$\mu = 4.9646 * (\frac{T_1}{T_2})^{4.0752} \quad (2)$$

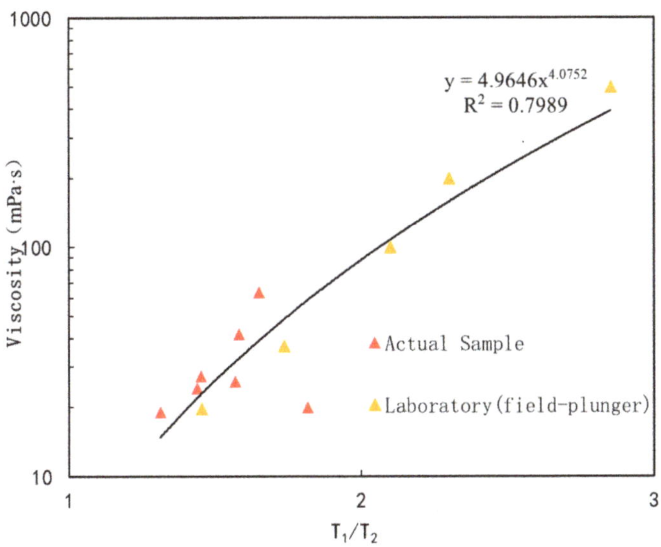

Figure 6. Field oil-bearing rock viscosity vs. T_1/T_2 cross plot.

3.3.3. Discrimination Diagram for Crude Oil Viscosity Based on Two-Dimensional (T_1–T_2) Maps

The chemical composition of heavy oil, particularly the asphaltene and resin content, is relatively high. These components have complex molecular structures and contain a large number of aromatic and heterocyclic structures, which have a significant effect on the relaxation time of NMR. Due to the decrease in the apparent hydrogen index caused by asphaltenes in the oil sample during the NMR recording, we classified the oil into three categories: light oil, medium oil, and heavy oil, in order to improve the accuracy of the fitting formula as much as possible. Based on the analysis of the relationship between crude oil viscosity and T_1, T_2, a discrimination diagram for crude oil viscosity was constructed using two-dimensional (T_1–T_2) maps from both laboratory experiments and field cores. The laboratory artificial samples, natural rock samples, and field full-diameter-core measurements all show a pattern in which oil viscosity increases as T_2 decreases. As the heavy oil signals and heavy hydrocarbon components increase, there is a rapid increase in the T_1/T_2 ratio (Figure 7). This discrimination method can qualitatively distinguish light, medium, and heavy oils (the T_1 and T_2 data used in the graph are all oil peak readings).

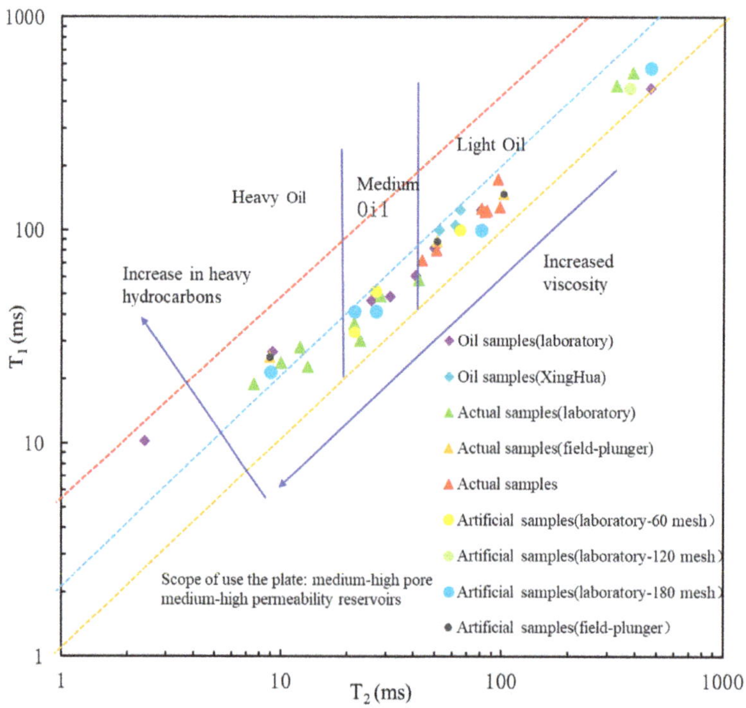

Figure 7. Discrimination diagram for crude oil viscosity based on T_1 and T_2 measurements.

3.3.4. Fitting Formulas Based on Crude Oil Viscosity Categories

After categorizing the experimental data by crude oil viscosity, fitting formulae were derived for each category. Figure 8 shows the cross plot of viscosity versus T_1/T_2 divided by crude oil viscosity, with different formulae derived for light, medium, and heavy oils (Figure 8).

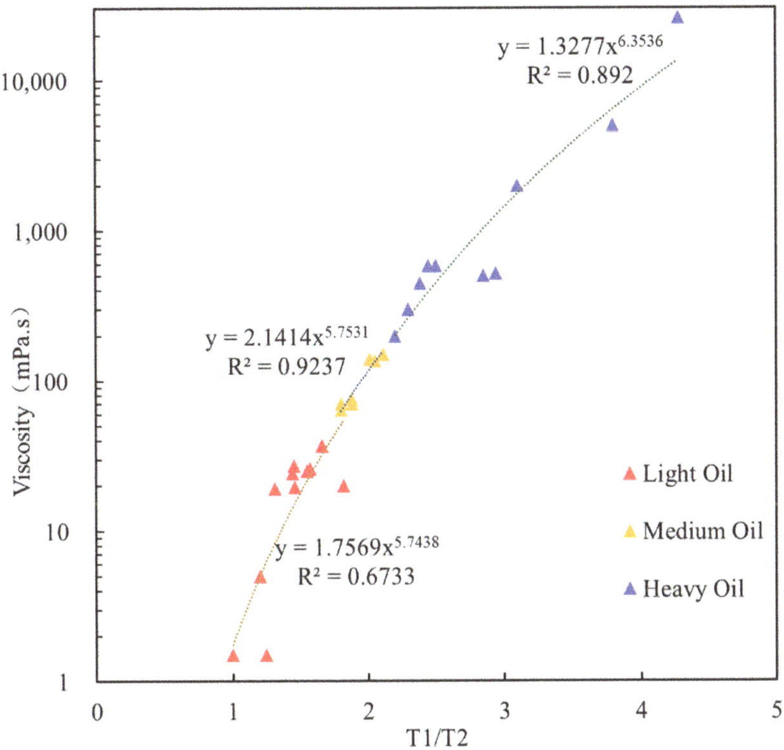

Figure 8. Viscosity vs. T_1/T_2 cross plot divided by crude oil viscosity.

For light oils, the relationship between viscosity and two-dimensional NMR is given by Formula (3):

$$\mu = 1.7577 * (\frac{T_1}{T_2})^{5.7428} \tag{3}$$

For medium oils, the relationship is described by Formula (4):

$$\mu = 2.1414 * (\frac{T_1}{T_2})^{5.7531} \tag{4}$$

For heavy oils, the relationship is given by Formula (5):

$$\mu = 1.3277 * (\frac{T_1}{T_2})^{6.3536} \tag{5}$$

The main parameters in Equations (1)–(5) have the following meanings: u represents the viscosity of crude oil; T_1 and T_2 represent longitudinal and transverse relaxation peaks.

3.3.5. Classification of Crude Oil Viscosity Levels

The viscosity measurements in this experiment were carried out under standard temperature and pressure conditions (20 °C, 0.101 mPa·s). Based on experimental conditions, the crude oil viscosity levels in the high-porosity-and-permeability sandstone of the North China Oilfield are classified as shown in Table 4 for reference (Table 4).

Table 4. Classification standards for crude oil viscosity levels.

Crude Oil Types	μ (mPa·s)	T_2 (ms)	Viscosity Formulas for Crude Oil
Heavy oil	$\mu \geq 150$ mPa·s	$T_2 < 20$ ms	$\mu = 1.3277 * (\frac{T1}{T2})^{6.3536}$
Medium oil	50 mPa·s < μ < 150 mPa·s	20 ms < T_2 < 40 ms	$\mu = 2.1414 * (\frac{T1}{T2})^{5.7531}$
Light oil	$\mu \leq 50$ mPa·s	$T_2 > 40$ ms	$\mu = 1.7577 * (\frac{T1}{T2})^{5.7428}$

This classification helps to understand the distribution of different types of crude oil in the field and provides a basis for further exploration and production strategies.

4. Applied Research

The BQ formation is a group of oil exploration wells in the Huabei oil field region. Figure 9 shows the comprehensive interpretation of the logging data from layer 21 of East Section 3 in BQ-20 well (Figure 9), which shows that the acoustic transit time is 293.0 μs/m, the resistivity is 9.7 Ω-m, the density is 2.19 g/cm^3, the porosity is 25.2%, and the permeability is 657.8 mD. These characteristics classify it as a typical medium-to-high-porosity-and-permeability sandstone reservoir, conventionally interpreted as an oil-bearing formation.

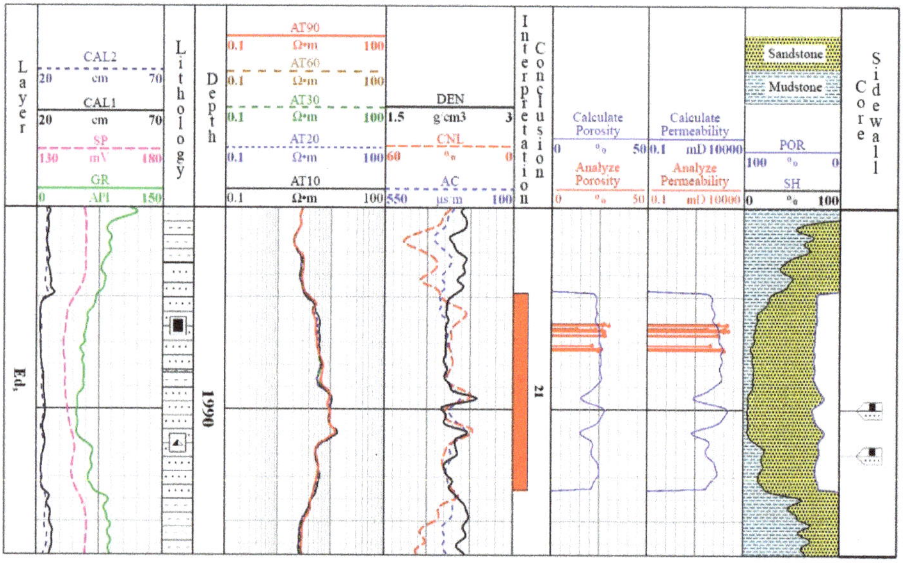

Figure 9. Comprehensive log interpretation diagram for the 21st layer of the East Third Section in Well BQ-20.

To validate the reliability of the crude oil viscosity discrimination chart and Nuclear Magnetic Resonance (NMR) viscosity calculation formulae, these tools were applied to practical scenarios in the medium-to-high-porosity sandstone reservoirs of the North China Oilfield. Figure 10 shows that the results of the crude oil viscosity type identified by the viscosity discrimination chart are consistent with the known viscosity types (Figure 10).

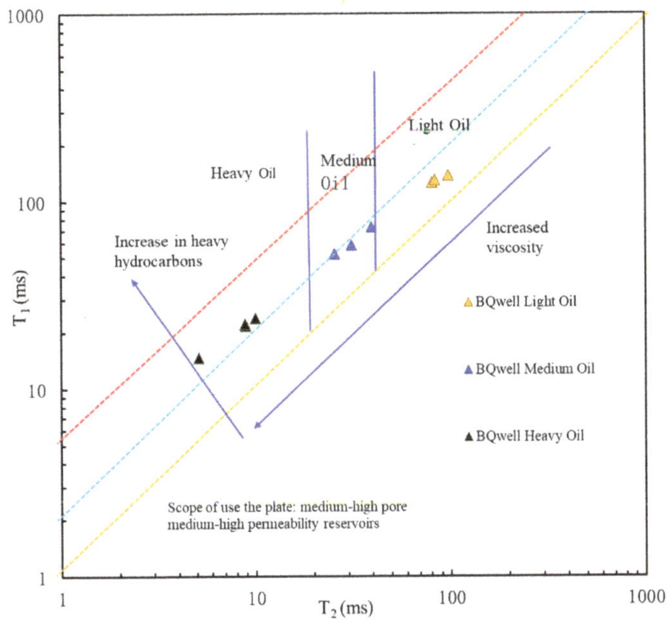

Figure 10. Discrimination chart for crude oil viscosity.

The TPI is determined by a rock pyrolyzer and represents the ratio of the total amount of hydrocarbons produced by pyrolysis from an oil-bearing rock sample under specified conditions to the total organic carbon content. It is also referred to as the oil quality factor. This parameter demonstrates a negative correlation with viscosity and is useful in assessing the viscosity of crude oil. According to the crude oil viscosity discrimination chart and the corresponding NMR viscosity calculation formula, the NMR-calculated crude oil viscosity is negatively correlated with the Total Petroleum Index (TPI), which is consistent with the actual viscosity trend relative to the TPI. The analysis of Table 5, which compares the geochemical pyrolysis and NMR-calculated viscosity for the BQ well group (Table 5), and Table 6, which presents the viscosity error analysis for the same group (Table 6), shows that the NMR-calculated crude oil viscosity is in close agreement with the measured viscosity for the light and medium crudes. Specifically, the average absolute error is 3.36, with an average relative error of 0.14, for light crude oil and 5.07, with an average relative error of 0.04, for medium crude oil, indicating high measurement accuracy.

Table 5. Comparison of geochemical pyrolysis and NMR-calculated viscosity for BQ well group.

Layer Number/Well	Midpoint of Layer (m)	TPI	Measured Viscosity (mPa·s)	Crude Oil Type	Visual Identification	T_1/T_2	NMR Viscosity (mPa·s)
35/BQ-25	2193.6	0.52	25.99	Light oil	Light oil	1.57	23.48
36/BQ-25	2197.2	0.59	16.13	Light oil	Light oil	1.41	12.79
37/BQ-25	2198	0.57	27.35	Light oil	Light oil	1.55	22.12
14/BQ-10	1893.6	0.47	70	Medium oil	Medium oil	1.80	63.08
15/BQ-10	1897.2	0.44	75.6	Medium oil	Medium oil	1.87	80.07
16/BQ-10	1898	0.39	148	Medium oil	Medium oil	2.05	134.25
21/BQ-20	1989	0.31	502.3	Heavy oil	Heavy oil	2.44	390.7
22/BQ-20	1992	0.32	445	Heavy oil	Heavy oil	2.38	331.74
23/BQ-20	1999	0.32	581	Heavy oil	Heavy oil	2.50	450.11
24/BQ-20	2001	0.24	1268	Heavy oil	Heavy oil	2.87	1084.49

Table 6. Viscosity error analysis for BQ well group.

Layer Number/Well	Midpoint of Layer (m)	Crude Oil Type	Measured Viscosity (mPa·s)	NMR Viscosity (mPa·s)	Absolute Error	Relative Error
35/BQ-25	2193.6	Light oil	25.99	23.48	2.51	0.10
36/BQ-25	2197.2	Light oil	16.13	12.79	2.34	0.15
37/BQ-25	2198	Light oil	27.35	22.12	4.23	0.16
14/BQ-10	1893.6	Medium oil	70	63.08	6.92	0.10
15/BQ-10	1897.2	Medium oil	75.6	80.07	−4.47	−0.06
16/BQ-10	1898	Medium oil	148	134.25	13.75	0.09
21/BQ-20	1989	Heavy oil	502.3	390.7	111.60	0.22
22/BQ-20	1992	Heavy oil	445	331.74	113.26	0.25
23/BQ-20	1999	Heavy oil	581	450.11	130.89	0.23
24/BQ-20	2001	Heavy oil	1268	1084.49	183.51	0.14

However, for heavy crudes, the NMR viscosity calculation formula shows a smaller but noticeable error compared to the measured viscosity, with an average absolute error of 134.32 and an average relative error of 0.21. This lower measurement accuracy is attributed to the high concentration of heavy hydrocarbon components and wax content in heavy crude oil. In addition, the two-dimensional NMR spectrum is less sensitive to T_1/T_2 variations, which further affects the accuracy of the formula calculation.

5. Conclusions

This study addresses the challenge of limited reservoir data for early-stage crude oil viscosity determination in oil field development. Utilizing nuclear magnetic resonance (NMR) logging technology, a novel method using T_1 and T_2 for crude oil viscosity calculation is proposed. Using the BQ well group in the North China Oilfield as a case study, a comprehensive multidimensional analysis of NMR experimental data was conducted, leading to the development of a robust crude oil viscosity calculation model. The key findings include the following:

(1) As crude oil viscosity increases, there is a tendency for the T_1 and T_2 measurements of laboratory crude oil samples, artificially saturated rock samples containing crude oil, natural rock samples, and full-diameter-core measurements obtained at the well site to decrease, with a corresponding shift in the oil peak positions in the 2D NMR T_1–T_2 spectrum.

(2) Components such as resins and asphaltenes have a significant influence on crude oil viscosity, which is clearly evident in the 1D T_2 and 2D T_1–T_2 spectra, where increasing hydrocarbon content correlates with a marked increase in the T_1/T_2 ratios.

(3) Through an analysis of the experimental data, a discriminant chart for crude oil properties in high-porosity-and-permeability sandstone reservoirs in the North China Oilfield was developed, and specific relationships between crude oil viscosity and T_1–T_2 were fitted. The comparison with the actual well data confirmed the applicability of the crude oil property discriminant chart and the NMR viscosity calculation formula.

(4) A crude oil viscosity calculation method using 2D NMR was proposed as an alternative to pyrolysis for non-destructive assessment, thereby reducing development costs.

In summary, the results of this study are applicable to low-resistivity, high-porosity, and high-permeability water-bearing reservoirs and provide critical technical support for the characterization of fluid properties and production potential in heavy oil reservoirs.

Author Contributions: Conceptualization, W.Z. and W.C.; methodology, J.S.; software, W.Y. (Weigao Yu); validation, S.W. and W.Y. (Weigao Yu); formal analysis, S.L.; investigation, W.Z.; resources, W.Z.; data curation, W.Y. (Wenkai Yang); writing—original draft preparation, W.Y. (Wenkai Yang); writing—review and editing, W.Y. (Wenkai Yang); visualization, H.D.; supervision, J.S.; project administration, W.Z.; funding acquisition, W.Z. All authors have read and agreed to the published version of the manuscript.

Funding: This research received no external funding.

Data Availability Statement: The original contributions presented in this study are included in the article. Further inquiries can be directed to the corresponding author.

Conflicts of Interest: Authors Wei Zhang, Li Si, Shaqoqing Wang, Wenyuan Cai, Weigao Yu, Hongxia Dai were employed by the company PetroChina Logging Co., Ltd., North China Branch. The remaining authors declare that the research was conducted in the absence of any commercial or financial relationships that could be construed as a potential conflict of interest.

ʀeferences

. Liu, Z.; Zou, L.; Qiao, J.; Shi, C.; Liu, X. The influence of the viscosity of crude oil on liquid-dynamic noise characteristics in the volute shell of oil transfer pump. *Measurement* **2022**, *197*, 111285. [CrossRef]
. Sekar, M.; Ramesh, P.T.; Palanivelu, E. Combined Analysis of Heavy Crude Oil Viscosity and Stress Acting on the Buried Oil Pipelines. *J. Pipeline Syst. Eng. Pract.* **2021**, *12*, 04020059. [CrossRef]
. Al-Marhoun, M.; Nizamuddin, S.; Raheem, A.A.; Ali, S.S.; Muhammadain, A. Prediction of crude oil viscosity curve using artificial intelligence techniques. *J. Pet. Sci. Eng.* **2012**, *86–87*, 111–117. [CrossRef]
. Al-Maamari, R.S.; Houache, O.; Abdul-Wahab, S.A. New Correlating Parameter for the Viscosity of Heavy Crude Oils. *Energy Fuels* **2006**, *20*, 2586–2592. [CrossRef]
. Akkurt, R.; Vinegar, H.J.; Tutunjian, P.N.; Guillory, A.J. NMR logging of natural gas reservoirs. *Log. Anal.* **1996**, *37*, 33–41.
. Mardon, D.; Prammer, M.G.; Coates, G.R. Characterization of light hydrocarbon reservoirs by gradient-NMR well logging. *Magn. Reson. Imaging* **1996**, *14*, 769–777. [CrossRef]
. Logan, W.D.; Horkowitz, J.P.; Laronga, R.; Cromwell, D.W. Practical application of NMR logging in carbonate reservoirs. *SPE Reserv. Eval. Eng.* **1998**, *1*, 438–448. [CrossRef]
. Freedman, R. Advances in NMR Logging. *J. Pet. Technol.* **2006**, *58*, 60–66. [CrossRef]
. Zittel, R.J.; Beliveau, D.; O'Sullivan, T.; Mohanty, R.; Miles, J. Reservoir Crude Oil Viscosity Estimation From Wireline NMR Measurements—Rajasthan, India. *SPE Res. Eval. Eng.* **2008**, *11*, 545–553. [CrossRef]
0. Xie, R.; Xiao, L. Advanced fluid-typing methods for NMR logging. *Pet. Sci.* **2011**, *8*, 163–169. [CrossRef]
1. Gruber, F.K.; Venkataramanan, L.; Habashy, T.M.; Freed, D.E. Estimation of petrophysical and fluid properties using integral transforms in nuclear magnetic resonance. *J. Magn. Reson.* **2013**, *228*, 104–115. [CrossRef] [PubMed]
2. Minagawa, H.; Nishikawa, Y.; Ikeda, I.; Miyazaki, K.; Takahara, N.; Sakamoto, Y.; Komai, T.; Narita, H. Characterization of sand sediment by pore size distribution and permeability using proton nuclear magnetic resonance measurement. *J. Geophys. Res. Solid Earth* **2008**, *113*, B07210. [CrossRef]
3. Tan, M.; Zou, Y.; Zhou, C. A new inversion method for (T_2, D) 2D NMR logging and fluid typing. *Comput. Geosci.* **2013**, *51*, 366–380. [CrossRef]
4. Bauer, K.; Kulenkampff, J.; Henninges, J.; Spangenberg, E. Lithological control on gas hydrate saturation as revealed by signal classification of NMR logging data. *J. Geophys. Res. Solid Earth* **2015**, *120*, 6001–6017. [CrossRef]
5. Xie, R.; Xiao, L. The (T_2, D) NMR Logging Method For Fluids Characterization. *Chin. J. Geophys.* **2009**, *52*, 1092–1100. [CrossRef]
6. He, Z.; Ni, J. Measurement results and study of NMR and viscosity experiments on thick oil samples. *J. Oil Gas (J. Jianghan Pet. Inst.)* **2006**, *28*, 82–84.
7. Kadkhodaie, A.; Rezaee, R.; Kadkhodaie, R. An effective approach to generate drainage representative capillary pressure and relative permeability curves in the framework of reservoir electrofacies. *J. Pet. Sci. Eng.* **2019**, *176*, 1082–1094. [CrossRef]
8. Li, G.; Liu, J.; Zhang, G.; Liu, S. Application of nuclear magnetic resonance logging technology in the evaluation of thick oil reservoirs in the Bohai Sea. *J. Oil Gas* **2009**, *31*, 64–67.
9. Tan, M.; Wu, H.; Zhang, J.; Wang, K.; Mao, K.; Li, B.; Li, C. Influencing mechanics and correction method of nuclear magnetic resonance measurement in igneous rocks reservoir. *J. Pet. Sci. Eng.* **2022**, *208*, 109648. [CrossRef]
:0. Dlubac, K.; Knight, R.; Song, Y.; Bachman, N.; Grau, B.; Cannia, J.; Williams, J. Use of NMR logging to obtain estimates of hydraulic conductivity in the High Plains aquifer, Nebraska, USA. *Water Resour. Res.* **2013**, *49*, 1871–1886. [CrossRef]
:1. Zhou, S.; Deng, Y.; Wang, H.; Pan, Z.; Yan, D. Connectivity and permeability of Zhuhai tight sandstone heterogeneous reservoirs from western Pearl River Mouth Basin (China) by nuclear magnetic resonance. *Appl. Geochem.* **2022**, *143*, 105381. [CrossRef]
:2. Maliva, R.G.; Clayton, E.A.; Missimer, T.M. Application of advanced borehole geophysical logging to managed aquifer recharge investigations. *Hydrogeol. J.* **2009**, *17*, 1547–1556. [CrossRef]
:3. Plastino, A.; Gonçalves, E.C.; da Silva, P.N.; Carneiro, G.; Azeredo, R.B.d.V. Combining classification and regression for improving permeability estimations from 1H NMR relaxation data. *J. Appl. Geophys.* **2017**, *146*, 95–102. [CrossRef]
:4. Liu, Y.; Yao, Y.; Liu, D.; Zheng, S.; Sun, G.; Chang, Y. Shale pore size classification: An NMR fluid typing method. *Mar. Pet. Geol.* **2018**, *96*, 591–601. [CrossRef]
:5. Hu, F.; Zhou, C.; Li, C.; Xu, H.; Zhou, F.; Si, Z. Water spectrum method of NMR logging for identifying fluids. *Pet. Explor. Dev.* **2016**, *43*, 268–276. [CrossRef]
:6. Muhammad, A.; Azeredo, R.B.d.V. 1H NMR spectroscopy and low-field relaxometry for predicting viscosity and API gravity of Brazilian crude oils—A comparative study. *Fuel* **2014**, *130*, 126–134. [CrossRef]

27. Yan, W.; Sun, J.; Cheng, Z.; Li, J.; Sun, Y.; Shao, W.; Shao, Y. Petrophysical characterization of tight oil formations using 1D and 2D NMR. *Fuel* **2017**, *206*, 89–98. [CrossRef]
28. Yan, W.; Sun, J.; Golsanami, N.; Li, M.; Cui, L.; Dong, H.; Sun, Y. Evaluation of wettabilities and pores in tight oil reservoirs by a new experimental design. *Fuel* **2019**, *252*, 272–280. [CrossRef]
29. Shi, Y.; Cai, W.; Liu, G.; Yuan, C.; Zhang, W.; Chuang, S.; Li, S.; Wang, S. Characteristics of two-dimensional nuclear magnetic resonance (NMR) mapping and evaluation method for pore fluids in shale oil reservoirs with full-diameter cores. *China Oil Explor.* **2023**, *28*, 132–144.

energies

MDPI

Article

Productivity Model for Multi-Fractured Horizontal Wells with Complex Fracture Networks in Shale Oil Reservoirs Considering Fluid Desorption and Two-Phase Behavior

Xin Liu, Ping Guo *, Junjie Ren, Zhouhua Wang and Hanmin Tu

State Key Laboratory of Oil and Gas Reservoir Geology and Exploitation, Southwest Petroleum University, Chengdu 610500, China; 202221000804@stu.swpu.edu.cn (X.L.); renjunjie1900@126.com (J.R.); wangzhouhua@126.com (Z.W.); t.hanmin@swpu.edu.cn (H.T.)

* Correspondence: guopingswpi@vip.sina.com; Tel.: +86-18628144040

Abstract: Shale oil reservoirs are characterized by extremely low porosity and permeability, necessitating the utilization of multi-fractured horizontal wells (MFHWs) for their development. Additionally, the complex phase behavior and desorption effect of two-phase fluids make the fluid flow characteristics of shale oil reservoirs exceptionally intricate. However, there are no productivity models for MFHWs in shale oil reservoirs that incorporate the complex hydraulically fractured networks, the oil–gas desorption effect, and the phase change of oil and gas. In this study, we propose a novel productivity model for MFHWs in shale oil reservoirs that incorporates these complex factors. The conformal transformation, fractal theory, and pressure superposition principle are used to establish and solve the proposed model. The proposed model has been validated by comparing its predicted results with the field data and numerical simulation results. A detailed analysis is conducted on the factors that influence the productivity of shale oil wells. It is found that the phase behavior results in a significant 33% reduction in well productivity, while the fluid desorption leads to a significant 75% increase in well productivity. In summary, the proposed model has demonstrated promising practical applicability in predicting the productivity of MFHWs in shale oil reservoirs.

Keywords: productivity model; shale oil reservoir; adsorption and desorption; phase behavior; tree-shaped fracture; multi-stage hydraulic fracture; complex fracture network; low-permeability reservoir

1. Introduction

In recent years, with the depletion of conventional resources, unconventional energy has been occupying an increasingly important position in the world's energy [1]. With the breakthrough in hydraulic fracturing technology, the development of shale oil and gas has become the focus of research in global energy development [2–6].

Shale oil reservoirs are distinguished by their extremely low porosity and permeability, with the majority of shale oil and gas being stored through adsorption mechanisms. Studies have revealed that the proportion of movable hydrocarbons in some shale oil reservoirs does not exceed 30% [7], which poses a significant obstacle to the development of shale oil resources. The use of multi-fractured horizontal wells (MFHWs) has been proven to be an effective method for increasing drainage area and mitigating wellbore contamination in the development of shale oil and gas reservoirs, leading to a substantial increase in production [8,9]. Therefore, there is a growing emphasis among researchers on the application of MFHWs in low- and ultra-low-permeability reservoirs.

Since the early 1970s, the configuration of wells in the petroleum industry has evolved through three primary stages: transitioning from vertical wells to fractured vertical wells, then progressing from fractured vertical wells to horizontal wells, and ultimately advancing from horizontal wells to MFHWs [10]. Wu et al. [11,12] transformed three-dimensional anisotropic reservoirs into equivalent isotropic reservoirs based on tensor theory and established the productivity equations of vertical fracture wells and multi-branch horizontal wells in three-dimensional anisotropic reservoirs by considering the coupling of natural fractures and artificial fractures. Chen et al. [13] proposed a comprehensive steady-state productivity equation for multi-fractured vertical wells operating under constant pressure drop conditions, taking into account stress-sensitive permeability, threshold pressure gradient, asymmetry in fracture angle, length, and conductivity. Wei et al. [14] developed a productivity equation for fractured vertical wells in gas reservoirs under stable seepage conditions, considering the influence of water saturation on various seepage mechanisms, and conducted an analysis of the influencing factors. In recent years, the development of a productivity model for MFHWs has emerged as a research hotspot. Soleimani et al. [15] used the method of 'process conversion flowing pressure correction' to convert the modified backpressure test process into an isochronous test, which was used to determine the productivity of horizontal wells in anisotropic tight gas formations. Lou et al. [16] developed a productivity equation for hydraulic fracturing wells that incorporates the desorption and diffusion of coalbed methane, based on the Langmuir equation and Fick's law. Clarkson and Williams-Kovacs [17] employed a support vector machine based on statistical theory and kernel function to develop a productivity model for MFHWs. Song et al. [18] derived a productivity model for MFHWs in water-bearing tight gas reservoirs, taking into account the impact of the threshold pressure gradient. Qi and Zhu [19] developed a productivity equation for MFHWs in shale gas reservoirs, taking into account diffusion, slippage, desorption, and absorption phenomena. Jiang et al. [20] developed a productivity model for MFHWs in shale oil reservoirs, taking into account the starting pressure gradient and stress sensitivity. However, the model did not consider the adsorption and desorption processes of shale oil, which are important aspects to consider when understanding production behavior in such reservoirs. Some scholars [21–23] proposed various productivity models for MFHWs in different reservoirs. Although some scholars have proposed non-steady-state productivity models for MFHWs in shale oil reservoirs and predicted the transient production rate of MFHWs [24–27], there is a lack of research on the steady-state productivity model for MFHWs in shale oil reservoirs.

Accurate production forecasting for MFHWs is crucial for the efficient development of shale oil reservoirs [28]. The productivity of shale low-permeability reservoirs is affected by reservoir transformation such as fracturing or drilling [29,30]. Among the influencing factors of MFHWs in shale oil reservoirs, hydraulic fracturing can increase the fluid discharge area and permeability of the matrix, thereby enhancing the well productivity. However, during pressure drop, shale oil will desorb in the reservoir. Additionally, fluid phase change during the seepage process will significantly impact the well productivity. Therefore, it is highly challenging to comprehensively consider hydraulic fracture characteristics, shale oil–gas desorption, and fluid phase change when predicting the productivity of MFHWs.

Some researchers [31–34] verified that the desorption characteristics of shale gas can be effectively described by the Langmuir isotherm theory, which was supported by experimental data. In the case of shale oil, a number of scholars [35–37] proposed a modified matrix permeability model to account for shale oil desorption. Li et al. [38] developed a comprehensive chemical reaction model that accounts for the interplay between shale oil desorption and fluid phase behavior. The simulation results show strong agreement with field data. However, accurately predicting well productivity remains challenging due to the complex geological conditions of shale oil reservoirs, variations in fracture parameters, and changes in fluid phase. Additionally, existing models do not consider the phase transition characteristics of shale oil and rely solely on single-phase oil or gas in simulation calculations [39]. As the reservoir pressure decreases and the formation pressure falls below the bubble point, dissolved gas is released, leading to changes in the fluid composition and migration [40]. Liu et al. [41] proposed a new equation of state suitable for confined fluids, which demonstrated improved matching with type I and IV adsorption isotherms. Therefore, it is crucial to consider the phase behavior of shale oil.

Alharthy et al. [42] found that the phase behavior has a significant impact on the initial oil production. Researchers have observed, through hydraulic fracturing model testing and numerical studies of the fracture network, that hydraulic fractures in fracture rock layers exhibit significant interruption, distortion, and bifurcation due to lithology and stress influences [43,44]. This leads to the development of a complex fracture network system with tree-like fractal characteristics (as shown in Figure 1). The formation fluid flows from the matrix to the tree-shaped fractures and subsequently from the fracture to the wellbore. These fractures consist of numerous branch networks, significantly enhancing the complexity of the pore medium and altering fluid seepage characteristics. In contrast to traditional productivity models for MFHWs, this fractal-like tree-shaped fracture model is more appropriate for describing fluid seepage in real reservoirs with complex fractures.

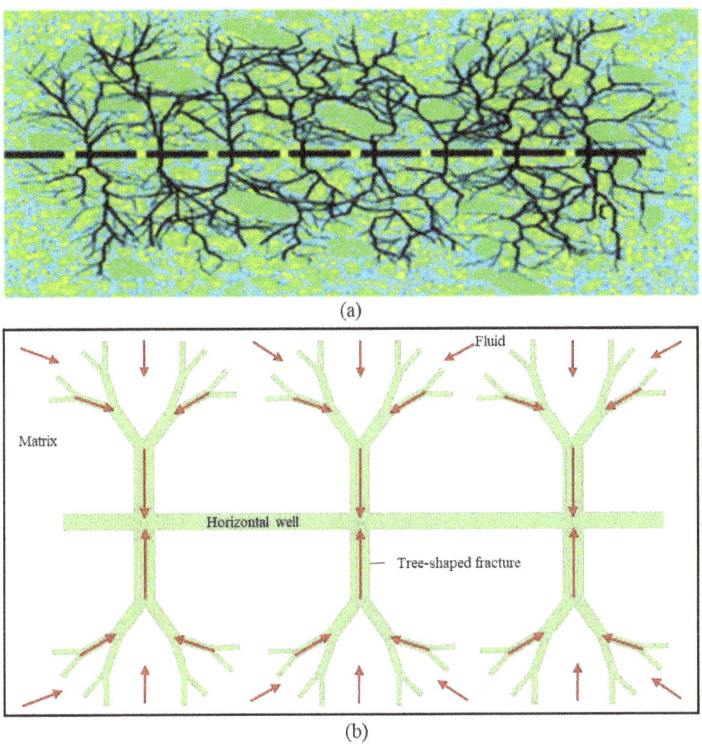

Figure 1. Horizontal well with tree-shaped fracture networks (**a**) Artificial fracture propagation law, (**b**) Tree-shaped fracture network model.

To the best of our knowledge, there are currently no productivity models for MFHWs in shale oil reservoirs that incorporate the complex, tree-shaped, hydraulically fractured networks, the oil–gas desorption effect, and the phase change of oil and gas. In this work, we have developed a novel productivity model for MFHWs in shale oil reservoirs that incorporates these complex factors and have investigated the effects of these factors on well productivity.

2. Mathematical Models

2.1. Model Assumptions

In order to develop a mathematical model for predicting the productivity of MFHWs, taking into account shale oil and gas desorption, phase change, and tree-shaped fracture characteristics, the following assumptions are made:

(1) The shale oil reservoir is a finite, vertically sealed formation with a thickness of h, and the reservoir exhibits homogeneity and isotropy.
(2) The horizontal well, with a length of L, is centrally situated within the reservoir and runs parallel to both the upper and lower boundaries of the reservoir.
(3) The N hydraulic fractures, each with a half-length of x_f, fully penetrate the reservoir while maintaining uniform spacing between them.
(4) The fluid flows from the formation into the fracture, and then it flows from the fracture into the wellbore. The production of the MFHW is the sum of the production from each fracture.
(5) The fluid flow in shale oil reservoirs is considered to be isothermal two-phase flow, governed by Darcy's law.

2.2. *Seepage Model for Oil and Gas Flows in Shale Matrix*

2.2.1. Model Establishment

Based on the principles of material balance [45], the steady-state continuity equation for shale oil and gas, taking into account desorption, is formulated as [46]

$$\nabla(\rho v) + q_d = 0, \tag{1}$$

where ρ is the density of shale oil or shale gas, $\mathrm{kg/m^3}$; q_d is the desorption amount of shale oil and gas desorbed from the matrix, $\mathrm{kg/(m^3 \cdot s)}$; and v is the velocity of shale oil or shale gas, m/s.

Based on the Langmuir isothermal adsorption theory and Fick's law of diffusion, the desorption amount of shale gas in Equation (1) can be obtained by [46]

$$q_{dg} = -GD_mF_s(c_m - c(p)) = -GD_mF_s\left(\frac{\rho_s M_g}{V_{std}} \cdot \frac{V_L p_e}{p_L + p_e} - \frac{\rho_s M_g}{V_{std}} \cdot \frac{V_L p}{p_L + p}\right), \tag{2}$$

where q_{dg} is the desorption amount of shale gas, $\mathrm{kg/(m^3 \cdot s)}$; G is the geometric factor, dimensionless; D_m is the diffusion coefficient, $\mathrm{m^2/s}$; F_s is the shape factor of the matrix block, $1/\mathrm{m^2}$; c_m is the adsorption concentration in the matrix block under the initial reservoir pressure p_e, $\mathrm{kg/m^3}$; $c(p)$ is the adsorption concentration in the matrix block under the reservoir pressure p, $\mathrm{kg/m^3}$; ρ_s is the density of the shale matrix, $\mathrm{kg/m^3}$; M_g is the molar mass of the gas, kg/mol; V_{std} is the molar volume of the gas under standard conditions, $\mathrm{m^3/mol}$; V_L is the Langmuir volume, $\mathrm{m^3/kg}$; p_L is the Langmuir pressure, MPa; p_e is the initial reservoir pressure, MPa; and p is the reservoir pressure, MPa.

The desorption amount of shale oil in Equation (1) can be determined using the following equation [38]:

$$q_{do} = -0.01k_{de}TOC\Delta p, \tag{3}$$

where q_{do} is the desorption amount of shale oil, $\mathrm{kg/(m^3 \cdot s)}$; k_{de} is the desorption amount per unit organic carbon content and pressure drop, $\mathrm{kg/(m^3 \cdot s \cdot MPa)}$; TOC is the organic carbon content of shale, %; and Δp is the pressure drop, MPa.

During the process of developing shale oil reservoirs through MFHWs, each fracture induces two-dimensional elliptical flow in the plane, forming conjugate isobaric ellipses and hyperbolic streamline clusters converging with the fracture endpoint as foci. By using the conformal transformation, the seepage region of a single fracture is transformed into a drainage tunnel with a width of π.

The point $Z_1 = x + iy$ in the Z_1-plane can be mapped to the point $Z_2 = u + iw$ in the Z_2-plane by the conformal transformation $Z_2 = f(Z_1)$.

According to Darcy's law, the two-dimensional steady-state continuity Equation (1) in the Z_1-plane is formulated as

$$\nabla(\rho v) + q_\mathrm{d} = \nabla\left(\rho \frac{K}{\mu} \nabla p\right) + q_\mathrm{d} = \rho \frac{K}{\mu} \nabla^2 p + q_\mathrm{d} = 0, \tag{4}$$

where K is the absolute permeability, mD; μ is the fluid viscosity, cp; and ∇^2 is the Laplace operator, which is expressed as

$$\nabla^2 = \frac{\partial^2}{\partial x^2} + \frac{\partial^2}{\partial y^2}. \tag{5}$$

The Laplace operator in the Z_1-plane (i.e., Equation (5)) can be mapped to the Z_2-plane, with specific details provided as follows:

According to the partial derivative chain rule, one can deduce that

$$\frac{\partial}{\partial x} = \frac{\partial u}{\partial x}\frac{\partial}{\partial u} + \frac{\partial w}{\partial x}\frac{\partial}{\partial w}, \tag{6}$$

$$\frac{\partial}{\partial y} = \frac{\partial u}{\partial y}\frac{\partial}{\partial u} + \frac{\partial w}{\partial y}\frac{\partial}{\partial w}, \tag{7}$$

$$\frac{\partial^2}{\partial x^2} = \frac{\partial^2 u}{\partial x^2}\frac{\partial}{\partial u} + \frac{\partial^2 w}{\partial x^2}\frac{\partial}{\partial w} + \left(\frac{\partial u}{\partial x}\right)^2\frac{\partial^2}{\partial u^2} + \left(\frac{\partial w}{\partial x}\right)^2\frac{\partial^2}{\partial w^2} + 2\frac{\partial u}{\partial x}\frac{\partial w}{\partial x}\frac{\partial^2}{\partial u \partial w}, \tag{8}$$

$$\frac{\partial^2}{\partial y^2} = \frac{\partial^2 u}{\partial y^2}\frac{\partial}{\partial u} + \frac{\partial^2 w}{\partial y^2}\frac{\partial}{\partial w} + \left(\frac{\partial u}{\partial y}\right)^2\frac{\partial^2}{\partial u^2} + \left(\frac{\partial w}{\partial y}\right)^2\frac{\partial^2}{\partial w^2} + 2\frac{\partial u}{\partial y}\frac{\partial w}{\partial y}\frac{\partial^2}{\partial u \partial w}. \tag{9}$$

Substituting Equations (8) and (9) into Equation (5) yields that

$$\begin{aligned}\nabla^2 &= \frac{\partial^2}{\partial x^2} + \frac{\partial^2}{\partial y^2} \\ &= \left[\left(\frac{\partial u}{\partial x}\right)^2 + \left(\frac{\partial u}{\partial y}\right)^2\right]\frac{\partial^2}{\partial u^2} + \left[\left(\frac{\partial w}{\partial x}\right)^2 + \left(\frac{\partial w}{\partial y}\right)^2\right]\frac{\partial^2}{\partial w^2} + \left(\frac{\partial^2 u}{\partial x^2} + \frac{\partial^2 u}{\partial y^2}\right)\frac{\partial}{\partial u} \\ &+ \left(\frac{\partial^2 w}{\partial x^2} + \frac{\partial^2 w}{\partial y^2}\right)\frac{\partial}{\partial w} + 2\left(\frac{\partial u}{\partial x}\frac{\partial w}{\partial x} + \frac{\partial u}{\partial y}\frac{\partial w}{\partial y}\right)\frac{\partial^2}{\partial u \partial w}\end{aligned}. \tag{10}$$

According to the theory of complex variable functions, the conformal transformation must satisfy the Cauchy–Riemann condition:

$$\frac{\partial u}{\partial x} = \frac{\partial w}{\partial y} \quad , \quad \frac{\partial u}{\partial y} = -\frac{\partial w}{\partial x}. \tag{11}$$

According to the Cauchy–Riemann condition (11), one can derive that

$$\frac{\partial^2 u}{\partial x^2} + \frac{\partial^2 u}{\partial y^2} = 0, \tag{12}$$

$$\frac{\partial^2 w}{\partial x^2} + \frac{\partial^2 w}{\partial y^2} = 0, \tag{13}$$

$$|f'(Z_1)|^2 = \left(\frac{\partial u}{\partial x}\right)^2 + \left(\frac{\partial u}{\partial y}\right)^2 = \left(\frac{\partial w}{\partial x}\right)^2 + \left(\frac{\partial w}{\partial y}\right)^2. \tag{14}$$

By substituting Equations (11)–(14) into Equation (10), we can obtain that

$$\nabla^2 = \frac{\partial^2}{\partial x^2} + \frac{\partial^2}{\partial y^2} = |f'(Z_1)|^2\left(\frac{\partial^2}{\partial u^2} + \frac{\partial^2}{\partial w^2}\right). \tag{15}$$

By substituting Equation (15) into Equation (4), we derive the two-dimensional steady-state continuity equation in the Z_2-plane:

$$\rho\frac{K}{\mu}|f'(Z_1)|^2\left(\frac{\partial^2 p}{\partial u^2} + \frac{\partial^2 p}{\partial w^2}\right) + q_{\mathrm{d}} = 0. \tag{16}$$

Equation (16) can be rewritten as

$$\rho\frac{K}{\mu}\left(\frac{\partial^2 p}{\partial u^2}+\frac{\partial^2 p}{\partial w^2}\right)+\frac{q_{\mathrm{d}}}{|f'(Z_1)|^2}=0. \tag{17}$$

The difference between the steady-state continuity equation in the Z_1-plane (i.e., Equation (1)) and that in the Z_2-plane (i.e., Equation (17)) lies in the fact that the desorption source term in Equation (17) is multiplied by a coefficient $1/|f'(Z_1)|^2$. The coefficient $1/|f'(Z_1)|^2$ can be determined as follows:

The conformal transformation in Figure 2 for the seepage region of a single fracture is

$$Z_1 = x_f \cosh Z_2. \tag{18}$$

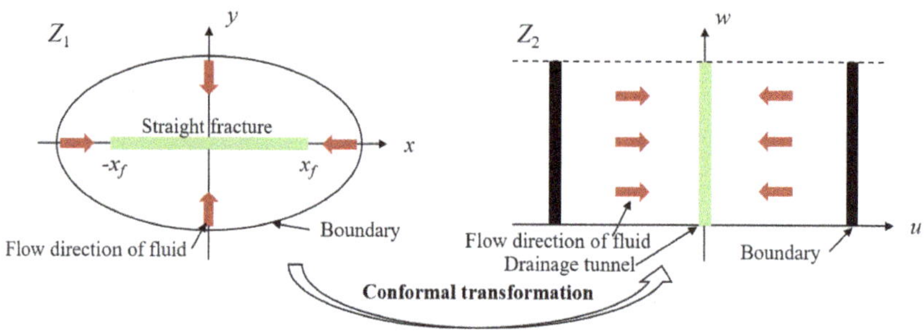

Figure 2. Conformal transformation diagram for the seepage region of a fracture.

The transformed coordinate correspondence can be obtained from Equation (18) as follows:

$$\begin{cases} x = x_f chu \cdot \cos w \\ y = x_f shu \cdot \sin w \end{cases}. \tag{19}$$

According to Equation (19), one can obtain that

$$\frac{x^2}{x_f^2 ch^2 u}+\frac{y^2}{x_f^2 sh^2 u}=1, \tag{20}$$

$$\frac{x^2}{x_f^2 \cos^2 w}-\frac{y^2}{x_f^2 \sin^2 w}=1, \tag{21}$$

where

$$u = arcch\frac{1}{\sqrt{2}}\left[\frac{x_f^2+x^2+y^2}{x_f^2}+\sqrt{\left(\frac{x_f^2+x^2+y^2}{x_f^2}\right)^2-4\frac{x^2}{x_f^2}}\right]^{0.5}, \tag{22}$$

$$w = \arccos\frac{1}{\sqrt{2}}\left[\frac{x_f^2+x^2+y^2}{x_f^2}-\sqrt{\left(\frac{x_f^2+x^2+y^2}{x_f^2}\right)^2-4\frac{x^2}{x_f^2}}\right]^{0.5}. \tag{23}$$

According to Equations (14) and (22), $|f'(Z_1)|^2$ can be obtained by

$$|f'(Z_1)|^2=\left(\frac{\partial u}{\partial x}\right)^2+\left(\frac{\partial u}{\partial y}\right)^2, \tag{24}$$

where

$$\frac{\partial u}{\partial x}=\frac{\left\{-\left[\frac{8x}{x_f^2}-\frac{4x(x^2+x_f^2+y^2)}{x_f^4}\right]/\left\{4\left[\frac{(x^2+x_f^2+y^2)^2}{x_f^4}-\frac{4x^2}{x_f^2}\right]^{0.5}\right\}-\frac{x}{x_f^2}\right\}}{\left\{\begin{array}{l}2\left\{\frac{(x^2+x_f^2+y^2)}{2x_f^2}+\left[\frac{(x^2+x_f^2+y^2)^2}{x_f^4}-\frac{4x^2}{x_f^2}\right]^{0.5}/2\right\}^{0.5}\times\left\{\left\{\frac{(x^2+x_f^2+y^2)}{2x_f^2}+\left[\frac{(x^2+x_f^2+y^2)^2}{x_f^4}-\frac{4x^2}{x_f^2}\right]^{0.5}/2\right\}^{0.5}-1\right\}^{0.5}\\ \times\left\{\left\{\frac{(x^2+x_f^2+y^2)}{2x_f^2}+\left[\frac{(x^2+x_f^2+y^2)^2}{x_f^4}-\frac{4x^2}{x_f^2}\right]^{0.5}/2\right\}^{0.5}+1\right\}^{0.5}\end{array}\right\}}, \tag{25}$$

$$\frac{\partial u}{\partial y}=\frac{\left\{\frac{y}{x_f^2}+\left[y(x^2+x_f^2+y^2)\right]/\left\{x_f^4\left[\frac{(x^2+x_f^2+y^2)^2}{x_f^4}-\frac{4x^2}{x_f^2}\right]^{0.5}\right\}\right\}}{\left\{\begin{array}{l}2\left\{\frac{(x^2+x_f^2+y^2)}{2x_f^2}+\left[\frac{(x^2+x_f^2+y^2)^2}{x_f^4}-\frac{4x^2}{x_f^2}\right]^{0.5}/2\right\}^{0.5}\times\left\{\left\{\frac{(x^2+x_f^2+y^2)}{2x_f^2}+\left[\frac{(x^2+x_f^2+y^2)^2}{x_f^4}-\frac{4x^2}{x_f^2}\right]^{0.5}/2\right\}^{0.5}-1\right\}^{0.5}\\ \times\left\{\left\{\frac{(x^2+x_f^2+y^2)}{2x_f^2}+\left[\frac{(x^2+x_f^2+y^2)^2}{x_f^4}-\frac{4x^2}{x_f^2}\right]^{0.5}/2\right\}^{0.5}+1\right\}^{0.5}\end{array}\right\}}. \tag{26}$$

Considering Equations (22)–(26), we can obtain that

$$|f'(Z_1)|^2=\frac{1}{x_f^2\cdot\left[cosh(u)^2-cos(w)^2\right]}. \tag{27}$$

The oil phase exhibits linear flow in a channel flow along the u direction, which can be described as

$$\frac{\rho_o K K_{ro}}{\mu_o}\left(\frac{d^2p}{du^2}\right)+\frac{q_{do}}{|f'(Z_1)|^2}=0, \tag{28}$$

where ρ_o is the oil phase density, kg/m^3; K_{ro} is the relative permeability of oil phase; and μ_o is the oil viscosity, cp.

When the formation pressure falls below the bubble point, it results in oil–gas two-phase seepage in the reservoir. The continuity equation for the gas phase is

$$\frac{\rho_g K K_{rg}}{\mu_g}\left(\frac{d^2p}{du^2}\right)+\frac{q_{dg}}{|f'(Z_1)|^2}=0, \tag{29}$$

where ρ_g is the gas phase density, kg/m^3; K_{rg} is the relative permeability of gas phase; and μ_g is the gas viscosity, cp.

Adding Equations (28) and (29) results in

$$\frac{K}{du}\left[\left(\frac{\rho_g K_{rg}}{\mu_g}+\frac{\rho_o K_{ro}}{\mu_o}\right)\frac{dp}{du}\right]+\frac{Q_d}{|f'(Z_1)|^2}=0, \tag{30}$$

where

$$Q_d=q_{dg}+q_{do}. \tag{31}$$

Introducing the pseudo-pressure function $\varphi(p)=\int\left(\rho_g K_{rg}/\mu_g+\rho_o K_{ro}/\mu_o\right)\mathrm{d}p$, the steady-state continuity equation in the Z_2-plane for two-phase flows is described as

$$\frac{d^2\varphi}{du^2}+\frac{Q_d}{|f'(Z_1)|^2K}=0. \tag{32}$$

2.2.2. Model Solution

In the Z_2-plane, the mass flow rate of the two-phase fluid extracted from the fracture is

$$q_t=2\pi Kh\frac{d\varphi}{du}, \tag{33}$$

where $q_t = \rho_g q_g + \rho_o q_o$ is the mass flow rate of the two-phase fluid, kg/d; q_g is the volume flow rate of gas, m^3/d; q_o is the volume flow of oil, m^3/d; and h is the thickness of the formation, m.

Solving Equation (32) with boundary condition (33) can produce the pseudo-pressure solution

$$\varphi(p) = \frac{q_t u}{2\pi Kh} + \frac{Q_d}{2K} x_f^2 \left(\sinh u^2 - u^2 \cos w^2\right) + C, \quad (34)$$

where C is a constant.

If the fracture center is located at the coordinate $(0, y_0)$ instead of the coordinate (0,0), Equations (22) and (23) can be rewritten as

$$u = arcch \frac{1}{\sqrt{2}} \left[\frac{x_f^2 + x^2 + (y - y_o)^2}{x_f^2} + \sqrt{\left(\frac{x_f^2 + x^2 + (y - y_o)^2}{x_f^2}\right)^2 - 4\frac{x^2}{x_f^2}} \right]^{0.5}, \quad (35)$$

$$w = \arccos \frac{1}{\sqrt{2}} \left[\frac{x_f^2 + x^2 + (y - y_o)^2}{x_f^2} - \sqrt{\left(\frac{x_f^2 + x^2 + (y - y_o)^2}{x_f^2}\right)^2 - 4\frac{x^2}{x_f^2}} \right]^{0.5}. \quad (36)$$

According to the superposition principle of potential, the pseudo-pressure at the j-th fracture induced by N fractures can be calculated as follows:

$$\varphi\left(0, y_j\right) = \sum_{i=1}^{N} \left[\begin{array}{l} \frac{q_{ti}}{2\pi Kh} \cdot arcch\sqrt{1 + \frac{(y_j - y_i)^2}{x_f^2}} + \frac{Q_d}{2K} x_f^2 \cdot \sinh\left(arcch\sqrt{1 + \frac{(y_j - y_i)^2}{x_f^2}} \right)^2 \\ -\left(arcch\sqrt{1 + \frac{(y_j - y_i)^2}{x_f^2}} \right)^2 \cos\left(\frac{\pi}{2}\right)^2 \end{array} \right] + C. \quad (37)$$

The pseudo-pressure at the outer boundary $(0, r_e)$ is given as

$$\varphi(0, r_e) = \sum_{i=1}^{N} \left[\begin{array}{l} \frac{q_{ti}}{2\pi Kh} \cdot arcch\sqrt{1 + \frac{(r_e - y_i)^2}{x_f^2}} + \frac{Q_d}{2K} x_f^2 \cdot \sinh\left(arcch\sqrt{1 + \frac{(r_e - y_i)^2}{x_f^2}} \right)^2 \\ -\left(arcch\sqrt{1 + \frac{(r_e - y_i)^2}{x_f^2}} \right)^2 \cos\left(\frac{\pi}{2}\right)^2 \end{array} \right] + C. \quad (38)$$

With the assistance of Equation (39), subtracting Equation (37) from Equation (38) results in Equation (40):

$$arcch\sqrt{1 + x^2} = \ln\left(x + \sqrt{1 + x^2}\right), \quad (39)$$

$$\begin{array}{l} \varphi_e - \varphi_{fj} = \frac{1}{2\pi Kh} \sum\limits_{i=1}^{N} q_{ti} \ln \left[\frac{\frac{r_e - y_i}{x_f} + \sqrt{1 + \frac{(r_e - y_i)^2}{x_f^2}}}{\left|\frac{y_j - y_i}{x_f}\right| + \sqrt{1 + \frac{(y_j - y_i)^2}{x_f^2}}} \right] + \frac{Q_d}{2K} x_f^2 \sum\limits_{i=1}^{N} \left\{ \sqrt{\frac{(r_e - y_i)^2}{x_f^2}} \cdot \sqrt{\frac{(r_e - y_i)^2}{x_f^2} + 2} \right. \\ -\sqrt{\frac{(y_i - y_j)^2}{x_f^2}} \cdot \sqrt{\frac{(y_i - y_j)^2}{x_f^2} + 2} - \cos\left(\frac{\pi}{2}\right)^2 \ln \left[\left[\frac{(r_e - y_i)^2}{x_f^2} + 1\right] + \sqrt{\left[\frac{(r_e - y_i)^2}{x_f^2} + 1\right]^2 - 1} \right] \\ \left. + \cos\left(\frac{\pi}{2}\right)^2 \ln \left[\frac{(y_i - y_j)^2}{x_f^2} + 1 + \sqrt{\left[\frac{(y_i - y_j)^2}{x_f^2} + 1\right]^2 - 1} \right] \right\} \end{array}. \quad (40)$$

2.3. Seepage Model for Oil and Gas Flows in Tree-Shaped Fractures

2.3.1. Characteristic Parameters of Tree-Shaped Fractures

Based on fractal theory [47], a tree-shaped fracture network, consisting of a combination of primary fractures and multiple branch fractures as shown in Figure 3, is used to better characterize the multi-fracture network of MFHWs in shale oil reservoirs. Some characteristic parameters, which are listed in Table 1, can be used to describe the tree-shaped fracture.

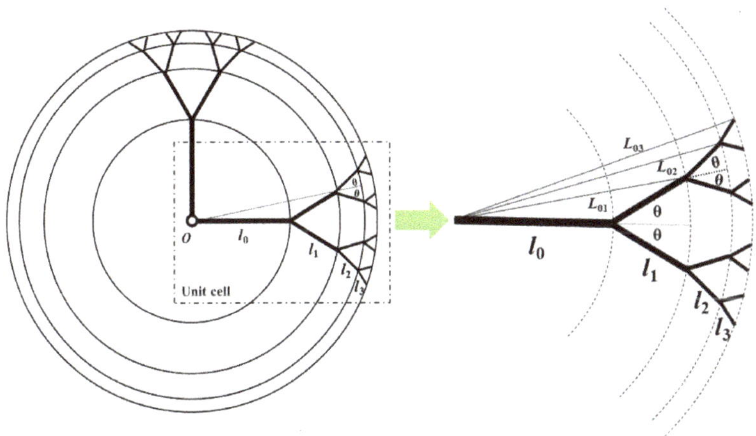

Figure 3. Diagram of tree-shaped fracture network.

Table 1. Characteristic parameters of tree-shaped fractures.

Characteristic Parameter	Symbol	Characteristic Parameter	Symbol
Fracture characteristic length at *i*-level	$L_{0,i}$	Branch angle of crack	θ
Fracture length at *i*-level	l_i	Length ratio of adjacent cracks	a
Fracture width at *i*-level	w_i	Width ratio of adjacent cracks	b
Fracture height at *i*-level	h_i	Height ratio of adjacent cracks	c
Initial fracture length	l_0	Fracture branch grade	n
Initial fracture width	w_0	Fracture branch number	m
Initial fracture height	h_0	Rectangular fracture permeability	k_{slit}
Effective width of tree-shaped crack	w_f	Fracture effective porosity	ϕ_f
Total volume of tree-shaped crack	V_f	Tree-shaped fracture permeability	k_f
Total length of tree-shaped crack	L_f	Correctional fracture permeability	K_f
Equivalent area of tree-shaped crack	A_f		

Based on the triangular cosine theorem, the characteristic length from the tree-shaped fracture inlet at 0 level to the *i*-level fracture outlet is derived as

$$L_{0,i} = \sqrt{L_{0,i-1}^2 + l_i^2 + 2L_{0,i-1} \cdot l_i \cos\theta}. \quad (41)$$

The relationship between the fracture half-length and the characteristic length is described as

$$x_f = L_{0,i}. \quad (42)$$

The total fracture volume can be determined by adding together the individual volumes of fractures across all levels and is calculated assuming a fracture branch number of 2 as follows:

$$V_f = l_0 w_0 h_0 + l_0 w_0 h_0 \cdot 2abc + \cdots + l_0 w_0 h_0 (2abc)^n = l_0 w_0 h_0 \cdot \frac{1-(2abc)^{n+1}}{1-2abc}. \quad (43)$$

The effective width of a tree-shaped crack is

$$w_f = w_0 (2b)^n. \quad (44)$$

The total length of tree-shaped cracks can be obtained by

$$L_f = l_0 + l_0 a + \cdots + l_0 (a)^n = l_0 \cdot \frac{1-a^{n+1}}{1-a}. \quad (45)$$

The equivalent area of tree-shaped cracks is

$$A_f = \frac{V_f}{L_f} = w_0 h_0 \cdot \frac{1-(2abc)^{n+1}}{1-2abc} \cdot l_0 \cdot \frac{1-a}{1-a^{n+1}}. \quad (46)$$

2.3.2. Tree-Shaped Fracture Permeability

If the cross-section of the fracture is assumed to be a rectangle with a width of w, the fracture permeability is

$$k_{slit} = \frac{w^2}{12}. \quad (47)$$

Assuming that the seepage of oil and gas in rectangular fractures conforms to Darcy's law, the seepage equation of each phase is

$$\frac{K_{ro}\rho_o}{\mu_o}\Delta p = \frac{12Q_o}{w^3 h}\Delta l, \quad (48)$$

$$\frac{K_{rg}\rho_g}{\mu_g}\Delta p = \frac{12Q_g}{w^3 h}\Delta l, \quad (49)$$

where Q_o is the mass flow rate of oil phase, kg/d; and Q_g is the mass flow rate of gas phase, kg/d.

The sum of Equations (48) and (49) yields the two-phase seepage equation:

$$\Delta\varphi = \frac{12q_t}{w^3 h}\Delta l, \quad (50)$$

where

$$\Delta\varphi = \left(\frac{K_{rg}\rho_g}{\mu_g} + \frac{K_{ro}\rho_o}{\mu_o}\right)\Delta p, \quad (51)$$

$$q_t = Q_g + Q_o. \quad (52)$$

The seepage equation for the i-level fracture is

$$\Delta\varphi_i = \frac{12q_t}{m_i w_i^3 h_i} l_i. \quad (53)$$

The total pressure drop from the fracture outlet to the bottom of the well for the tree-shaped fracture is

$$\Delta\varphi_f = \frac{12q_t l_o}{w_o^3 h_o} \cdot \frac{1-\left(\frac{a}{2b^3 c}\right)^{n+1}}{1-\frac{a}{2b^3 c}}. \quad (54)$$

A complex, tree-shaped fracture can be transformed into a rectangular fracture with a half-length of $L_{0,i}$, a width of w_f, and a height of h after determining their characteristic parameters using fractal theory (Figure 4).

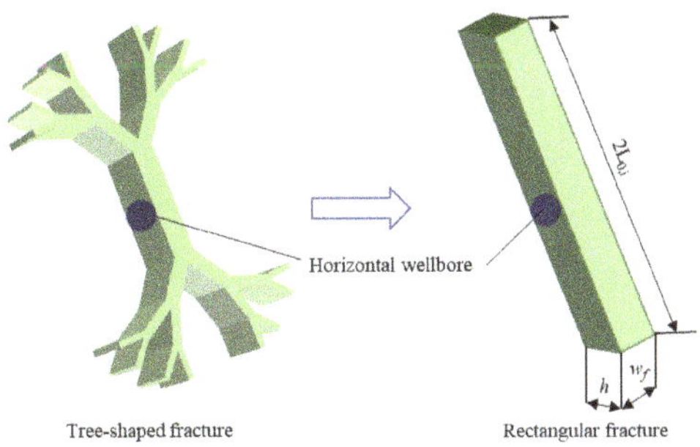

Figure 4. Transformation diagram of tree-shaped fractures and rectangular fractures.

According to Darcy's law, the total mass flow rate of fluid through rectangular fractures is determined by

$$q_t = Ak_f \frac{\Delta \varphi_f}{L_{0,i}}. \tag{55}$$

The effective permeability of the tree-shaped fracture is obtained by combining Equations (41), (46), (54) and (55):

$$k_f = \frac{w_0^2}{12} \cdot \frac{\sqrt{L_{0,n-1}^2 + l_n^2 + 2L_{0,n-1} l_n \cos\theta}}{l_0} \cdot \frac{1 - a^{n+1}}{1 - a} \cdot \frac{1 - 2abc}{1 - (2abc)^{n+1}} \cdot \frac{1 - \frac{a}{2b^3 c}}{1 - \left(\frac{a}{2b^3 c}\right)^{n+1}}. \tag{56}$$

Similarly, the effective permeability of the tree-shaped fracture with a fracture branch number of m is

$$k_f = \frac{w_0^2}{12} \cdot \frac{\sqrt{L_{0,i-1}^2 + l_i^2 + 2L_{0,i-1} l_i \cos\theta}}{l_0} \cdot \frac{1 - a^{n+1}}{1 - a} \cdot \frac{1 - mabc}{1 - (mabc)^{n+1}} \cdot \frac{1 - \frac{a}{mb^3 c}}{1 - \left(\frac{a}{mb^3 c}\right)^{n+1}}. \tag{57}$$

After the hydraulic fracturing operation of the horizontal well, it is imperative to inject sand into the fractures to prevent closure. Assuming a fracture porosity of ϕ_f after sand injection, the effective permeability of the fracture is [39]

$$K_f = k_f \phi_f. \tag{58}$$

2.3.3. Oil–Gas Two-Phase Seepage Model

It is assumed that the fluid flow within the fracture follows Darcy's law, with no consideration for the influence of gravity. The seepage of oil and gas within the fracture can be considered as a plane radial flow with a radius of $h/2$ and a thickness of w_f. The productivity equation for radial flow within the fracture plane is as follows:

$$\varphi\left(p_f\right) - \varphi\left(p_{wf}\right) = \frac{q_t}{2\pi K_f w_f} \cdot \ln \frac{h}{2r_w}, \tag{59}$$

where

$$q_t = \rho_g q_g + \rho_o q_o = \rho_{gsc} q_{gsc} + \rho_{osc} q_{osc}. \tag{60}$$

2.4. Productivity Model for MFHWs

The formation fluid flows from the matrix into the fractures and then flows from the fractures towards the wellbore. The detailed schematic is shown in Figure 5.

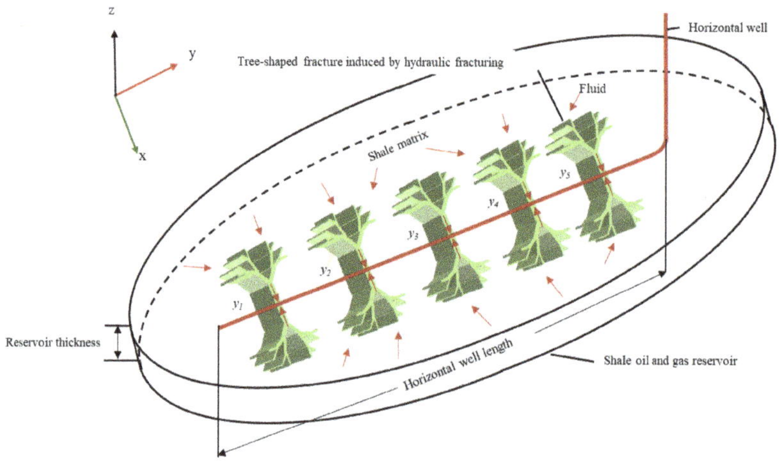

Figure 5. Multiple tree-shaped fractures distributed along a horizontal well.

The total pressure drop from the matrix to the wellbore is equal to the sum of the pressure drop from the matrix to the fracture and that from the fracture to the wellbore. Therefore, adding together Equations (40) and (59) yields the productivity model of the jth fracture:

$$
\begin{aligned}
\varphi(p_e)-\varphi\left(p_{wf}\right) = {} & \frac{1}{2\pi Kh}\sum_{i=1}^{N} q_{ti}\ln\left[\frac{\frac{r_e-y_i}{x_f}+\sqrt{1+\frac{(r_e-y_i)^2}{x_f^2}}}{\left|\frac{y_j-y_i}{x_f}\right|+\sqrt{1+\frac{(y_j-y_i)^2}{x_f^2}}}\right]+\frac{Q_d}{2K}x_f^2\sum_{i=1}^{N}\left\{\sqrt{\frac{(r_e-y_i)^2}{x_f^2}}\cdot\sqrt{\frac{(r_e-y_i)^2}{x_f^2}+2}\right. \\
& -\sqrt{\frac{(y_i-y_j)^2}{x_f^2}}\cdot\sqrt{\frac{(y_i-y_j)^2}{x_f^2}+2}-\cos\left(\frac{\pi}{2}\right)^2\ln\left[\left[\frac{(r_e-y_i)^2}{x_f^2}+1\right]+\sqrt{\left[\frac{(r_e-y_i)^2}{x_f^2}+1\right]^2-1}\right] \\
& \left.+\cos\left(\frac{\pi}{2}\right)^2\ln\left[\frac{(y_i-y_j)^2}{x_f^2}+1+\sqrt{\left[\frac{(y_i-y_j)^2}{x_f^2}+1\right]^2-1}\right]\right\}+\frac{q_{tj}}{2\pi K_f w_f}\cdot\ln\frac{h}{2r_w}
\end{aligned}
\quad (61)
$$

In addition, the total productivity of the MFHW is equal to the sum of the productivity of each fracture:

$$q_t=\sum_{i=1}^{N} q_{ti}. \quad (62)$$

By substituting Equations (2), (3), (44), (58) and (61) into Equation (62), N sets of productivity equations can be obtained, and the total mass productivity of the MFHW can be obtained by solving the simultaneous equations.

2.5. Solution for Two-Phase Pseudo-Pressure

Based on the definition of two-phase pseudo-pressure, calculating this parameter requires determining the relative permeabilities (i.e., K_{ro} and K_{rg}) of oil and gas, as well as establishing the functional relationship between oil–gas physical parameters (i.e., μ_o, μ_g, ρ_o, and ρ_g) and pressure (p). The correlation between oil–gas physical parameters (i.e., μ_o,

μ_g, ρ_o, and ρ_g) and pressure (p) can be determined through pressure–volume–temperature (PVT) fitting using the Peng–Robinson (PR) state equation (i.e., Equation (63)). Additionally, the relationship between gas-phase molar percentage (V_1) and pressure (p) is established through a two-phase flash program.

$$p = \frac{RT}{V-b} - \frac{a\alpha(T)}{V(V+b)+b(V-b)}, \tag{63}$$

where R is the molar gas constant, which is equal to 8.31 $\text{MPa} \cdot \text{cm}^3/(\text{mol} \cdot \text{K})$; a, b are molecular gravitational and repulsive coefficients; V is molecular volume; T is temperature, K; and $\alpha(T)$ is the temperature function.

According to the steady-state seepage theory, the relationship between K_{ro}/K_{rg} and pressure (p) can be obtained by Equation (64). Based on this, the relationship between the relative permeabilities (i.e., K_{ro} and K_{rg}) and pressure can be easily derived with the assistance of relative permeability curves.

$$\frac{K_{ro}}{K_{rg}} = \frac{\rho_g L_1 \mu_o}{\rho_o V_1 \mu_g}, \tag{64}$$

where L_1 is the liquid-phase molar percentage, %; and V_1 is the gas-phase molar percentage, %.

The two-phase pseudo-pressure in Equation (65) can be calculated by the steps outlined in Figure 6. The integral of the pseudo-pressure should be divided into two segments, with the bubble point pressure serving as the dividing point.

$$\varphi(p_e) - \varphi\left(p_{wf}\right) = \int_{p_b}^{p_e} \left(\frac{\rho_o K_{ro}}{\mu_o}\right) dp + \int_{p_{wf}}^{p_b} \left(\frac{\rho_g K_{rg}}{\mu_g} + \frac{\rho_o K_{ro}}{\mu_o}\right) dp, \tag{65}$$

where p_b is the bubble point pressure, MPa.

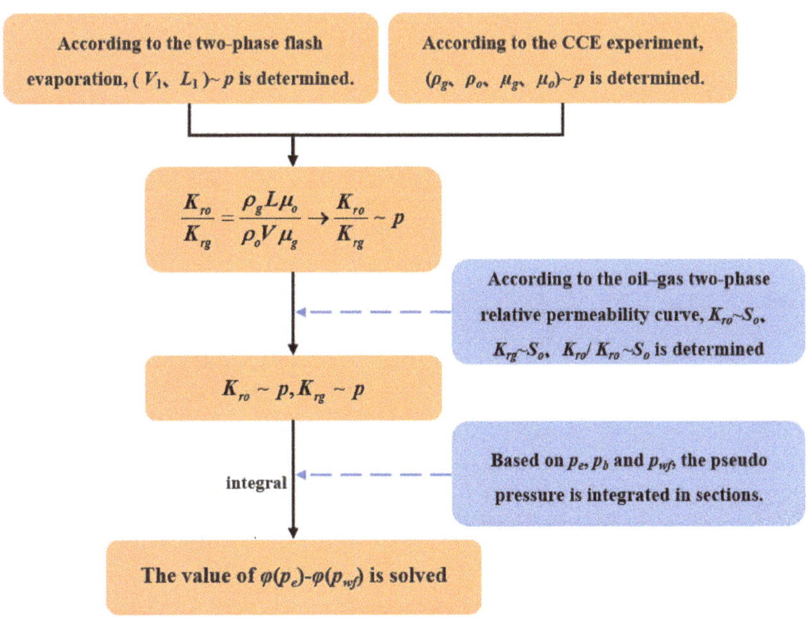

Figure 6. Flowchart for calculating two-phase pseudo-pressure.

2.6. Solution for Productivity Model

Given the presence of N fractures in the horizontal well, Equations (61) and (62) form a system of N linear equations with N unknowns, i.e., $q_{ti}(1 \le i \le N)$, which can be expressed in matrix form as follows:

$$\begin{bmatrix} E_{11} + \frac{Kh}{K_f w_f}\ln\frac{h}{2r_w} & \cdots & E_{1N} \\ \vdots & \vdots & \vdots \\ E_{N1} & \cdots & E_{NN} + \frac{Kh}{K_f w_f}\ln\frac{h}{2r_w} \end{bmatrix} \begin{bmatrix} q_{t1} \\ \vdots \\ q_{tN} \end{bmatrix} = \begin{bmatrix} 2\pi Kh(\varphi_e - \varphi_{wf}) - (F_{11} + \cdots + F_{1N} - G_{11} - \cdots - G_{1N})\pi x_f^2 h Q_d \\ \vdots \\ 2\pi Kh(\varphi_e - \varphi_{wf}) - (F_{N1} + \ldots + F_{NN} - G_{N1} - \ldots - G_{NN})\pi x_f^2 h Q_d \end{bmatrix}, \tag{66}$$

$$E = \begin{bmatrix} \ln\left(\frac{r_e - y_1}{x_f} + \sqrt{1 + \frac{(r_e - y_1)^2}{x_f^2}}\right) & \cdots & \ln\left(\dfrac{\frac{r_e - y_N}{x_f} + \sqrt{1 + \frac{(r_e - y_N)^2}{x_f^2}}}{\left|\frac{y_1 - y_N}{x_f}\right| + \sqrt{1 + \frac{(y_1 - y_N)^2}{x_f^2}}}\right) \\ \vdots & \vdots & \vdots \\ \ln\left(\dfrac{\frac{r_e - y_1}{x_f} + \sqrt{1 + \frac{(r_e - y_1)^2}{x_f^2}}}{\left|\frac{y_N - y_1}{x_f}\right| + \sqrt{1 + \frac{(y_N - y_1)^2}{x_f^2}}}\right) & \cdots & \ln\left(\frac{r_e - y_N}{x_f} + \sqrt{1 + \frac{(r_e - y_N)^2}{x_f^2}}\right) \end{bmatrix}, \tag{67}$$

$$F = \begin{bmatrix} \begin{matrix} \frac{r_e - y_1}{x_f} \cdot \sqrt{\frac{(r_e - y_1)^2}{x_f^2} + 2 - \cos\left(\frac{\pi}{2}\right)^2} \cdot \\ \ln\left[\frac{(r_e - y_1)^2}{x_f^2} + 1 + \sqrt{\left[\frac{(r_e - y_1)^2}{x_f^2} + 1\right]^2 - 1}\right] \end{matrix} & \cdots & \begin{matrix} \frac{r_e - y_N}{x_f} \cdot \sqrt{\frac{(r_e - y_N)^2}{x_f^2} + 2 - \cos\left(\frac{\pi}{2}\right)^2} \cdot \\ \ln\left[\frac{(r_e - y_N)^2}{x_f^2} + 1 + \sqrt{\left[\frac{(r_e - y_N)^2}{x_f^2} + 1\right]^2 - 1}\right] \end{matrix} \\ \vdots & \vdots & \vdots \\ \begin{matrix} \frac{r_e - y_1}{x_f} \cdot \sqrt{\frac{(r_e - y_1)^2}{x_f^2} + 2 - \cos\left(\frac{\pi}{2}\right)^2} \cdot \\ \ln\left[\frac{(r_e - y_1)^2}{x_f^2} + 1 + \sqrt{\left[\frac{(r_e - y_1)^2}{x_f^2} + 1\right]^2 - 1}\right] \end{matrix} & \cdots & \begin{matrix} \frac{r_e - y_N}{x_f} \cdot \sqrt{\frac{(r_e - y_N)^2}{x_f^2} + 2 - \cos\left(\frac{\pi}{2}\right)^2} \cdot \\ \ln\left[\frac{(r_e - y_N)^2}{x_f^2} + 1 + \sqrt{\left[\frac{(r_e - y_N)^2}{x_f^2} + 1\right]^2 - 1}\right] \end{matrix} \end{bmatrix} \tag{68}$$

$$G = \begin{bmatrix} 0 & \cdots & \begin{matrix} \left|\frac{y_1 - y_N}{x_f}\right| \cdot \sqrt{\frac{(y_1 - y_N)^2}{x_f^2} + 2 - \cos\left(\frac{\pi}{2}\right)^2} \cdot \\ \ln\left[\frac{(y_1 - y_N)^2}{x_f^2} + 1 + \sqrt{\left[\frac{(y_1 - y_N)^2}{x_f^2} + 1\right]^2 - 1}\right] \end{matrix} \\ \vdots & \vdots & \vdots \\ \begin{matrix} \left|\frac{y_N - y_1}{x_f}\right| \cdot \sqrt{\frac{(y_N - y_1)^2}{x_f^2} + 2 - \cos\left(\frac{\pi}{2}\right)^2} \cdot \\ \ln\left[\frac{(y_N - y_1)^2}{x_f^2} + 1 + \sqrt{\left[\frac{(y_N - y_1)^2}{x_f^2} + 1\right]^2 - 1}\right] \end{matrix} & \cdots & 0 \end{bmatrix} \tag{69}$$

where E, F, and G are $N \times N$ square matrices; E_{ij} is the value in the ith row and jth column of matrix E; F_{ij} is the value in the ith row and jth column of matrix F; and G_{ij} is the value in the ith row and jth column of matrix G.

Solving the system of linear equations allows for determining the production rate (q_{ti}) of each fracture, subsequently enabling the calculation of the productivity of MFHWs. The solution procedure for the proposed productivity model is illustrated in Figure 7, which provides a comprehensive summary of the process.

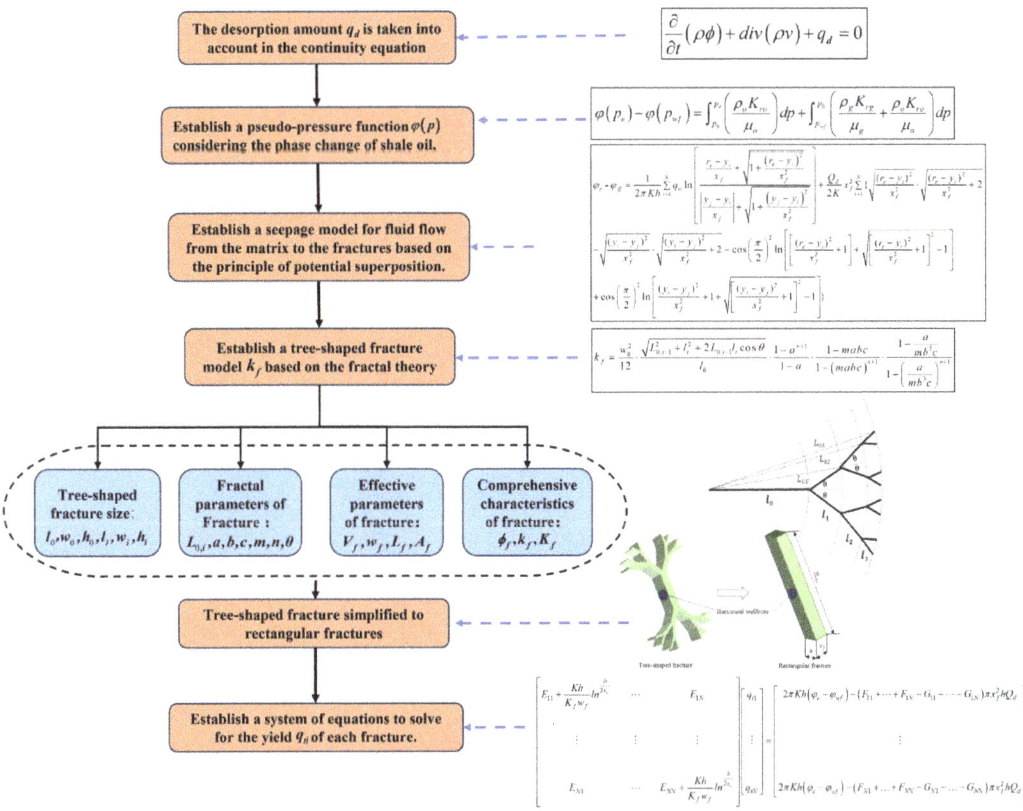

Figure 7. Flow chart of solution procedure for the proposed productively model.

3. Model Verification

In this section, the proposed productivity model is validated by comparing it with field data and a numerical model. We only have access to field data from two shale oil wells, namely the PA-1 well and the YY-1 well, in China and are unable to obtain additional field data due to confidentiality concerns. The YY-1 well is a newly drilled well, and as of now, no fluid PVT test has been carried out. Given that the YY-1 well is situated within the same block as the pre-existing PA-1 well, it is hypothesized that the fluid properties of YY-1 are similar to those of PA-1. The proposed productivity model is used to calculate the productivity of the two actual shale oil wells. Through conducting PVT tests on fluid samples extracted from the PA-1 well, precise correlations between thermodynamic parameters and pressure have been established in order to derive accurate two-phase pseudo-pressure. The proposed productivity model is utilized to predict the inflow performance relationship (IPR) curve, which is then compared with field data to validate the model.

3.1. PVT Fitting

The composition of the fluid sample obtained from the PA-1 well is detailed in Table 2. The PR state equation is used to fit the PVT testing data of the fluid sample. As shown in Figure 8, the predicted parameters, such as fluid density and gas deviation factor, are in good agreement with the experimental values. In addition, it is interesting that there is an inflection point in the oil phase density–pressure curve in Figure 8a. This occurs because when the pressure is greater than the bubble point pressure (36.17 MPa), the fluid consists of single-phase oil. The higher the pressure, the greater the density of the oil phase becomes. When the pressure is less than the bubble point pressure (36.17 MPa), gas phase deaeration occurs as the pressure decreases. The content of heavy components in crude oil gradually increases, leading to a gradual increase in density.

Table 2. Molar composition of sample.

Component	Molar Composition, %	Molecular Weight, g/mol
CO_2	0.112	44.01
N_2	0.280	28.01
C_1	56.798	16.04
C_2	16.716	30.07
C_3	5.765	44.1
iC_4	1.415	58.12
nC_4	1.571	58.12
iC_5	0.942	72.15
nC_5	0.468	72.15
C_6	0.403	86.18
C_7	1.304	100.2
C_8	1.959	114.23
C_9	1.420	128.26
C_{10}	1.175	142.29
C_{11+}	9.672	248.18

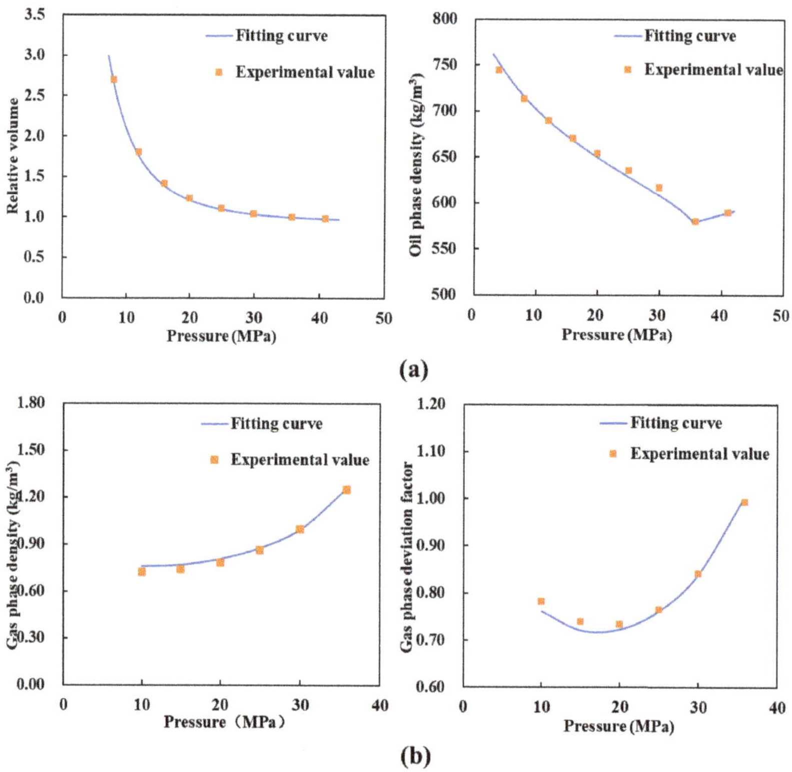

Figure 8. Comparison of the calculated PVT parameters with experimental results (**a**) CCE experimental fitting results, (**b**) differential liberation experimental fitting results.

3.2. Productivity Calculation

Based on the basic parameters of the PA-1 well and YY-1 well listed in Table 3, the proposed productivity model is employed to calculate the IPR curves of the PA-1 well and YY-1 well. The comparisons between the predicted results and the field data are shown in Figures 9 and 10, which indicate that the predicted results are consistent with the field data. Therefore, the proposed productivity model is validated as feasible for accurately calculating the productivity of shale oil wells. The predicted oil and gas productivities of the PA-1 well are 192.87 m^3/d and 199,600 m^3/d, respectively. The predicted oil and gas productivities of the YY-1 well are 19.75 m^3/d and 22,700 m^3/d, respectively.

Table 3. Basic parameters of the PA-1 well and YY-1 well.

Parameter	PA-1	YY-1	Parameter	PA-1	YY-1
Fracture number, N	21	31	Bubble point pressure, P_b, MPa	36.17	36.17
Horizontal well length, L, m	817	1400	Reservoir pressure, P_e, MPa	50.14	32.19
Control radius, r_e, m	500	750	Matrix permeability, K, mD	0.025	0.016
Well radius, r_{wf}, m	0.06985	0.06985	Reservoir thickness, h, m	63.5	20
0-level fracture length, l_0, m	22	28	Oil density, ρ_{osc}, kg/m^3	786.2	786.2
0-level fracture width, w_0, mm	0.005	0.005	Gas density, ρ_{gsc}, kg/m^3	0.9608	0.9608
Length ratio, a	0.5	0.5	Reservoir temperature, T, °C	66.5	66.5
Width ratio, b	0.5	0.5	Langmuir volume, V_L, m^3/kg	5.82×10^{-3}	5.82×10^{-3}
Height ratio, c	1	1	Langmuir pressure, P_L, MPa	4.45	4.45
Branch angle, θ	30°	30°			
Branch number, m	2	2			
Fracture branch level, n	3	3			

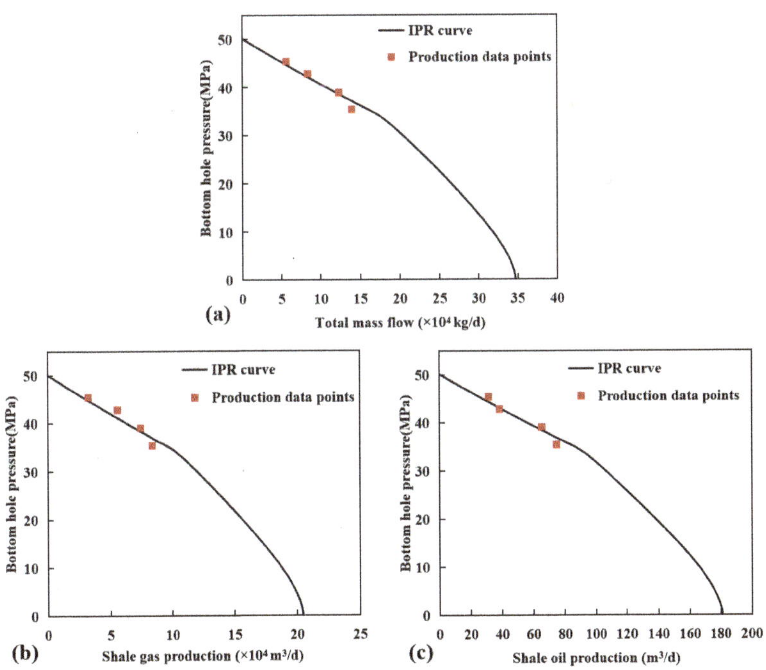

Figure 9. Comparison of the calculated IPR curves with field data of the PA-1 well for (**a**) total mass flow rate q_t, (**b**) oil production rate q_o, and (**c**) gas production rate q_g.

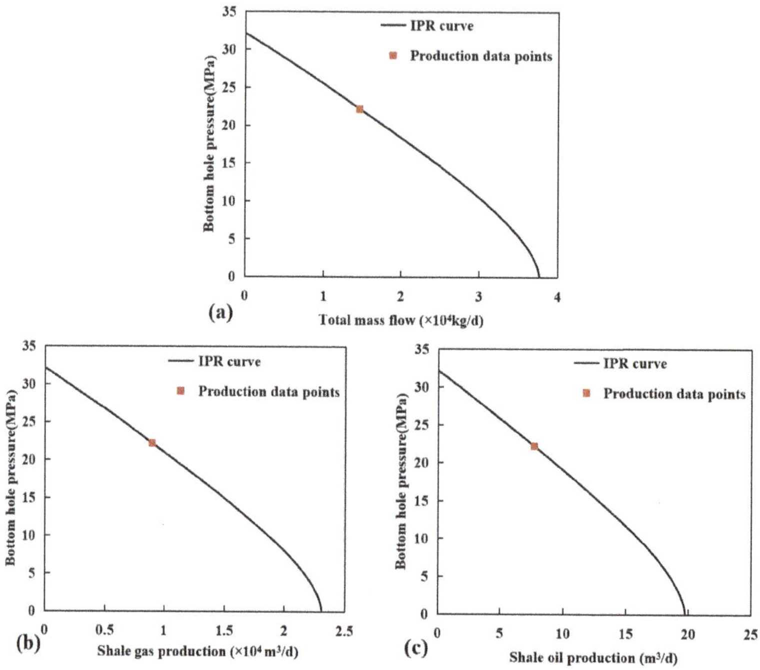

Figure 10. Comparison of the calculated IPR curves with field data of the YY-1 well for (**a**) total mass flow rate q_t, (**b**) oil production rate q_o, and (**c**) gas production rate q_g.

3.3. Numerical Simulation Verification

To validate the accuracy of the proposed productivity model, a numerical model of a MFHW is established using CMG. By assuming the same parameters as those of the YY-1 well presented in Table 3, the production rate is obtained using CMG. A conceptual model of a MFHW is generated by CMG (Figure 11). The production well maintains consistent bottom hole flowing pressure during operation. When the pressure wave propagates to the outer boundary, the average formation pressure and well production at different times are recorded. The average formation pressure and constant bottom hole flowing pressure are employed to obtain the IPR curve based on the proposed productive model. The comparison of the calculated IPR curves with CMG numerical solutions is shown in Figure 12. It is observed that numerical solutions agree well with the calculated IPR curves under different average formation pressures, indicating the reliability of the model.

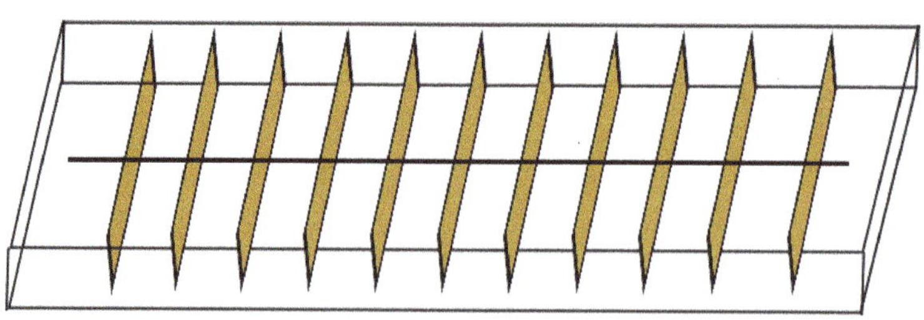

Figure 11. Conceptual model of a MFHW generated by CMG.

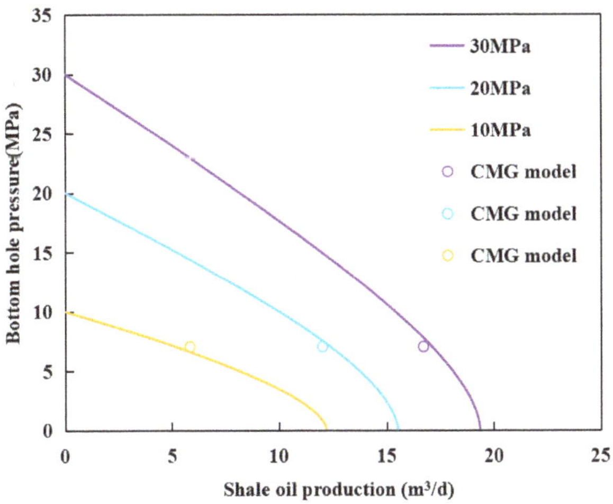

Figure 12. Comparison of the calculated IPR curves with CMG numerical solutions under different average formation pressures.

4. Results and Analyses

4.1. The Impact of Shale Oil Phase Behavior on Well Productivity

Figure 13 shows the effect of the phase behavior on the IPR curve of shale oil wells. As depicted in Figure 13, the IPR curves demonstrate a significant disparity when considering the phase behavior versus not considering it. Without accounting for phase change, a decrease in bottom hole pressure results in a linear increase in production rate. However, factoring in the phase change leads to the production rate initially exhibiting a linear increase with decreasing bottom hole pressure; subsequently transitioning from single-phase flow to two-phase flow leads to a sharp decline in production rate. As shown in Figure 13, the phase behavior results in a 33% reduction in the well productivity. The productivity of shale oil wells is notably reduced when accounting for phase behavior, as compared to the productivity without considering phase behavior.

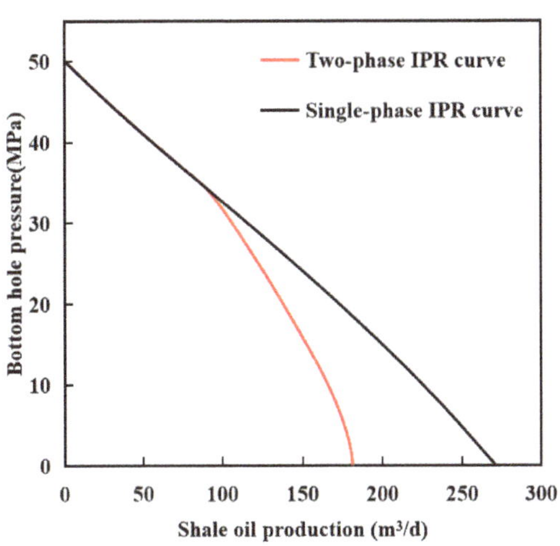

Figure 13. The relationship between shale oil production and bottom hole pressure with and without considering the phase behavior.

4.2. The Impact of Fracture Spacing on Well Productivity

Figure 14 shows the effect of fracture spacing on IPR curves of shale oil wells. It is seen from Figure 14 that as the fracture spacing increases, the production rate of shale oil wells increases, and the lower the bottom hole pressure, the more pronounced this increase in production rate becomes. The reason is that as the fracture spacing decreases, the interference between hydraulic fractures becomes stronger, resulting in a decrease in production rate.

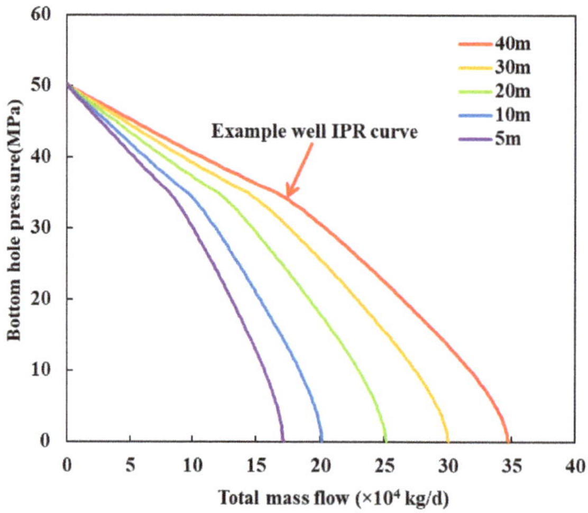

Figure 14. Effect of fracture spacing on IPR curves of shale oil wells.

4.3. The Impact of Tree-Shaped Fracture Characteristics on Well Productivity

Figure 15 shows the influence of length ratio a at different branch levels n on fracture characteristic length, fracture effective permeability, and well productivity. It can be observed from Figure 15a,b that with an increase in the length ratio and branch level, there is an increase in the fracture characteristic length, accompanied by a corresponding decrease in fracture permeability. This observation suggests that an increase in fracture characteristic length leads to a reduction in fracture conductivity, which can be elucidated through Equation (57). Figure 15c demonstrates a positive correlation between the length ratio and well productivity, indicating that as the length ratio increases, so does the well productivity. Furthermore, it is evident that higher branch levels result in a greater increase in well productivity with an increase in the length ratio.

Figure 16 shows the influence of width ratio b at different branch levels n on fracture effective width, fracture characteristic length, fracture effective permeability, and well productivity. The fracture effective width w_f increases as the width ratio b increases. For $b \geq 0.5$, w_f is greater than the initial width w_0, whereas for $b \leq 0.5$, w_f is less than the initial width w_0 (Figure 16a). The fracture characteristic length is independent of the width ratio and slowly increases with the elevation of the branch level (Figure 16b). The fracture effective permeability increases with an increase in the width ratio b, while it decreases with an increase in the branch level (Figure 16c). As the width ratio increases, the well productivity gradually increases and reaches a stable value. When $n = 0$, the fracture exhibits a linear configuration without any branching, and the productivity remains constant regardless of the width ratio of adjacent cracks. When $n = 1$, the fracture demonstrates high permeability and no starting pressure gradient within the fracture. Under these conditions, the production also remains unaffected by changes in the width ratio of adjacent cracks. Compared to the conventional straight fractures, tree-shaped fractures have the potential to enhance well productivity, although their influence is limited (Figure 16d).

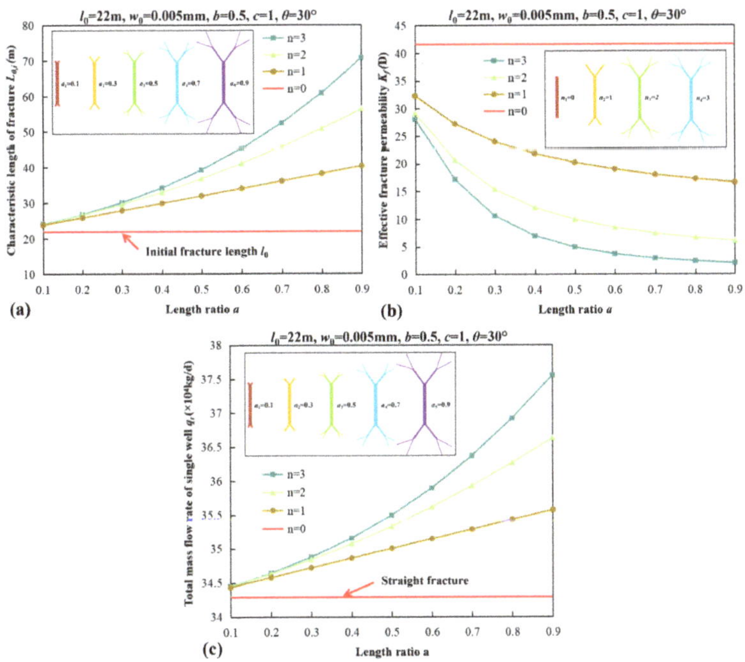

Figure 15. The influence of length ratio a at different branch levels n on (**a**) fracture characteristic length $L_{0,i}$, (**b**) fracture effective permeability K_f, and (**c**) well productivity q_t.

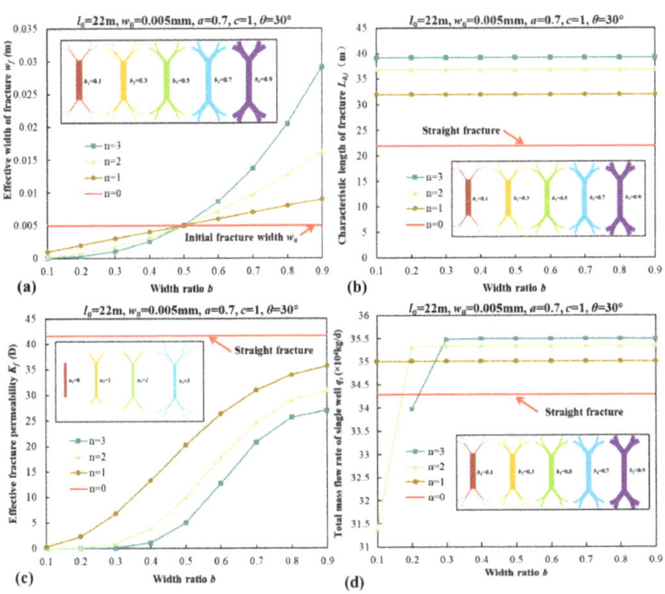

Figure 16. The influence of width ratio b at different branch levels n on (**a**) fracture effective width w_f, (**b**) fracture characteristic length $L_{0,i}$, (**c**) fracture effective permeability K_f, and (**d**) well productivity q_t.

Figure 17 shows the influence of branch angle θ at different branch levels n on fracture characteristic length, fracture effective permeability, and well productivity. The smaller the branch angle, the larger the fracture characteristic length, and the greater the fracture effective permeability. It is evident that well productivity decreases as the branch angle increases; however, there is minimal impact of the branch angle on well productivity.

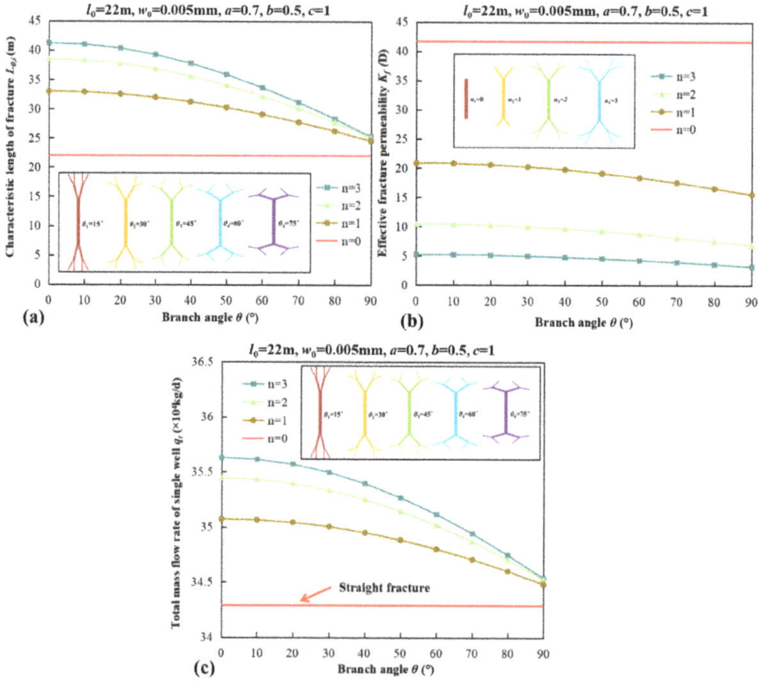

Figure 17. The influence of branch angle θ at different branch levels n on (**a**) fracture characteristic length $L_{0,i}$, (**b**) fracture effective permeability K_f, and (**c**) well productivity q_t.

4.4. The Impact of Oil and Gas Desorption on Well Productivity

Figure 18 shows the desorption amounts of shale oil and gas change with the pressure. It is observed from Figure 18 that as the pressure decreases, the desorption amount of oil increases linearly, while the desorption amount of gas initially increases slowly and then rapidly.

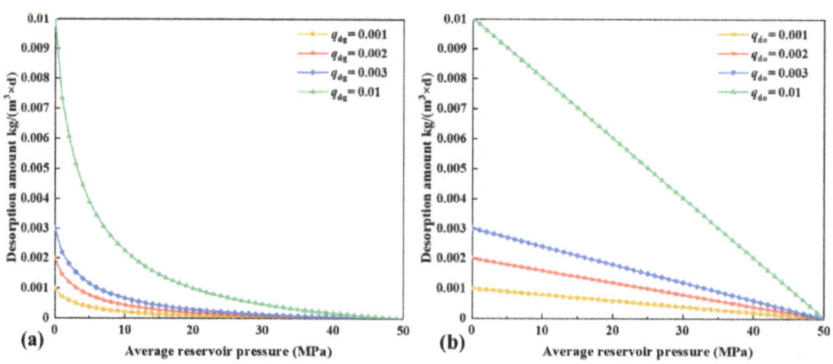

Figure 18. (**a**) Desorption amount of shale gas q_{dg} and (**b**) desorption amount of shale oil q_{do} change with pressure.

Figure 19 shows the influence of the maximum desorption amount on IPR curves of shale oil wells. It is evident that fluid desorption has a great influence on well productivity. As the maximum desorption amount increases, there is a corresponding increase in well productivity. As shown in Figure 19, when the maximum desorption amount increases from 0.001 kg/(m^3·d) to 0.01 kg/(m^3·d), there is a significant 75% increase in well productivity. Additionally, as the bottom hole pressure decreases, the impact of desorption on well productivity becomes more pronounced. This can be attributed to the fact that the desorption amounts of shale oil and gas increase with decreasing pressure.

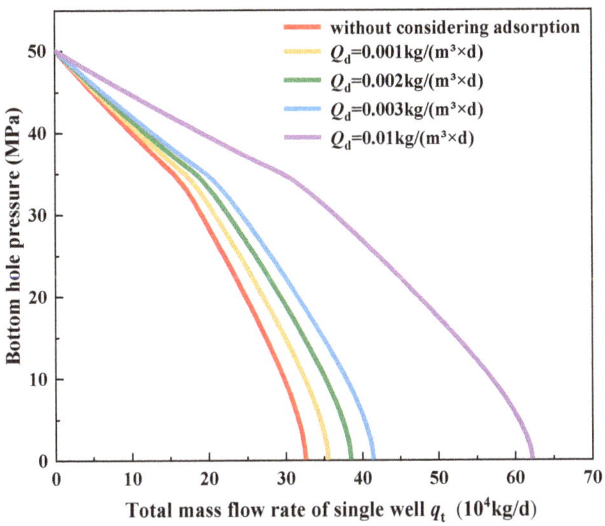

Figure 19. The influence of the maximum desorption amount on IPR curves of shale oil wells.

5. Conclusions

In this study, a novel productivity model is proposed for MFHWs in shale oil reservoirs to accurately predict well productivity. This model takes into account the complex, tree-shaped, hydraulically fractured networks, as well as the oil–gas desorption effect and phase change of oil and gas, ensuring a comprehensive understanding of the production process. The proposed model has been validated through comparison with field data from an actual shale oil well. The effects of key parameters on IPR curves are obtained and analyzed. Several important conclusions are drawn as follows:

(1) Compared to previous productivity models, this model considers the effects of fluid desorption, hydraulic fracture networks, and oil–gas phase behavior, which is more consistent with the actual development of shale oil reservoirs. The predicted results of the proposed model show strong agreement with the field data from an actual shale oil well, confirming the reliability of the proposed model.

(2) As the pressure decreases, the oil phase undergoes a phase change into a two-phase mixture of oil and gas, leading to an increase in flow resistance. Consequently, this phase behavior of oil and gas results in a significant decrease in well production. As the fracture spacing decreases, the interference between hydraulic fractures becomes stronger, leading to a reduction in well productivity.

(3) The characteristic parameters of hydraulic fracture networks have a substantial influence on well productivity. The increase in length ratio a, the decline in branch angle θ, and the increase in branch level n result in an increase in well productivity. As the width ratio b increases, the well productivity gradually increases and reaches a stable value.

(4) As the pressure decreases, the desorption amount of oil exhibits a linear increase, whereas the desorption amount of gas demonstrates an initial slow increase followed by a rapid escalation. As the maximum desorption amount increases, there is a significant enhancement in well productivity, highlighting the crucial role of fluid desorption in improving well performance.

Author Contributions: Conceptualization, X.L. and P.G.; Methodology, X.L. and J.R.; Software, X.L.; Validation, X.L.; Resources, J.R. and Z.W.; Writing—original draft, X.L.; Writing—review & editing, X.L., P.G., J.R., Z.W. and H.T.; Visualization, X.L.; Supervision, P.G., J.R., Z.W. and H.T.; Project administration, P.G.; Funding acquisition, P.G. All authors have read and agreed to the published version of the manuscript.

Funding: This research received no external funding.

Data Availability Statement: The original contributions presented in the study are included in the article, further inquiries can be directed to the corresponding author/s.

Conflicts of Interest: The authors declare no conflict of interest.

References

1. Zou, C.N.; Yang, Z.; Cui, J.W.; Zhu, R.K.; Hou, L.H.; Tao, S.Z.; Yuan, X.J.; Wu, S.T.; Lin, S.H.; Wang, L. Formation mechanism geological characteristics and development strategy of nonmarine shale oil in China. *Pet. Explor. Dev.* **2013**, *40*, 15–27. [CrossRef]
2. Bažant, Z.P.; Salviato, M.; Chau, V.T.; Viswanathan, H.; Zubelewicz, A. Why fracking works. *J. Appl. Mech.* **2014**, *81*, 101010 [CrossRef]
3. Peplow, M. The great gas gold rush. *Nature* **2017**, *550*, 26–28. [CrossRef] [PubMed]
4. Hughes, J.D. A reality check on the shale revolution. *Nature* **2013**, *494*, 307–308. [CrossRef] [PubMed]
5. Li, J.H.; Li, Y.J.; Sang, Q.; Gong, H.J.; Xu, L.; Zhang, H.Y.; Arif, M.; Dong, M.Z.; Cui, C.Z. Quantification of seepage characteristics in shale oil reservoirs: A triple medium model-driven approach. *Geoenergy Sci. Eng.* **2024**, *242*, 213225. [CrossRef]
6. Qu, J.; Tang, Z.; Lei, G.; Wu, Q.; Liao, Q.; Ning, F. A novel threshold pressure gradient model and its influence on production simulation for shale oil reservoirs. *Energy Fuels* **2024**, *38*, 11644–11661. [CrossRef]
7. Zou, C.; Zhao, Q.; Dong, D.; Yang, Z.; Qiu, Z.; Liang, F.; Wang, N.; Huang, Y.; Duan, A.; Zhang, Q.; et al. Geological characteristics main challenges and future prospect of shale gas. *Nat. Gas Geosci.* **2017**, *2*, 273–288. [CrossRef]
8. Lei, Q.; Xu, Y.; Cai, B.; Guan, B.; Wang, X.; Bi, G.Q.; Li, H.; Li, S.; Ding, B.; Fu, H.F.; et al. Progress and prospects of horizontal well fracturing technology for shale oil and gas reservoirs. *Pet. Explor. Dev.* **2022**, *49*, 191–199. [CrossRef]
9. Zhang, D.; Zhang, L.; Zhao, Y.; Guo, J. A composite model to analyze the decline performance of a multiple fractured horizontal well in shale reservoirs. *J. Nat. Gas Sci. Eng.* **2015**, *26*, 999–1010. [CrossRef]
10. Wang, C.; Li, Z.P.; Li, H.; Wei, Q.; Bai, R.T. A new method to calculate the productivity of fractured horizontal gas wells considering non-Darcy flow in the fractures. *J. Nat. Gas Sci. Eng.* **2015**, *26*, 981–991. [CrossRef]
11. Wu, J.; Liu, Y.; Yang, H. New method of productivity equation for multibranch horizontal well in three-dimensional anisotropic oil reservoirs. *J. Energy Resour. Technol.* **2012**, *134*, 032801. [CrossRef]
12. Wu, J.; Liu, Y.; Yang, H. A new analytical solution of the productivity equation for a vertical fractured well in 3D anisotropic oil reservoirs. *Pet. Sci. Technol.* **2014**, *32*, 433–441. [CrossRef]
13. Chen, Z.M.; Liao, X.W.; Zhao, X.L.; Lyu, S.B.; Zhu, L.T. A comprehensive productivity equation for multiple fractured vertical wells with non-linear effects under steady-state flow. *J. Pet. Sci. Eng.* **2017**, *149*, 9–24. [CrossRef]
14. Wei, B.C.; Nie, X.R.; Zhang, Z.H.; Ding, J.C.; Shayireatehan, R.Y.Z.; Ning, P.Z.; Deng, D.T.; Cao, Y. Productivity equation of fractured vertical well with gas–water co-production in high-water-cut tight sandstone gas reservoir. *Processes* **2023**, *11*, 3123 [CrossRef]
15. Soleimani, R.; Jahanpeyma, Y.; Salehian, M. Analysis of horizontal well productivity in tight gas formations and its sensitivity to reservoir properties. *J. Pet. Explor. Prod. Technol.* **2019**, *9*, 1237–1244. [CrossRef]
16. Lou, Y.; Song, H.Q.; Yang, J.S.; Huang, X.H.; Dong, H. Productivity equation of fractured well in CBM reservoirs. *J. Nat. Gas Sci Eng.* **2013**, *11*, 39–45. [CrossRef]
17. Clarkson, C.R.; Williams-Kovacs, J.D. Modeling two-phase flowback of multifractured horizontal wells completed in shale. *SPE J.* **2013**, *18*, 795–812. [CrossRef]
18. Song, H.Q.; Liu, Q.P.; Yang, D.W.; Yu, M.X.; Lou, Y.; Zhu, W.Y. Productivity equation of fractured horizontal well in a water-bearing tight gas reservoir with low-velocity non-Darcy flow. *J. Nat. Gas Sci. Eng.* **2014**, *18*, 467–473. [CrossRef]
19. Qi, Q.; Zhu, W.Y. The three-zone composite productivity model for a multi-fractured horizontal shale gas well. *IOP Conf. Ser. Earth Environ. Sci.* **2018**, *121*, 052051. [CrossRef]
20. Jiang, B.M.; Mu, L.F.; Yan, Y.Q.; Li, Q.; Wu, Z.B.; Chen, J.Y.; Zhang, Q.S. Calculation method of productivity for shale oil in volumetric fractured horizontal wells. *ACS Omega* **2022**, *7*, 20495–20504. [CrossRef]

21. Wang, Q.; Wan, J.; Mu, L.; Shen, R.; Jurado, M.J.; Ye, Y. An analytical solution for transient productivity prediction of multi-fractured horizontal wells in tight gas reservoirs considering nonlinear porous flow mechanisms. *Energies* **2020**, *13*, 1066. [CrossRef]
22. Yao, S.S.; Wang, X.Z.; Yuan, Q.W.; Guo, Z.X.; Zeng, F. Production analysis of multifractured horizontal wells with composite models: Influence of complex heterogeneity. *J. Hydrol.* **2020**, *583*, 124542. [CrossRef]
23. Zhang, H.; Sheng, J.J. Complex fracture network simulation and optimization in naturally fractured shale reservoir based on modified neural network algorithm. *J. Nat. Gas Sci. Eng.* **2021**, *95*, 104232. [CrossRef]
24. Lu, M.J.; Qian, Q.; Zhong, A.H.; Yang, F.; He, W.J.; Li, M. High-precision flow numerical simulation and productivity evaluation of shale oil considering stress sensitivity. *Fluid Dyn. Mater. Process.* **2024**, *20*, 2281–2300. [CrossRef]
25. Guan, Q.S.; Chen, C.W.; Pu, X.G.; Wan, Y.G.; Xu, J.; Zeng, H.W.; Jia, C.; Gao, H.H.; Yang, W.; Peng, Z.S. Production performance analysis of a continental shale oil reservoir in Bohai Bay basin. *Petroleum* **2024**, *10*, 294–305. [CrossRef]
26. Yan, Z.; Wang, F.; Liu, Y.; Wang, P. A coupled matrix-fracture productivity calculation model considering low-velocity non-Darcy flow in shale reservoirs. *Fuel* **2024**, *357*, 129845. [CrossRef]
27. Zhang, Q.; Wang, W.D.; Su, Y.L.; Chen, W.; Lei, Z.D.; Li, L.; Hao, Y.M. A semi-analytical model for coupled flow in stress-sensitive multi-scale shale reservoirs with fractal characteristics. *Pet. Sci.* **2024**, *21*, 327–342. [CrossRef]
28. Wang, Z.P.; Zhu, W.Y.; Yue, M.; Gao, Y.; Zhao, G.J.; Wang, H.Q. A method to predict the production of fractured horizontal wells in low/ultra-low permeability reservoirs. *Chin. J. Eng.* **2012**, *34*, 750–754.
29. Li, Q.; Li, Q.C.; Wang, F.L.; Wu, J.J.; Wang, Y.L. The carrying behavior of water-based fracturing fluid in shale reservoir fractures and molecular dynamics of sand-carrying mechanism. *Processes* **2024**, *12*, 2051. [CrossRef]
30. Li, Q.C.; Li, Q.; Han, Y. A numerical investigation on kick control with the displacement kill method during a well test in a deep-water gas reservoir: A case study. *Processes* **2024**, *12*, 2090. [CrossRef]
31. Wang, J.G.; Hu, B.W.; Wu, D.; Dou, F.K.; Wang, X.L. A multiscale fractal transport model with multilayer sorption and effective porosity effects. *Transp. Porous Media* **2019**, *129*, 25–51. [CrossRef]
32. Lane, H.S.; Lancaster, D.E.; Watson, A.T. Characterizing the role of desorption in gas production from Devonian shales. *Energy Sources* **1991**, *13*, 337–359. [CrossRef]
33. Bumb, A.C.; McKee, C.R. Gas-well testing in the presence of desorption for coalbed methane and Devonian shale. *SPE Form. Eval.* **1988**, *3*, 179–185. [CrossRef]
34. Jenkins, C.D.; Boyer, C.M. Coalbed- and shale-gas reservoirs. *J. Pet. Technol.* **2008**, *60*, 92–99. [CrossRef]
35. Guo, C.; Xu, J.; Wu, K. Study on gas flow through nano pores of shale gas reservoirs. *Fuel* **2015**, *143*, 107–117. [CrossRef]
36. Xiong, H.; Devegowda, D.; Huang, L.L. Water bridges in clay nanopores: Mechanisms of formation and impact on hydrocarbon transport. *Langmuir* **2020**, *36*, 723–733. [CrossRef]
37. Singh, H.; Javadpour, F.; Ettehadtavakkol, A.; Darabi, H. Nonempirical apparent permeability of shale. *SPE Reserv. Eval. Eng.* **2014**, *17*, 414–424. [CrossRef]
38. Li, M.; Su, Y.L.; Dong, M.Z.; Zhou, L.H.; Li, Y.J.; Pu, X.G.; Li, S.; Han, W.Z.; Wang, W.D. A numerical study of fluids desorption and phase behavior in shale oil reservoir using a chemical reaction model. *J. Pet. Sci. Eng.* **2021**, *196*, 108050. [CrossRef]
39. Mo, C.P.; Zhang, G.D.; Zhang, Z.W.; Yuan, Y.H.; Yang, S. Productivity calculation model for fractured horizontal wells considering phase variation. *J. Pet. Sci. Eng.* **2022**, *209*, 109811. [CrossRef]
40. Liu, X.Y.; Zhang, D.X. A review of phase behavior simulation of hydrocarbons in confined space: Implications for shale oil and shale gas. *J. Nat. Gas Sci. Eng.* **2019**, *68*, 102901. [CrossRef]
41. Liu, X.Y.; Li, X.; Zhang, D.X. A statistical thermodynamics-based equation of state and phase equilibrium calculation for confined hydrocarbons in shale reservoirs. *J. Nat. Gas Sci. Eng.* **2022**, *102*, 104579. [CrossRef]
42. Alharthy, N.; Teklu, T.; Kazemi, H.; Graves, R.; Hawthorne, S.; Braunberger, J.; Kurtoglu, B. Enhanced oil recovery in liquid–rich shale reservoirs: Laboratory to field. *SPE Reserv. Eval. Eng.* **2018**, *21*, 137–159. [CrossRef]
43. Yang, Y.; Yang, Y.M.; Ma, S.; Ju, Y.; Guo, Y.C.; Yu, Y. Crack propagation induced by hydraulic fracturing in low permeability rocks using CT technology. *J. Min. Saf. Eng.* **2013**, *30*, 739–743.
44. Chen, M. Re-orientation and propagation of hydraulic fractures in shale gas reservoir. *J. China Univ. Pet.* **2013**, *37*, 88–94.
45. Deng, J.; Zhu, W.Y.; Liu, J.X.; Zhang, Z.; Ma, Q.; Zhang, M.; Deng, K.; Ma, L. Productivity prediction model of shale gas considering stress sensitivity. *Nat. Gas Geosci.* **2013**, *24*, 456–460.
46. Zhu, W.Y.; Yue, M.; Song, H.Q.; Yang, J.S.; Wang, Y.B.; Yu, M.X. Productivity model of gas flow in CBM fractured reservoirs considering desorption and diffusion. *J. Basic Sci. Eng.* **2013**, *21*, 953–960.
47. Xu, P.; Yu, B.M.; Yun, M.J.; Zou, M.Q. Heat conduction in fractal tree-like branched networks. *Int. J. Heat Mass Transf.* **2006**, *49*, 3746–3751. [CrossRef]

energies

MDPI

Article

Analysis on Correlation Model Between Fracture Network Complexity and Gas-Well Production: A Case in the Y214 Block of Changning, China

Zhibin Gu [1], Bingxiao Liu [1], Wang Liu [1], Lei Liu [1], Haiyu Wei [2], Bo Yu [2], Lifei Dong [2], Pinzhi Zhong [2,*] and Hun Lin [3]

1 Sichuan Changning Natural Gas Development Co., Ltd., Chengdu 610000, China; 15320759184@163.com (Z.G.); 19823175647@163.com (B.L.); l_wang@petrochina.com.cn (W.L.); a15524811816@163.com (L.L.)
2 College of Civil Engineering, Chongqing Three Gorges University, Wanzhou, Chongqing 404120, China; 19823015562@163.com (H.W.); 18683318298@163.com (B.Y.); lfdong2012@sina.com (L.D.)
3 Department of Safety Engineering, Chongqing University of Science & Technology, Chongqing 401331, China; 2017031@cqust.edu.cn
* Correspondence: zpz944@163.com

Abstract: The fracture network of the Y214 block in the Changning area of China is complex, and there are significant differences in the productivity of different shale gas wells. However, traditional machine learning models have problems such as missing key parameters, poor fitting effects and low prediction accuracy, which make it difficult to effectively evaluate the impact of crack network complexity on productivity. Therefore, the Pearson correlation coefficient was used to analyze the correlation between evaluation parameters, such as mineral content, horizontal stress difference, natural fractures and gas production. Combined with the improved particle swarm optimization (IPSO) algorithm and support vector machine (SVM) algorithm, a fracture network index (FNI) model was proposed to effectively evaluate the complexity of fracture networks, and the model was verified by comparing it with the performance evaluation results from the other two traditional models. Finally, the correlation between the fracture network index and the actual average daily gas production of different fracturing sections was calculated and analyzed. The results showed that the density of natural fractures was the key factor in controlling gas production (the Pearson correlation coefficient was 0.39), and the correlation between other factors was weak. In the process of fitting the actual data, the coefficient of determination, R^2, of the IPSO-SVM-FNI model training set increased by 8% and 24% compared with the two traditional models, and the fitting effect was greatly improved. In the prediction process based on actual data, the R^2 of the IPSO-SVM-FNI model test set was improved by 22% and 20% compared with the two traditional models, and the prediction accuracy was also significantly improved. The fracture index was concentrated, and its main distribution range was in the range of [0.2, 0.8]. The fracturing section with a higher FNI showed higher average daily gas production, and there was a significant positive correlation between fracture network complexity and gas production. Indeed, the research results provide some ideas and references for the evaluation of fracturing effects in shale reservoirs.

Keywords: friability; improved particle swarm optimization; support vector machine; fracture network complexity; natural gas production

1. Introduction

The global petroleum industry has been gradually moving toward unconventional energy sources. Shale gas has become a new standard for unconventional natural gas development in North America [1–3]. The United States has adopted new fracturing technology in many old oil fields to revitalize them [4,5]. In the total natural gas production in the United States, the proportion of shale gas increased from 5% in 2007 to more than 60% in 2023 [6]. Drawing on the success of shale oil and gas reform abroad, China has successfully established three national shale gas development areas: Fuling, Zhaotong and Changning [7]. As one of these development areas, after more than ten years of exploration and practice, the Changning area has proposed a development route to promote the concept of integrating geology and engineering and continuously improved their horizontal well fracture network technology [8–10]. By the end of 2023, the annual shale gas production in China's Changning region reached 52.13×10^8 m^3 [11]. However, in the Changning area, block Y214 still shows significant differences in productivity among shale gas wells under the same fracturing conditions [12]. The complexity of the fracture network morphology leads to differences in production capacity [13,14]. Fracture networks can directly affect the path of gas flow, limiting the productivity rate of shale gas wells and thereby reducing natural gas production [15]. Therefore, studying the impact of fracture network complexity on shale gas well productivity is critical to improving the development efficiency of this block.

Early scholars mainly used the brittleness index evaluation method to evaluate the compressibility of shale reservoirs [16–18]. However, this method is too simple to effectively characterize the ability of shale reservoirs to form complex fracture networks. To this end, Chong et al. [19] summarized the success of fracking transformation cases in the United States over the past 20 years and analyzed the ability to obtain shale gas from the fracturing section in the process of compressible hydraulic fracturing. Tang [20], Wang [21] and Liu et al. [22] constructed a compressibility evaluation model through an analytical hierarchy process to evaluate the influence of different geological and engineering factors on the effects of shale fracturing. Zhao et al. [23,24] calculated and analyzed the compressibility index of different shale reservoirs through a harmonic average method, proposed an evaluation standard that can scientifically characterize the compressibility grade of shale reservoirs and constructed the fracture network fracturing theory of shale reservoirs. Shi et al. [25] proposed a compressibility evaluation method based on the gray correlation method, which effectively solved the problem of the sensitivity of small samples, but the study did not consider the evaluation index of natural fractures. Wang et al. [26] constructed a shale reservoir fracturability evaluation model based on rock mechanics tests and comprehensively analyzed the correlation effect of geological and engineering factors on the fracture network, but the model ignored the influence of natural fractures on the formation of the fracture network. Joseph et al. [27] employed the Fault Upgrade Method (FUM) to demonstrate the impact of the distribution of complex fracture networks on different geological features and fracturing development data. Although the above evaluation methods can effectively integrate a variety of evaluation parameters, they are still prone to large errors when there is a lack of data.

The machine learning method can deal with different complex problems with limited data samples, and this method has already achieved preliminary results in the field of natural gas exploration and development. Lin [28] and Zhu [29] evaluated the hydraulic fracturing effect of shale reservoirs based on their previous research ideas and an artificial neural network model, but the model had problems such as a poor fitting effect and low prediction accuracy when dealing with nonlinear factors. Han et al. [30] combined the gray correlation method with the support vector machine algorithm to quantitatively evaluate the influence of shale reservoir fracture network complexity on gas production by using machine learning methods, but the traditional SVM algorithm still faces a bottleneck in hyperparameter selection when dealing with complex geological conditions and still has the problem of being prone to long local optimal and training times.

At present, there are the following problems in research on compressibility evaluation methods for shale reservoirs at home and abroad: firstly, the existing compressibility evaluation methods lack a quantitative analysis of natural fractures and secondly, traditional machine learning methods have problems such as a poor fitting effect and low prediction accuracy. Therefore, based on the on-site fracturing data of shale gas wells in the Y214 block in the Changning area, this paper comprehensively considers evaluation parameters, such as the mineral content, the horizontal stress difference and natural fractures, and uses the Pearson correlation coefficient to analyze the influence of each key factor on gas production. Combined with the IPSO algorithm and SVM algorithm, an FNI model with a small error, good fitting and high accuracy was proposed, and the accuracy of the model was verified by comparing it with the performance evaluation results of the other two traditional models. Finally, the proposed model was used to calculate and analyze the correlation between the fracture network index of different fracturing sections and the actual average daily gas production, which provides a scientific basis for improving the development efficiency of the block.

2. Data and Methods

2.1. Data

In this section, we shall take block Y214 in the Changning area as the research subject, conduct a systematic analysis of the geological characteristics and fracturing development data of various shale gas wells and explore the impact of geological and engineering factors on the productivity of shale gas wells.

2.1.1. Database Sample

The main geological and engineering factors used in this study are mineral content, horizontal stress difference coefficient, natural fracture density, number of perforations and construction displacement. These are used to figure out how the complexity of the fracture network affects the productivity of shale gas wells. These parameters directly influence the creation and propagation of fractures during the fracturing process, thereby impacting gas flow and overall output. The data are presented in Table 1.

Table 1. Comparison of geological characteristics and fracturing development data of block Y214 in Changning area, China.

Well Number	Fracturing Section	Mineral Content (%)	Horizontal Stress Difference Coefficient	Natural Fracture Density (Strip/Meter)	Number of Perforations	Construction Displacement (m^3/min)	Average Daily Gas Production (10^4 m^3/d)
Y202	1	63.4	0.102	1.78	3	10	1.7375
	2	64.2	0.102	1.78	3	11	1.7395
	…	…	…	…	…	…	…
	19	63.8	0.102	1.78	3	11	1.7598
Y203	1	69.1	0.12	0.2	3	8.8	1.5448
	2	70.9	0.12	0.2	6	8.8	1.5494
	…	…	…	…	…	…	…
	25	71.8	0.13	0.2	6	8.6	1.5524
Y205	1	77.9	0.12	0.14	3	12	8.0422
	2	76.3	0.12	0.14	8	14	8.6083
	…	…	…	…	…	…	…
	28	76.9	0.12	0.14	6	16	9.7609
Y206	1	82.9	0.14	0.2	3	15.5	1.2354
	2	82.9	0.14	0.2	6	15	1.5489
	…	…	…	…	…	…	…
	34	83.7	0.14	0.2	7	18	3.1289
Y207	1	81.8	0.166	0.7	3	13.5	1.1281
	2	81.8	0.166	0.7	6	15.5	1.2084
	…	…	…	…	…	…	…
	25	71.3	0.140	1.1	7	18	4.7098
Y215	1	70.6	0.140	1.1	7	18.5	5.6797
	2	69.0	0.140	1.1	7	19	5.8198
	…	…	…	…	…	…	…
	24	69.8	0.140	1.1	7	10	9.9027
Y251	1	65.5	0.132	0.2	3	18	2.1094
	2	64.8	0.132	0.2	10	18	2.1694
	…	…	…	…	…	…	…
	26	69.1	0.136	0.2	10	17	2.6734

2.1.2. Pearson Correlation Analysis

In order to clarify the influence of a single geological engineering factor on the average daily gas production, the linear correlation between the mineral content, the horizontal stress difference coefficient, the natural crack density, the perforation number and the construction displacement was selected to be quantitatively studied [31,32]. The results are shown in Figure 1.

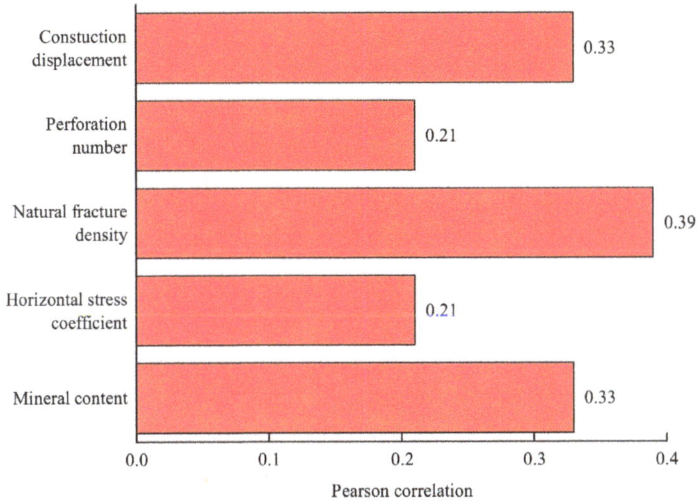

Figure 1. Pearson correlation analysis results.

Figure 1 illustrates that, among the effects of the five geological and engineering factors on the average daily gas production, the correlation with natural fracture density was the most significant (the correlation coefficient was 0.39). The correlation between mineral content and construction displacement was second only to natural crack density (with a correlation coefficient of 0.33). In contrast, the link between the horizontal stress coefficient and the number of perforations was weak (the correlation coefficient was 0.21). Studies have shown that during fracturing development, the natural fracture density is a crucial factor in controlling gas production.

However, Pearson correlation analysis was only used to study the linear correlation between single factors and gas production and failed to capture the more complex nonlinear interactions between geoengineering factors. Therefore, based on the influence law of the above single factors, combined with the proposed IPSO-SVM-FNI model, the nonlinear relationship between multiple factors is analyzed in depth.

2.1.3. Data Normalization

The geotechnical engineering factors have different units and data ranges. If the data are directly substituted into the model for calculation, it may cause some of the output values to be large and thus dominate the analysis of the results. This work normalizes the data selection equation to address this problem [33].

For positive indicators, take

$$S = \frac{P - P_{\min}}{P_{\max} - P_{\min}} \tag{1}$$

For the inverse indicator, take

$$S = \frac{P_{\max} - P}{P_{\max} - P_{\min}} \tag{2}$$

where S stands for the normalized value, P stands for the original value, $P_{\max}$ stands for the maximum value in the original database and $P_{\min}$ represents the minimum value in the original database.

2.2. Methods

2.2.1. Support Vector Machine

The SVM is known to be a valuable technique for supervised learning tasks, extensively utilized for its efficacy in addressing diverse classification and regression problems [34]. Its strong generalization capability across diverse datasets, along with its effectiveness in handling both linear and nonlinear problems, makes the SVM a popular choice in machine learning [35]. This research employs the SVM algorithm to determine the weight coefficients (w_1~w_5) of several geological engineering elements affecting gas production, which are utilized to develop the FNI model used for assessing the influence of fracture networks on production capacity during the fracturing process.

When there is a linear relationship between various geological engineering factors, the sample dataset can be divided into

$$(x_i, y_i), i = [1, n] \tag{3}$$

where x_i is the input vector value, $x_i \in R^d$, and y_i is the output vector value, $y_i \in [-1, 1]$.

The optimal classification surface is divided according to the sample:

$$f(x) = \omega \cdot x + b = 0 \tag{4}$$

The decision function for the linearly related data classification problem that can be obtained from the above formula is as follows:

$$f(x) = \mathrm{sgn}(\omega \cdot x_i + b) \tag{5}$$

The optimal classification surface can only partially separate the samples when there is a nonlinear relationship between various geological engineering factors. The slack factor can solve the problem of nonlinear data classification to a certain extent by reflecting nonlinear characteristics. Therefore, this paper adds the slack factor and the penalty coefficient to the objective function, and the objective function is given in Equation (6):

$$\min_{\omega,b,\xi} \frac{\|\omega\|^2 + 2C \cdot \sum_{i=1}^{m} \xi_i}{2} \tag{6}$$

The constraints are

$$y_i(\omega \cdot x_i + b) \geq 1 - \xi_i, i = 1, 2, \ldots, m, \xi_i = \max(0, 1 - y_i(\omega \cdot x_i + b)) \tag{7}$$

where ξ_i is the relaxation factor and C is the penalty coefficient.

The above formula can be used to convert the problem of obtaining an optimal classification surface into a quadratic programming problem with optimization parameters. Equation (8) is as follows:

$$\begin{cases} \max Q(\alpha) = \left\{ \sum\limits_{i=1}^{m} \alpha_i - \frac{1}{2} \sum\limits_{i=1}^{m} \sum\limits_{j=1}^{m} (\alpha x y)_i \cdot (\alpha x y)_j \right\}, \\ s.t.\ 0 \leq \alpha_i \leq C, \\ \sum\limits_{i=1}^{m} y_i \alpha_i = 0, \end{cases} \quad (8)$$

where $Q(\alpha)$ is the objective function and α_i and α_j are Lagrange multipliers.

According to the Kuhn–Tucker criterion [36], α_i must satisfy the following condition:

$$\alpha_i [y_i(\omega \cdot x + b) - 1 + \xi_i] = 0 \quad (9)$$

Substituting Equation (9) into Equation (8) gives the following function:

$$f(x) = \text{sgn}\left\{ \sum_{i=1}^{m} [(\alpha y)_i \cdot K(x, x_i)] + b \right\} \quad (10)$$

where m is the number of support vectors.

The radial basis function and kernel function are substituted into Equation (10) to obtain the final kernel function:

$$K(x, x_i) = \exp(-1) \cdot (g) \cdot \|x - x_i\|^2 \quad (11)$$

where g is the kernel function and $g > 0$.

2.2.2. Improved Particle Swarm Optimization

Kennedy et al. [37] created a particle swarm optimization (PSO) algorithm that simulates the foraging behavior of migratory bird groups. In comparison to alternative machine learning techniques, the PSO algorithm is straightforward and easy to use, needing less parameter matching to address basic optimization challenges. However, when solving complex optimization problems, the inertia factor, ω, is usually a fixed value or linearly decreasing, which leads to slow convergence in the early stages of the algorithm, and the algorithm is prone to becoming stuck in a local optimum in the later stages.

This work presents an IPSO approach to overcome these constraints. First, a nonlinear decreasing inertia factor is introduced to ensure strong global search capability in the early stages while accelerating convergence in the later stages. Second, adaptive mutation mechanisms are added to increase the population's diversity and prevent premature particle convergence. Finally, by dynamically adjusting the weight parameters, the adaptability to different optimization problems is enhanced, thereby improving the search efficiency and accuracy of the global optimal solution. The position and velocity equations for the IPSO algorithm are as follows:

$$f_i(h+1) = c_1 r_1 \cdot \left[p_{optimal,i} - x_i(h) \right] + c_2 r_2 \cdot \left[p_{optimal,g} - x_i(h) \right] + \omega \cdot f_i(h) \quad (12)$$

$$x_i(h+1) = x_i(h) + f_i(h+1) \quad (13)$$

where ω is the inertia factor and $f_i(h)$ is the velocity of particle i at time h. $x_i(h)$ represents the position of particle i at time h. The particle moves toward its local optimal location and its global optimal position with the help of learning factors called c_1 and c_2, respectively. The acceleration weights r_1 and r_2, which are randomly selected from the interval [0, 1], give the particle's motion some stochasticity. Furthermore, the local optimal position of particle i is denoted by $p_{optimal,i}$, while the global optimal position for the entire swarm is represented by $p_{optimal,g}$.

The algorithm's capacity for optimization and the rate of convergence are largely dependent on the value of the inertia factor, ω [38]. In this study, ω is designed to gradually decrease with the number of iterations, enhancing both exploration and convergence. The decay formula is as follows:

$$\omega = (\omega_0 - \omega_t) \cdot \left(\frac{t}{T}\right)^2 + \omega_t \tag{14}$$

where ω_0 is the initial inertia factor, ω_t is the inertia factor at the current iteration and t is the current iteration number. T is the maximum number of iterations.

In order to improve the diversity of the population, the adaptive mutation factor is used to randomly perturb the particles trapped in the local optimal solution so as to improve the accuracy of the model. The specific formula is as follows:

$$x_i(t+1) = x_i(t+1) + \sigma \cdot N(0,1) \tag{15}$$

where σ is the amplitude of the variation, which is related to the current population fitness distribution. N (0, 1) is a standard normally distributed random number.

Introducing Equations (14) and (15) into the traditional PSO algorithm can reduce the randomness of the learning process to a certain extent, thereby accelerating the convergence speed of PSO and improving its global search performance.

2.2.3. Fracture Network Index Algorithm

The FNI is a crucial parameter for assessing the complexity of the fracture network formed during hydraulic fracturing. In order to more accurately reflect the impact of the fracture network on the production performance of shale gas wells, the FNI calculation formula integrates several key geological and engineering factors. Equation (16) is as follows:

$$F_n = S_W \cdot w_1 + S_{K_h} \cdot w_2 + S_{N_f} \cdot w_3 + S_{N_p} \cdot w_4 + S_Q \cdot w_5 \tag{16}$$

where F_n is the fracture network index, W is the mineral content, K_h is the horizontal stress difference coefficient, N_f is the natural fracture density, Np is the number of perforations, Q is the construction displacement, S_W~S_Q are the normalized values of each factor, respectively, and w_1~w_5 are the weight coefficients of each factor, respectively.

These five factors were chosen based on their critical role in fracture formation and propagation. A significant association exists between the mineral composition of underground rocks and the brittleness of shale gas reservoirs. An increased mineral content correlates with a heightened probability of subsurface cracks and an accelerated rate of shale gas flow [39]. Yang et al. [40] have shown that a large horizontal stress difference will increase the heterogeneity and complexity of the fracture network and promote its development. The natural crack density determines the initial conditions of the fracture network and the interactions between fractures, which is crucial to the complexity of the fracture network [16]. The number of perforations and the construction displacement affect the distribution of the starting point of the crack and the propagation rate and scale of the crack, respectively [41,42]. Therefore, these factors can be used as evaluation parameters for the fracture network index.

2.2.4. Model Parameters

The main parameters of the SVM-FNI model were set as follows: penalty coefficient (c) = 2 and kernel function parameter (g) = 0.4. The main parameters of the PSO-SVM-FNI model were as follows: individual weight coefficient (c_1 = 1.5) and population weight coefficient (c_2 = 1.7), maximum number of iterations (T = 50), population size (i = 5), penalty coefficient, c, and search interval for the kernel function [0.1, 100]. Based on the traditional model, the IPSO-SVM-FNI model introduces a nonlinear attenuation formula for the inertia factor. The ω parameter was set to an initial value of ω_{start} = 2 and a final value of ω_{end} = 0.4. Table 2 lists the parameters set for the three different models.

Table 2. Model parameters.

Parameter \ Model	SVM-FNI	PSO-SVM-FNI	IPSO-SVM-FNI
c	2	-	-
g	0.4	-	-
c_1	-	1.5	1.5
c_2	-	1.7	1.7
T	-	50	50
i	-	5	5
$f_{max}(h)$	-	100	100
$f_{min}(h)$	-	0.1	0.1
$x_{max}(h)$	-	100	100
$x_{min}(h)$	-	0.1	0.1
ω_{start}	-	-	2
ω_{end}	-	-	0.4

2.2.5. Model Evaluation Index

Four evaluation indexes—the determination coefficient (R^2), mean absolute error (MAE), mean absolute percentage error (MAPE), and mean bias error (MBE)—were chosen in this work to assess the model's performance in data fitting [43]. The following are the precise calculation formulas:

$$R^2 = \sum_{i=1}^{n} \frac{(\hat{p}_i - c)^2 - (p_i - \hat{p}_i)^2}{(\hat{p}_i - \overline{p})^2} \tag{17}$$

$$\text{MAE} = \frac{1}{n} \sum_{i=1}^{n} |p_i - \hat{p}_i| \tag{18}$$

$$\text{MAPE} = \frac{1}{n} \frac{\sum_{i=1}^{n} |p_i - \hat{p}_i|}{\sum_{i=1}^{n} |p_i|} \times 100\% \tag{19}$$

$$\text{MBE} = \sum_{i=1}^{n} \frac{(\hat{p}_i - p_i)}{n} \tag{20}$$

where n is the number of samples in the dataset, p_i is the actual value, $\hat{p}_i$ is the predicted value and $\overline{p}$ is the mean of the real values.

2.2.6. Model Operation Steps

This study integrates the IPSO method with the SVM algorithm to formulate an FNI model. IPSO-SVM-FNI models effectively manage the interplay among many geological and engineering variables to precisely forecast natural gas output. The research findings can offer theoretical support and an engineering foundation for optimizing fracturing design and predicting production performance. The calculation steps of the model are as follows:

(1) Designate 127 fracturing sections (about 70% of the total) to the training set and the remaining 54 fracturing sections (approximately 30% of the total) to the test set and normalize the data accordingly.

(2) Set parameters such as the population size, learning factor, inertia factor and position and optimize the population parameters using the weighted decay mechanism of the IPSO algorithm.

(3) Assess the efficacy of the IPSO-SVM algorithm utilizing the evaluation metrics of a sample test set. If the coefficient of determination, R^2, MAE, MAPE and MBE meet the set requirements, output the weight factors (w_1 to w_5) of the input vector to the output vector.

(4) Construct the FNI model using the weighting factors obtained using the IPSO-SVM algorithm. Calculate the fracture network index of different fracturing sections of shale gas wells, and perform a correlation analysis with the average daily gas production.

For comparison, the computational steps of the PSO-SVM-FNI and SVM-FNI models are consistent with those of the IPSO-SVM-FNI model. Both traditional models adopt the same process for data division, normalization and parameter optimization for training and testing. The overall workflow and calculation process in this paper are shown in Figure 2.

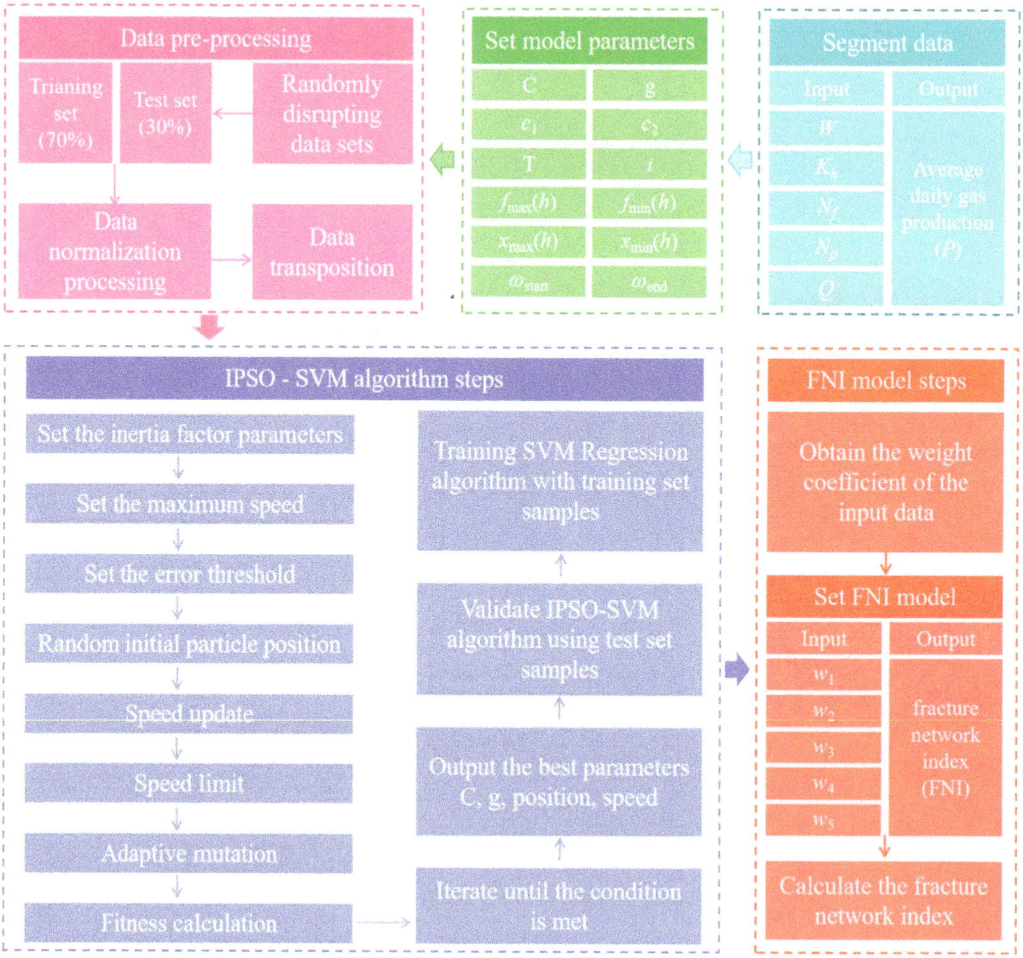

Figure 2. Workflow diagrams.

3. Results

3.1. The Weight Factor Calculation Results

This research employs the IPSO-SVM algorithm to determine the impact of geological engineering parameters on natural gas output. The weight coefficients (w_1 to w_5) are presented in Table 3.

Table 3. The weight coefficients of geological engineering factors in gas production in Y214 block.

Geotechnical Factor	Mineral Content (w_1)	Horizontal Stress Difference Coefficient (w_2)	Natural Fracture Density (w_3)	Number of Perforations (w_4)	Construction Displacement (w_5)
weight coefficient	0.0044	0.3839	0.4984	0.0959	0.0174

Table 3 shows that the natural fracture density has the highest weight coefficient (w_3 = 0.4954), which indicates that the natural fracture density is the geological factor that has the greatest impact on gas production. The higher the density of natural fractures, the more gas flow paths and the higher the gas production. The horizontal stress difference coefficient is the second most important factor, and its weight coefficient, w_2 = 0.3839, indicates that its contribution to gas production is also quite large. In the process of hydraulic fracturing, the existence of a horizontal stress difference will directly affect the formation and conductivity of fractures, which further determines the gas production performance of shale gas wells.

The weights for perforation number, construction displacement and mineral content were relatively small (w_4 = 0.0959, w_5 = 0.0174, w_1 = 0.0044), indicating that their influence on gas production was limited. However, in some studies, the above factors have a greater impact on gas production, which may be related to factors such as construction costs and reservoir type. Zhu et al. [44] found that increasing the number of perforations and construction displacement can help to increase gas production after fracturing, but this will also increase the construction costs. Cao et al. [45] pointed out that reservoirs with a higher mineral content usually have better fracturing effects, but too high a quartz or carbonate content may lead to a decrease in fracture conductivity, which may affect gas production performance. In summary, the specific role of these factors under different geological and engineering conditions still needs to be further explored.

3.2. Model Evaluation Index Results

The IPSO-SVM-FNI model is contrasted with and examined against the PSO-SVM-FNI and SVM-FNI models. Table 4 lists the results of the performance evaluation of the three models and shows the models numbered by performance.

Table 4. Comparison of the results of the performance evaluation of different models.

Model		① SVM-FNI	② PSO-SVM-FNI	③ IPSO-SVM-FNI
R^2	Training	0.7821	0.9012	0.9728
	Testing	0.7911	0.7778	0.9516
MAE	Training	1.5873	0.3797	0.1810
	Testing	2.2088	0.6307	0.2871
MAPE	Training	0.3869	0.1574	0.0658
	Testing	0.3715	0.2823	0.1312
MBE	Training	0.7187	0.0225	0.1334
	Testing	1.2717	0.2332	0.2066

Table 4 indicates that the R^2 values for the training and test sets of the model ③ are 0.9728 and 0.9516, respectively, accompanied by MAE values of 0.1810 and 0.2871, MAPE values of 0.0658 and 0.1312, and MBE values of 0.1334 and 0.2066. In comparison to model ② and model ①, model ③ enhanced the R^2 of the training set by 8% and 24%, respectively, while decreasing the MAE by 52% and 89%, the MAPE by 58% and 83%, and the MBE by 40% and 81%. In the test set, model ③ enhanced the R^2 by 22% and 20%, decreased the MAE by 54% and 87%, decreased the MAPE by 53% and 65%, and lowered the MBE by 11% and 84%, respectively. The results indicate that model ③ markedly surpasses the other two models in terms of optimization and overall performance.

Figure 3 illustrates the convergence trajectories of model ③ (black curve) and model ② (red curve). As can be seen from Figure 3, the black curve reaches the optimal solution in the 21st iteration, while the red curve reaches the optimal solution in the 35th iteration. The final convergence fitness value of the black curve (left Y-axis) is about 0.0084256, while the final fitness value of the red curve (right Y-axis) is about 0.0114766. The results show that model ③ converges faster and has higher accuracy.

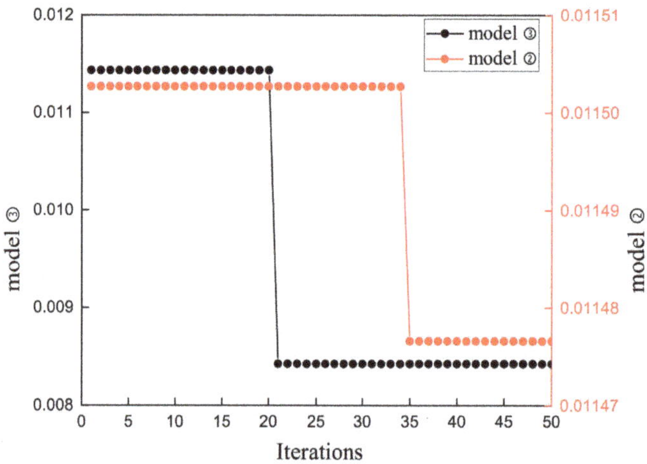

Figure 3. Comparison of the model fitness curves of model ③ and model ②.

The above results reflect the excellent nonlinear processing capacities of model ③. By combining the improved particle swarm optimization algorithm and the support vector machine algorithm, the complex nonlinear relationship between the geological and engineering factors can be efficiently captured, and the problem of traditional algorithms insufficiently fitting complex data can be overcome so as to achieve an accurate evaluation of the complexity of shale gas well fracture networks.

3.3. FNI Results

Based on the weight coefficients of the above key influencing factors, a seam index evaluation model for block Y214 was constructed. The correlation between the FNI value of each fracturing section of the shale gas well and the actual average daily gas production was calculated and analyzed, and the influence of fracture network complexity on productivity was quantitatively evaluated, as shown in Figure 4.

The results show that there is a significant positive correlation between the FNI and the average daily gas production, which is in line with the fracture network fracturing theory of shale reservoirs proposed by Zhao et al. [23,24]. The FNI values are mainly concentrated within the range of 0.2 to 0.8, and gas production increases notably as the FNI rises. When the FNI approaches or exceeds 0.7, the average daily gas production reaches approximately 8×10^4 m^3/d, demonstrating that higher FNI values correspond to greater gas production. Additionally, when the FNI is low (FNI < 0.2), gas production tends to remain small, typically around 2×10^4 m^3/d, indicating that simpler fracture networks result in weaker production capacity. In contrast, when the FNI is in the range of 0.4 to 0.6, gas production becomes more volatile, with significant fluctuations and strong dispersion. This suggests that the influence of varying geological conditions and engineering factors on gas production capacity becomes more complex within this range.

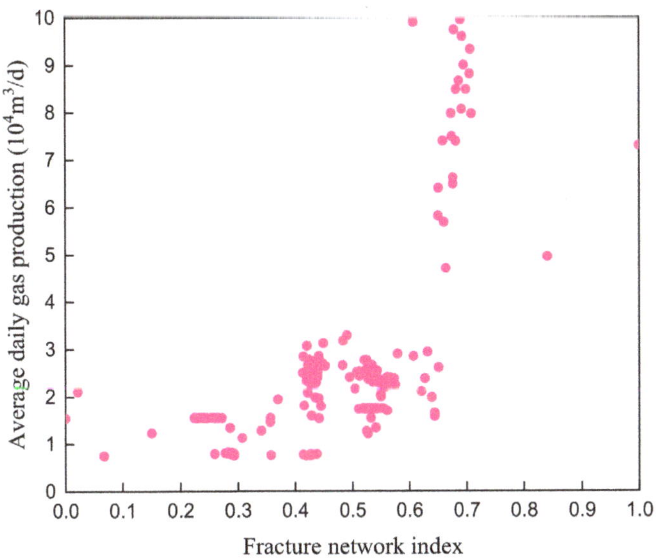

Figure 4. Correlation between the FNI of model ③ and the average daily gas production.

For the fracturing section with a low FNI (FNI < 0.2), the main factors are insufficient fracture development, poor reservoir geological conditions or unreasonable fracturing construction parameters. Under such conditions, the natural fractures are sparse or small, and the gas flow path is limited, resulting in poor reservoir connectivity and difficulty in effectively guiding deep gas in the wellbore. In order to improve the productivity performance of the low-fracture-network-index section, reservoir modification technology can be used to optimize the fracturing design; increase the amount of sand added, the amount of fracturing fluid and the construction displacement; and enhance the ductility and complexity of the fracture network.

For fracturing sections with a high FNI (FNI > 0.7), the high complexity of the fracture network and the significant increase in overall gas production may lead to potential problems such as excessive fracture development, which will lead to rapid pressure decay and increased production fluctuations, which will affect the production stability of gas wells. In order to solve the above problems, it is possible to limit the fracture propagation range in the high-FNI section to stop the fracture from entering the non-reservoir area. The construction parameters for the high-FNI section were refined and optimized to control the crack morphology and fluid distribution. A hierarchical production design was adopted to appropriately reduce the early production rate, prolong the production life of gas wells, and alleviate the problem of excessive production degradation.

4. Conclusions

Based on the on-site fracturing development of the Y214 block in the Changning area combined with the Pearson correlation coefficient, the IPSO-SVM algorithm was used to construct a complex FNI mode, which provided a multi-factor comprehensive evaluation method to quantify the fracture network complexity. The main conclusions are as follows:

(1) The linear relationship between individual geoengineering factors and the actual average daily gas production was analyzed through Pearson correlation. The natural crack

density, construction displacement and mineral content were significantly positively correlated with gas production, and their weight coefficients were 0.39, 0.33 and 0.33, respectively. However, this method is only suitable for studies on the linear effects of univariates and cannot capture the complex nonlinear interactions between multiple variables.

(2) In order to solve the problem of nonlinear interactions between complex geoengineering factors, an ISO-SVM-FNI model is constructed. By introducing an improved particle swarm optimization algorithm, the prediction accuracy and robustness of the model are significantly improved. The results show that the IPSO-SVM-FNI model performs well in capturing the multi-factor nonlinear relationship, and the R^2 of the training and test sets reaches 0.9728 and 0.9516, respectively, which is significantly better than that of the traditional PSO-SVM-FNI and SVM-FNI models, showing strong nonlinear processing capacities.

(3) There was a significant positive correlation between the FNI value and the actual average daily gas production. The FNI value is mainly concentrated in the range of 0.2 to 0.8, and the higher the FNI value, the greater the gas production of shale gas wells. When the FNI is close to 0.7 or higher, the average daily gas production is significantly increased, and conversely, when the FNI value is low, the gas production is significantly reduced, indicating that the complexity of the fracture network has a key impact on the capacity for gas production.

Author Contributions: Methodology, L.L.; Software, H.W. and B.Y.; Validation, L.D.; Investigation, W.L.; Data curation, B.L.; Writing—original draft, Z.G.; Writing—review & editing, P.Z. and H.L. All authors have read and agreed to the published version of the manuscript.

Funding: This research was funded by the following projects: (1) Chongqing Municipal Education Commission Science and Technology Research Project (key project): "Research on self-supporting effect and influencing factors of secondary fractures in shale gas reservoirs" (KJZD-K202201204); (2) Chongqing Social Science Planning Project: "Risk assessment of shale gas resources development and utilization in Chengdu-Chongqing economic circle" (2023SZ41); (3) Chongqing Basic Research and Frontier Exploration (surface project): "Self-supporting mechanism and conductivity evaluation of far-field branch fractures in shale gas reservoirs based on micro-contact mechanics" (CSTB2022NSCQ-MSX1135); (4) Chongqing Three Gorges Reservoir Bank Slope and Engineering Structure Disaster Prevention and Control Engineering Technology Research Center 2024 Open Fund: "Bridge Town Shale Oil Horizontal Well Productivity Prediction Model and Influencing Factors" (SXAPGC24YB02).

Data Availability Statement: The original contributions presented in the study are included in the article, further inquiries can be directed to the corresponding author.

Conflicts of Interest: Z.G., B.L., W.L. and L.L. were employed by Sichuan Changning Natural Gas Development Co., Ltd. The remaining authors declare that the research was conducted in the absence of any commercial or financial relationships that could be construed as a potential conflict of interest.

References

1. He, X.; Chen, G.; Wu, J.; Liu, Y.; Wu, S.; Zhang, J.; Zhang, X. Deep shale gas exploration and development in the southern Sichuan Basin: New progress and challenges. *Nat. Gas Ind.* **2022**, *42*, 24–34. [CrossRef]
2. She, C. New progress and development prospect of key technologies for drilling of deep shale gas wells in the Sichuan Basin. *Nat. Gas Ind.* **2024**, *44*, 1–9. [CrossRef]
3. Nie, H.; Dang, W.; Zhang, K.; Su, H.; Ding, J.; Li, D.; Liu, X.; Li, F.; Li, P.; Yang, S.; et al. Two decades of shale gas research & development in China: Review and prospects. *Nat. Gas Ind.* **2024**, *44*, 20–52. [CrossRef]
4. Eddine, A.; Peng, H.; Khetib, Y.; Ben, C. Natural gas flaring status in the Bakken shale play and potential remedial solutions. *Fuel* **2023**, *342*, 127807. [CrossRef]
5. Hou, L.; Xia, L.; Yu, Z.; Wu, S.; Zhao, H.; Lin, S. Key factors controlling the occurrence of shale oil and gas in the Eagle Ford Shale, the Gulf Coast Basin: Models for sweet spot identification. *J. Nat. Gas Sci. Eng.* **2021**, *94*, 104063. [CrossRef]
6. Alex, H.; Mark, Z. The relation between stimulated shear fractures and production in the Barnett Shale: Implications for unconventional oil and gas reservoirs. *Geophysics* **2019**, *84*, B461–B469. [CrossRef]

. Yong, R.; Chen, G.; Yang, X.; Huang, S.; Li, B.; Zheng, M.; Liu, W.; He, Y. Profitable development technology of the Changning Weiyuan National Shale Gas Demonstration Area in the Sichuan Basin and its enlightenment. *Nat. Gas Ind.* **2022**, *42*, 136–147. [CrossRef]

. Zhang, M.; Tang, Q.; Cao, C.; Lv, Z.; Zhang, T. Molecular and carbon isotopic variation in 3.5 years shale gas production from Longmaxi Formation in Sichuan Basin, China. *Mar. Pet. Geol.* **2018**, *89*, 27–37. [CrossRef]

. Han, C.; Chen, S.; Wang, F.; Li, W.; Yin, D.; Zhang, J.; Zhang, W.; Bai, Y. Data-driven hydraulic property analysis and prediction of two-dimensional random fracture networks. *Comput. Geotech.* **2024**, *171*, 106353. [CrossRef]

0. Wang, S.; Zhou, J.; Zhang, L.; Han, Z.; Kong, Y. Numerical insight into hydraulic fracture propagation in hot dry rock with complex natural fracture networks via fluid-solid coupling grain-based modeling. *Energy* **2024**, *295*, 131060. [CrossRef]

1. Tang, X.; Yang, H.; Zhu, H.; Dusseault, M.B.; Mclennan, J.D.; Li, K.; Xiao, J.; Zeng, B.; Zheng, M.; Lin, L.; et al. Investigation on Well Interferences of Infill Well Fracturing in Shale Gas Reservoirs Across Sichuan Basin. *Rock Mech. Rock Eng.* **2024**, 1–20. [CrossRef]

2. Shen, P.; Wu, J.; Fu, Y.; Zeng, B. Integrated dynamic evaluation of long lateral fracturing in shale gas wells: A case study on the Changning National Shale Gas Demonstration Area. *Nat. Gas Ind.* **2022**, *42*, 123–132. [CrossRef]

3. Wang, J.; Jia, A.; Wei, Y.; Wang, J.; Huang, X. A new method for evaluating tridimensional development effect of shale gas horizontal wells based on complex fracture network simulation: A case study of Longmaxi Formation shale gas in the southern Sichuan Basin. *Nat. Gas Ind.* **2022**, *42*, 175–189. [CrossRef]

4. Guo, J.; Zhao, Z.; Lu, Q.; Yin, C.; Chen, C. Research progress in key mechanical theories of deep shale network fracturing. *Nat. Gas Ind.* **2021**, *41*, 102–117. [CrossRef]

5. Wei, Y.; Wang, J.; Yu, W.; Dong, Y.; Miao, J. A smart productivity evaluation method for shale gas wells based on 3D fractal fracture network model. *Pet. Explor. Dev.* **2021**, *48*, 787–796. [CrossRef]

6. Guo, T.; Zhang, S.; Ge, H. A new method for evaluating ability of forming fracture network in shale reservoir. *Rock Soil Mech.* **2013**, *34*, 947–954. [CrossRef]

7. Xiao, J.; Li, Y.; Hou, Z.; Guo, Y.; Wang, L. A method for evaluating the brittleness of shale reservoirs. *Fault-Block Oil Gas Field* **2017**, *24*, 486–489. [CrossRef]

8. Zhang, C.; Xia, F.; Xia, Y.; Zhou, X. Comprehensive evaluation of tight sandstone reservoir compressibility based on analytic hierarchy process. *Drill. Prod. Technol.* **2021**, *44*, 61–64.

9. Chong, K.; Grieser, W.; Passman, A. A completions guide book to shale-play development: A review of successful approaches towards shale-play stimulation in the last two decades. In Proceedings of the Canadian Unconventional Resources and International Petroleum Conference, Calgary, AB, Canada, 19–21 October 2010. [CrossRef]

0. Tang, Y.; Xing, Y.; Li, L.; Zhang, B.H.; Jiang, S.X. Influence factors and evaluation methods of the gas shale fracability. *Earth Sci. Front.* **2012**, *19*, 356–363.

1. Wang, J.; Feng, M.; Yan, W.; Liu, S. Influence factors and evaluation methods for shale reservoir fracability in Jiaoshiba Area. *Fault-Block Oil Gas Field* **2016**, *23*, 216–220. [CrossRef]

2. Liu, X.; Wang, X.; Zhao, B.; Xiong, J.; Liang, L. Propagation of Hydraulic Fractures and Fracability Evaluation of Sandy Conglomerate Reservoirs. *Xinjiang Pet. Geol.* **2023**, *44*, 169–177. Available online: https://www.zgxjpg.com/CN/10.7657/XJPG20230206 (accessed on 26 November 2024).

3. Zhao, J.; Xu, W.; Li, Y.; Hu, J.Y.; Li, J.Q. A new method for fracability evaluation of shale gas reservoirs. *Nat. Gas Geosci.* **2015**, *26*, 1165–1172. Available online: http://www.nggs.ac.cn/CN/10.11764/j.issn.1672-1926.2015.06.1165 (accessed on 26 November 2024).

4. Zhao, J.; Ren, L.; Shen, C.; Li, Y. Latest research progresses in network fracturing theories and technologies for shale gas reservoirs. *Nat. Gas Geosci.* **2018**, *38*, 1–14. [CrossRef]

5. Shi, C.; Lin, B.; Xie, B.; Pan, Y.; Zhang, J. Evaluation method of shale reservoir fracability based on double sweet spots. *J. Shenzhen Univ. Sci. Eng.* **2024**, *41*, 183–191. [CrossRef]

6. Wang, X.; Liang, L.; Zhao, L.; Liu, X.; Qin, Z.; Li, W. Rock mechanics and fracability evaluation of the Lucaogou Formation oil shales in Jimusaer sag, Junggar Basin. *Oil Gas Geol.* **2019**, *40*, 661–668. [CrossRef]

7. Sherratt, J.; Haddad, A.S.; Wejzerowski, F.; Rafati, R. Optimising well orientation in hydraulic fracturing of naturally fractured shale gas formations. *J. Nat. Gas Sci. Eng.* **2021**, *94*, 104141. [CrossRef]

8. Lin, H.; Sun, X.; Song, X. A model for shale gas well production prediction based on improved artificial neural network. *Pet. Reserv. Eval. Dev.* **2023**, *13*, 467–473. [CrossRef]

9. Zhu, Q.; Lin, B.; Yang, G.; Wang, L.; Chen, M. Intelligent production optimization method for a low pressure and low productivity shale gas well. *Pet. Explor. Dev.* **2022**, *49*, 770–777. [CrossRef]

0. Han, S.; Che, M.; Su, W.; Xiao, M.; Wu, Z. Prediction Method and Application of Single Shale Gas Well Production in Weiyuan Block. Sichuan Basin. *Spec. Oil Gas Reserv.* **2022**, *29*, 141–149. [CrossRef]

1. Qi, N.; Li, X.; Wu, Z.; Wan, Y.; Wang, N.; Duan, G.; Wang, L.; Xiang, J.; Zhao, Y.; Zhan, H. Machine Learning-Based Research for Predicting Shale Gas Well Production. *Symmetry* **2024**, *16*, 600. [CrossRef]

2. Liu, Z. Method for selecting repeated fracturing wells in low-permeability sand stone reservoirs based on Pearson correlation coefficient. *Pet. Geol. Recovery Effic.* **2022**, *29*, 140–144. [CrossRef]

33. Rasouli, H.S.V.; Tokhmechi, B. A data processing algorithm proposed for identification of breakout zones in tight formations: A case study in Barnett gas shale. *J. Pet. Sci. Eng.* **2010**, *74*, 154–162. [CrossRef]
34. Radha, P.; Himansu, B.; Sibarama, P. A novel high order hesitant fuzzy time series forecasting by using mean aggregated membership value with support vector machine. *Inf. Sci.* **2023**, *626*, 494–523. [CrossRef]
35. Chen, Y.; Deng, X.; Wang, X.; He, Q.; Huang, D.; Cheng, L.; Zhang, Y.; Li, Q. Application of a PSO-SVM algorithm for predicting the TOC content of a shale gas reservoir: A case study in well Z in the Yuxi area. *Geophys. Prospect. Pet.* **2021**, *60*, 652–663. [CrossRef]
36. Ram, C.; Debdas, G. An erratum to "Extended Karush-Kuhn-Tucker Condition for Constrained Interval Optimization Problems and its Application in Support Vector Machines". *Inf. Sci.* **2021**, *559*, 309–313. [CrossRef]
37. Kennedy, J.; Eberhart, R. Particle swarm optimization. In Proceedings of the ICNN'95—International Conference on Neural Networks, Perth, WA, Australia, 27 November–1 December 1995; Volume 4, pp. 1942–1948. [CrossRef]
38. Wang, Z.; Sun, J.; Yin, C. A support vector machine based on an improved particle swarm optimization algorithm and its application. *J. Harbin Eng. Univ.* **2016**, *37*, 1728–1733. [CrossRef]
39. Sun, W.; Hou, M.; Liang, B.; Liu, Q.; Zhao, H. Study on the Fracture Extension Pattern of Shale Reservoir Fracturing under the Influence of Mineral Content. *ACS Omega* **2024**, *9*, 32525–32535. [CrossRef]
40. Yang, Y.; Xiao, L.; Yang, J. Influence mechanisms of plasticity and horizontal stress difference on the fracture propagation in plastic reservoir rocks: A 3D XFEM-based plastic fracturing model. *Geomech. Geophys. Geo-Energy Geo-Resour.* **2022**, *8*, 145. [CrossRef]
41. Zhang, Z.; Tang, T.Z.; Zhang, T.; Meng, W.; Li, L. Modeling of scale-dependent perforation geometrical fracture growth in naturally layered media. *Eng. Geol.* **2024**, *336*, 107499. [CrossRef]
42. Zhang, S.; Wu, F.; Huang, Y.; Wang, H.; Luo, R.; Yue, H.; Hou, B. Technological parameter optimization for improving the complexity of hydraulic fractures in deep shale reservoirs. *Nat. Gas Ind.* **2021**, *41*, 125–135. [CrossRef]
43. Jiang, R.; He, J.; Jiang, Y.; Fan, H. Establishment and application of Blasingame production decline analysis method for fractured horizontal well in shale gas reservoirs. *Acta Pet. Sin.* **2019**, *40*, 1503–1510. [CrossRef]
44. Zhu, H.; Song, Y.; Tang, X.; Li, K.; Xiao, J. Optimization of fracturing timing of infill wells in shale gas reservoirs: A case study on Well Group X1 of Fuling Shale Gas Field in the Sichuan Basin. *Nat. Gas Ind.* **2021**, *41*, 154–168. [CrossRef]
45. Cao, D.; Han, J.; Xiao, J.; Liu, Q.; Fu, F.; Liu, D. Method for evaluating the brittleness of shale minerals under the constraints of elastic characteristics. *Chin. J. Geophys.* **2023**, *66*, 4781–4791. [CrossRef]

energies

MDPI

Article

Numerical Study on Heat Leakage, Thermal Stratification, and Self-Pressurization Characteristics in Liquid Helium Storage Tanks

Jing Xu [1], Fa'an Liu [1], Jianguo Zhang [1], Chao Li [1], Qinghua Liu [1], Changjun Li [2,3], Wenlong Jia [2,3,*], Shixiong Fu [2] and Longjiang Li [2]

1 China Petroleum Engineering & Construction Corp. North China Company, Renqiu 062552, China
2 School of Petroleum Engineering, Southwest Petroleum University, Chengdu 610500, China; llj14799@163.com (L.L.)
3 CNPC Key Laboratory of Oil & Gas Storage and Transportation, Southwest Petroleum University, Chengdu 610500, China
* Correspondence: jiawenlong08@126.com

Abstract: During the operation of liquid-phase He-4 (LHe-4) storage tanks, heat leakage changes the thermophysical parameters and phase properties of the LHe-4 in the tanks, resulting in the thermal layering phenomenon. This phenomenon is characterized by the LHe-4 temperature gradient and pressure increase (self-pressurization) phenomena in the tanks. Based on the Layer-by-Layer model, a heat transfer model of a composite adiabatic structure with multilayer insulation and liquid nitrogen screen (LNCS) insulation was established, and the Neumann boundary heat flux of the thermal response model was determined. A numerical simulation model of the thermal response of a liquid helium storage tank was established. The spatial and temporal evolutions of the pressure distribution, natural convection characteristics, thermal stratification characteristics, and self-pressurization characteristics of the LHe-4 tank were investigated. Finally, the self-pressurization thermodynamic model of the LHe-4 storage tank was built based on the isothermal saturation and homogeneous model. It is shown that the predictive performance of the mLee model for the self-boosting characteristics (relative deviation of 14.32%) was significantly improved compared with that of the Lee model (relative deviation of 39.64%). The thermal stratification degree (TSD) of the tank increased with the operation time, with TSDs of 1.023, 1.028, and 1.036 at 1 h, 2 h, and 3 h, which exacerbated the self-pressurization of the tank. The wall surface in contact with the phase interface is a strong evaporation point, so the interfacial mass transfer rate maps show a pattern of high at both ends and low in the middle.

Keywords: helium-4; storage tank; heat leakage; thermal stratification; self-pressurization; numerical study

1. Introduction

He-4 has special properties, with a boiling point as low as 4.2 K, and it retains its liquid phase at temperatures close to 0 K [1]. LHe-4 undergoes a rare Lambda phase transition below 2.17 K [1]. Because of this, He-4 plays a pivotal role in cutting-edge scientific research fields such as space exploration, the defense and military industry, superconductivity, and medicine [2]. Being a rare and strategic resource [3], LHe-4 is used in infrared detectors, particle gas pedals, cancer treatment, and the low-temperature performance testing of experimental materials [4]. The liquid-phase storage of LHe-4 in deep cryogenic adiabatic tanks is the main He-4 storage method [2]. LHe-4 tanks are double-layer containers, with the inner tanks filled with LHe-4 and the outer tanks protecting the inner tanks and constituting a vacuum adiabatic space. The vacuum adiabatic space is alternately wrapped with radiant screens and spacers [2]. In contrast, a liquid nitrogen screen is installed to form a 77 K cold wall, thereby realizing the efficient adiabatic insulation of the LHe-4 storage tank.

The thermal stratification of tanks and the self-pressurization of tanks are due to heat transfer from the LHe-4 tank insulation structure. Thermal intrusion leads to natural convection driven by the density difference in the LHe-4 near the wall of the inner tank, which continuously migrates the heat intruded from the wall to the gas–liquid interface, and as a result, the heat spreads to the liquid-phase region, resulting in the phenomenon of thermal stratification. Furthermore, VHe-4 generated by evaporation accumulates in the gas-phase region, leading to the self-pressurization of the tank. The thermal stratification and self-pressurization of storage tanks not only lead to economic losses caused by the phase change of LHe-4, but the self-pressurization of storage tanks also poses a serious threat to the safe and efficient storage of LHe-4 [5]. This problem is common in the field of cryogenic storage. In the aerospace field, thermal response problems lead to up to 12% and 28% [6] losses of liquid hydrogen and liquid oxygen in liquid hydrogen and liquid oxygen tanks and up to 6% LNG loss during LNG transportation [7], and they lead to the self-pressurization and tumbling of LNG tanks. It is worth noting that the very low temperature and latent heat of the vaporization of LHe-4 (20.4 kJ/kg) lead to a higher evaporation rate and tank self-pressurization rate of LHe-4 during the thermal response than those of liquid hydrogen [8], liquid nitrogen, and LNG tanks, and the economic loss and safety risk due to the thermal response problems of LHe-4 tanks are higher than those of other tanks.

However, little research has been reported on the thermal stratification and self-pressurization of LHe-4 storage tanks. Hence, the engineering community lacks an in-depth understanding of the spatial and temporal evolution of the thermal stratification and self-pressurization characteristics of LHe-4 storage tanks. In recent years, scholars have used the finite volume method [9] to carry out the phase-change analysis of cryogenic storage tanks and have researched the thermal stratification [10], self-pressurization [11], and evaporation rate [12] in the thermal response process of cryogenic storage tanks.

Numerical simulation has become the main method to study the thermal response characteristics of cryogenic storage tanks. In the numerical simulation study of the thermal responses of cryogenic storage tanks, the Lee model [13] is the most widely used and successful model. Joseph et al. [14] investigated the effect of the thickness of the adiabatic layer on the pressure of the tanks and the thermal stratification characteristics of the liquid. Gursu et al. [15] investigated the self-augmentation rate and the gasification rate of a liquid hydrogen storage tank by using three different self-augmentation models. Kumar et al. [16] investigated the thermal stratification characteristics of cylindrical liquid hydrogen storage tanks under different height-to-diameter ratios based on the VOF model and Lee model. Tatom et al. [17] investigated the thermal stratification phenomenon of a rocket propellant liquid hydrogen storage tank when the bottom was heated, the side walls were heated, and the bottom side walls were heated simultaneously. Das et al. [18] investigated the natural convection and thermal stratification phenomena in water and liquid hydrogen storage tanks using numerical methods. Zhang et al. [19] investigated the thermal stratification and self-pressurization phenomena in liquid nitrogen storage tanks based on the VOF model and the Lee model with an evaporation–condensation frequency of 10^{-6} s^{-1}. Stewart et al. [20] analyzed the self-pressurization characteristics of the thermal response process of a liquid hydrogen storage tank based on the Schrage equation using Fluent 18.1 software. Kang et al. [21] and Gong et al. [22] investigated the thermal response characteristics of LNG and liquid hydrogen storage tanks. Gong et al. [22] developed a simplified model considering the thermal stratification zone to study the thermal response characteristics of liquefied petroleum gas (LPG) storage tanks under fire conditions. Liu et al. [23] investigated the thermal stratification characteristics of liquid hydrogen storage tanks by using the CFD method. Abdullah et al. [24] investigated the thermal response characteristics in LNG storage tanks by using the Lee model. Fu et al. [25] investigated the thermal response characteristics of liquid hydrogen storage tanks by using the VOF model with the Lee model. Fu et al. [25] investigated the thermal stratification characteristics of liquid hydrogen storage tanks using the VOF model and Lee model. Hands et al. [26] investigated the thermal stratification and rollover phenomenon of liquid nitrogen storage tanks using numerical methods. Lu et al. [27] investigated the thermal response in a low-temperature storage tank using the Lee model [13] and the VOF model. Li et al. [28] developed a numerical model of the thermal response of propellant tanks based on the VOF model and the Lee model by taking the capillary effect under microgravity into consideration.

Based on the existing research, many numerical studies have reported the thermal responses of cryogenic tanks, but the existing studies were mainly carried out on liquid hydrogen, liquid oxygen, liquid nitrogen, and LNG tanks, and there is no literature report on the study of the thermal response characteristics of LHe-4 tanks by using numerical methods, so the thermal stratification and self-pressurization characteristics of LHe-4 tanks are not yet clear. There are multiple challenges to address: Firstly, the LHe-4 tank employs a composite insulation structure comprising vacuum multilayer insulation and a liquid nitrogen screen to achieve efficient cryogenic preservation. Consequently, the primary task is to establish a heat transfer model for this insulation structure, enabling the precise determination of the heat flux density infiltrating into the LHe-4 tank. This serves as an essential boundary condition for the subsequent numerical model. Secondly, in the numerical investigation of the thermal response of LHe-4 storage tanks, a critical challenge lies in constructing a robust numerical model. This model should facilitate a quantitative analysis of both the thermal stratification within the tank and the rate of self-pressurization at the gas–liquid interface of the LHe-4. Addressing these core issues is imperative for a comprehensive understanding of the thermal behavior of the LHe-4 storage system.

To address the thermal response characteristics of LHe-4 storage tanks, first, the heat transfer model of the composite adiabatic structure with a vacuum multilayer and liquid nitrogen screen was established based on the Layer-by-Layer model [29], and the convective heat transfer boundary was used to obtain the Neumann boundary condition boundary of the thermal response model. Secondly, the numerical model of the thermal response of the LHe-4 tank was established based on the mass, kinetic energy, and energy conservation equations, VOF model, physical parameter equations, and Lee model, and the physical parameter equations and Lee model were loaded into the CFD solver using the UDF, based on the mLee model, combined with the typical LHe-4 storage tank self-pressurization engineering application data. Furthermore, the established numerical model was used to calculate and investigate the thermal stratification and self-pressurization characteristics of the LHe-4 storage tanks and their spatiotemporal evolution laws. Finally, a thermodynamic model was established based on the saturated homogeneous model to study the warming curves and self-pressurization curves of the tanks for long-term storage.

2. Heat Transfer Modeling of the Adiabatic Structure

A heat transfer model was developed to calculate the heat fluxes to provide the boundary conditions for the numerical modeling of the thermal response of the fluid domain of the tank. The heat leakage into the tank mainly includes the heat transfer from the vacuum multilayer adiabatic layer and that from the thermal bridges, such as supports and pipelines.

2.1. Assumptions

The following reasonable assumptions are put forward under the premise of ensuring the accuracy of the heat transfer calculation: (1) the spacer in the vacuum multilayer insulation structure has a small absorption effect on the radiation heat transfer, which is negligible; (2) the interlayer vacuum of the vacuum multilayer insulation structure is high, and the residual gas convection inside the multilayer insulation structure is not taken into account; (3) the expansion and delamination between the multilayer insulation materials are not taken into account; and (4) the vacuum multilayer insulation structure has a constant internal pressure, and the deterioration of the vacuum is not considered.

2.2. Heat Transfer Modeling

2.2.1. Heat Transfer in Composite Insulation Structures

For the LHe-4 tank in this study, we adopted a composite insulation structure combining vacuum multilayer insulation and liquid nitrogen screen insulation, and the liquid nitrogen screen was installed in the vacuum sandwich, close to the inner tank; therefore, according to the position of the liquid nitrogen screen, the vacuum multilayer insulation layer is divided into an inner insulation layer and an outer insulation layer. The inner and outer insulation layers are alternately arranged with several layers of insulation materials. Each layer of insulation material consists of one layer of radiation screen (double-sided aluminized polyester film) and one layer of spacer (polyester fiber paper).

Compared to the Lockheed Martin model [29], the Layer-by-Layer model [30] makes it easier to obtain the heat flux and temperature distribution [2] between two adjacent layers. Therefore, the Layer-by-Layer model was used for the heat transfer calculation of the adiabatic structure of the LHe-4 tank. The node distribution of the vacuum multilayer adiabatic structure of the LHe-4 tank is shown in Figure 1.

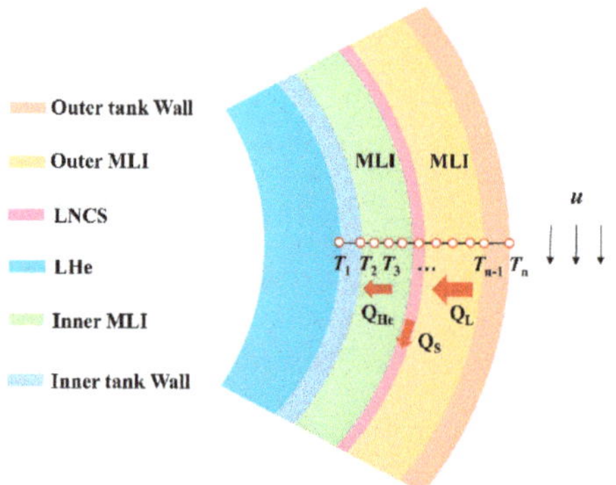

Figure 1. Nodal distribution of vacuum multilayer insulation structure of LHe-4 tanks.

The Layer-by-Layer model assumes that the heat transfer modes, such as the radiation heat transfer, residual gas heat conduction, and solid heat conduction, do not affect each other, and it calculates the heat transfer thermal resistance from the relationship between the heat transfer coefficients of each multilayer adiabatic material and then iteratively calculates according to the cold boundary conditions and the thermal boundary conditions to calculate the temperature distribution of the adiabatic structure [2]. Finally, the heat transfer coefficient and heat flux of the vacuum multilayer adiabatic material are then calculated separately. The relationship between the total heat transfer coefficient of the vacuum multilayer adiabatic structure and radiative heat transfer coefficient, residual gas heat transfer coefficient, and solid heat transfer coefficient is shown in Equation (1):

$$h_{\text{tot}} = h_{\text{rad}} + h_{\text{g-cond}} + h_{\text{s-cond}} \tag{1}$$

where h_{tot} represents the total heat transfer coefficient, W/(m^2·K); h_{rad} represents the radiant heat transfer coefficient, W/(m^2·K); $h_{\text{g-cond}}$ represents the residual gas heat transfer coefficient, W/(m^2·K); and $h_{\text{s-cond}}$ represents the solid heat transfer coefficient, W/(m^2·K).

Considering the heat transfer from the tank wall, the vacuum gap between the LNCS and the inner layer, and the vacuum gap between the outer tank wall and the outer layer, the tank wall–vacuum multilayer adiabatic layer–vacuum gap heat transfer model can be established, as shown in Equations (2)–(6):

$$Q_{\mathrm{iw}} = \frac{2\pi k_{\mathrm{iw}} L (T_2 - T_1)}{\ln(d_{\mathrm{iw,od}}/d_{\mathrm{iw,id}})} \tag{2}$$

$$Q_{\mathrm{rad}} = \sigma\left(T_i^4 - T_{i-1}^4\right)/\left(\frac{1}{\varepsilon_i} + \frac{1}{\varepsilon_{i-1}} - 1\right)\pi d_i L \tag{3}$$

$$Q_{\mathrm{gc}} = \frac{\gamma+1}{\gamma-1}\sqrt{\frac{R}{8\pi M T_g}}\, p_{\mathrm{abs}} \alpha c_{\mathrm{p}} c_{\mathrm{t}} (T_i - T_{i-1})\pi d_i L \tag{4}$$

$$Q_{\mathrm{sc}} = \frac{E_{\mathrm{C}} f_{\mathrm{s}} k_{\mathrm{s}}}{\Delta x}(T_i - T_{i-1})\pi d_i L \tag{5}$$

$$Q_{\mathrm{ow}} = \frac{2\pi k_{\mathrm{ow}} L (T_n - T_{n-1})}{\ln(d_{\mathrm{ow,od}}/d_{\mathrm{ow,id}})} \tag{6}$$

where Q_{iw} and Q_{ow} represent the heat passing through the wall surface of the inner tank and outer tank, W; Q_{rad}, Q_{gc}, and Q_{sc} represent the heat of the radiation heat transfer, residual gas heat conduction, and solid heat conduction in the multilayer adiabatic layer, W; T_1 and T_2 represent the temperatures of the inner wall of the inner tank and the outer wall of the inner tank, K; $T_{\mathrm{n}-1}$ and T_{n} represent the temperatures of the inner wall of the outer tank and the outer wall of the outer tank, K; k_{iw} and k_{ow} denote the thermal conductivities of the inner and outer tank walls, W/(m·K); L denotes the length of the cylindrical part of the inner tank, m; $d_{\mathrm{iw,od}}$ and $d_{\mathrm{iw,id}}$ denote the outer and inner diameters of the cylindrical part of the inner tank, m; $d_{\mathrm{ow,od}}$ and $d_{\mathrm{ow,id}}$ denote the outer and inner diameters of the cylindrical part of the outer tank, m; σ denotes the value of the Stephan–Boltzmann constant with a value of 5.675×10^{-8} W/(m^2·K^4); T_{i} and $T_{\mathrm{i}-1}$ denote the temperatures of the high- and low-temperature radiation screens, K; ε_{i} and $\varepsilon_{\mathrm{i}-1}$ denote the emissivities of the high- and low-temperature radiation screens, which are calculated in the temperature region of 4–300 K by using the formula in the literature [31]; d_{i} denotes the diameters of the positions of the radiation screens of the various layers, m; p_{abs} denotes the gas pressure of the adiabatic structure, with the unit of pascal; α denotes the thermal adaptation coefficient; M denotes the molar mass of the gas, g/mol; γ denotes the specific heat capacity ratio; c_{p} and c_{t} denote the temperature and pressure correction coefficients, which were calculated using the equations in the literature [32]; E_{C} denotes the empirical constants; f_{s} denotes the relative density of the spacer material concerning the solid material; Δx denotes the actual thickness of the spacer, m; k_{s} denotes the thermal conductivity of the spacer, W/(m·K).

The thermal resistance of the wall surface of the inner tank, the thermal resistance of the adiabatic layer, the thermal resistance of the vacuum interlayer, and the thermal resistance of the wall surface of the outer tank are connected in series to form the total heat transfer thermal resistance, as shown in Equation (7). The thermal resistance of the adiabatic layer includes the radiation heat transfer thermal resistance, solid thermal conductivity thermal resistance, and residual gas thermal conductivity thermal resistance, which are connected in parallel and are calculated using Equation (8). The vacuum sandwich thermal resistance is the result of the parallel connection of the radiation thermal resistance and residual gas thermal conductivity thermal resistance, which is calculated using Equation (9):

$$R_{\mathrm{tot}} = R_{\mathrm{iw}} + R_{\mathrm{MLI}} + R_{\mathrm{vac}} + R_{\mathrm{ow}} \tag{7}$$

$$\frac{1}{R_{\mathrm{MLI}}} = \frac{1}{R_{\mathrm{rad}}} + \frac{1}{R_{\mathrm{gc}}} + \frac{1}{R_{\mathrm{sc}}} \tag{8}$$

$$\frac{1}{R_{\mathrm{vac}}} = \frac{1}{R_{\mathrm{rad}}} + \frac{1}{R_{\mathrm{gc}}} \tag{9}$$

where R_{tot} represents the total heat transfer thermal resistance of the adiabatic layer, K/W; R_{MLI} represents the thermal resistance of the adiabatic layer, K/W; R_{rad}, R_{gc}, and R_{sc} represent the thermal resistances to radiant heat transfer, residual gas thermal conductivity, and solid thermal conductivity in the adiabatic structure, K/W; R_{vac} represents the thermal resistance of the vacuum interlayer, K/W; R_{iw} represents the thermal resistance of the wall surface of the inner can, K/W; and R_{ow} represents the thermal resistance of the wall surface of the outer can, K/W.

2.2.2. Heat Transfer in Thermal Bridges

The tank is installed with piping, such as an inlet pipe, outlet pipe, overflow pipe, and vent pipe, which are also the main heat transfer thermal bridges of the tank. There are nine thermal bridges between the support columns and support rods of the outer adiabatic layer and eight thermal bridges between the support rods of the inner adiabatic layer, among which, to reduce the heat flux, the support columns of the outer adiabatic layer are designed as an "S" structure to extend the thermal bridges, and the thermal resistances of the support rods of the inner and outer adiabatic layers are in a parallel relationship. Equations (10)–(13) show the calculation model of the heat flux and thermal resistance of the supporting thermal bridges of the inner and outer adiabatic layers:

$$Q_{os} = \frac{(T_{owo} - T_{LNCS})}{R_{os}} \tag{10}$$

$$Q_{is} = \frac{(T_{LNCS} - T_{iwo})}{R_{is}} \tag{11}$$

$$\frac{1}{R_{os}} = \frac{8k_{osup}A_{osup}}{L_{osup}} + \frac{\pi k_{ocyl}\left(d^2_{ocyl,od} - d^2_{ocyl,id}\right)}{4L_{ocyl}} \tag{12}$$

$$R_{is} = \frac{L_{isup}}{8k_{isup}A_{isup}} \tag{13}$$

where Q_{is} and Q_{os} denote the heat fluxes of the thermal bridge supported by the inner and outer insulation layers, W; T_{iwo} and T_{owo} denote the temperatures of the outer wall surface of the inner and outer tanks, K; T_{LNCS} denotes the liquid nitrogen screen temperature, K; R_{is} and R_{os} denote the thermal resistances of the inner and outer adiabatic layers to support the thermal bridge, K/W; k_{ocyl}, k_{osup}, and k_{isup} denote the coefficients of thermal conductivity of the support columns and rods located in the outer adiabatic layer and the support rods located in the inner adiabatic layer, W/(m·K); L_{ocyl}, L_{osup}, and L_{isup} denote the lengths of the support columns and support rods located in the outer insulation layer and the support rods located in the inner insulation layer, m; A_{isup} and A_{osup} denote the cross-sectional areas of the support rods located in the inner and outer insulation layers, m^2; and $d_{ocyl.id}$ and $d_{ocyl.od}$ denote the inner and outer diameters of the support columns in the outer insulation layer, m.

The pipeline thermal bridge of the LHe-4 tank starts at the wall surface of the outer tank and connects to the wall surface of the inner tank after passing through the liquid nitrogen screen. Considering that the liquid nitrogen screen is in contact with the pipeline thermal bridge and absorbs the heat from the pipeline thermal bridge, the pipeline thermal bridge is divided into the outer adiabatic layer pipeline thermal bridge and the inner adiabatic layer pipeline bridge, and Equations (14)–(17) are shown for the calculation model of the heat flux and thermal resistance of the pipeline thermal bridge:

$$Q_{op} = \frac{(T_{owo} - T_{LNCS})}{R_{op}} \tag{14}$$

$$Q_{ip} = \frac{(T_{LNCS} - T_{iwi})}{R_{ip}} \tag{15}$$

$$\frac{1}{R_{\text{op}}} = \sum_{\text{i}=1}^{4} \frac{\pi k_{\text{pi}}\left(d_{\text{pi,od}}^2 - d_{\text{pi,id}}^2\right)}{4L_{\text{opi}}} \tag{16}$$

$$\frac{1}{R_{\text{ip}}} = \sum_{i-1}^{4} \frac{\pi k_{\text{pi}}\left(d_{\text{pi,od}}^2 - d_{\text{pi,id}}^2\right)}{4L_{\text{ipi}}} \tag{17}$$

where Q_{op} and Q_{ip} indicate the heat fluxes of the thermal bridges of the inner and outer adiabatic layer pipelines, W; R_{ip} and R_{op} indicate the thermal resistances of the thermal bridges of the inner and outer adiabatic layer pipelines, K/W; i indicates the number of pipelines; k_{pi} indicates the thermal conductivity of the pipeline, W/(m·K); L_{ipi} and L_{opi} indicate the lengths of the thermal bridges of the inner and outer adiabatic layer pipelines, m; $d_{\text{pi.od}}$ indicates the inner and outer diameter of the thermal bridge of the pipeline, m; $d_{\text{pi. id}}$ and $d_{\text{pi.od}}$ indicate the inner and outer diameters of the thermal bridge of each pipeline, m.

2.3. Boundary Conditions

The Robin boundary, the convective heat transfer boundary condition, is used to solve the above heat transfer model to obtain the heat flux of the tank, and the expressions of the convective heat transfer boundary condition are shown in Equations (18) and (19):

$$-k_{\text{ow}} \left.\frac{\partial T}{\partial r}\right|_{\text{ow}} = h_{\text{c}}(T_{\text{owo}} - T_{\text{air}}) \tag{18}$$

$$\left.\frac{\partial T}{\partial r}\right|_{\text{ow}} = \frac{2}{d_{\text{owo}}} \frac{T_{\text{owo}} - T_{\text{owi}}}{\ln(d_{\text{owo}}/d_{\text{owi}})} \tag{19}$$

where k_{ow} denotes the thermal conductivity of the outer tank wall surface, W/(m·K); h_{c} denotes the heat transfer coefficient of the outer tank wall surface and the environment, W/(m^2·K); T_{air} denotes the ambient temperature, K; d_{owi} and d_{owo} denote the diameters of the inner and outer wall surfaces of the outer tank, m.

Equations (18) and (19) can be deduced from the column coordinates of the outer wall surface temperature of the outer tank, convective heat transfer coefficient, and ambient temperature of the function between the relationship shown in Equation (20):

$$T_{\text{owo}} = \frac{T_{\text{owi}} + \frac{h_{\text{c}} d_{\text{owo}} \ln(d_{\text{owo}}/d_{\text{owi}})}{2k_{\text{ow}}} T_{\text{air}}}{\left(\frac{h d_{\text{owo}} \ln(d_{\text{owo}}/d_{\text{owi}})}{2k_{\text{ow}}} + 1\right)} \tag{20}$$

During the operation of LHe-4 tanks, the tanks are usually installed in awnings or the corners of buildings [19]; therefore, natural convection boundary conditions were used to carry out the heat flux calculations of the tanks. At the same time, the heat fluxes of the storage tanks with forced convection boundary conditions were calculated for comparison. The surface heat transfer coefficient (h_{c}) was calculated using Equation (21):

$$h_{\text{c}} = \frac{\text{Nu} d_{\text{owo}}}{k_{\text{air}}} \tag{21}$$

where Nu is the Nusselt dimensionless number, the physical meaning of which is the dimensionless temperature gradient on the wall of the outer tank; k_{air} is the thermal conductivity of the air, W/(m·K).

According to the cylindrical structure of the LHe-4 tank, the natural convection and forced convection heat transfer modes on the outer wall of the tank are considered according to the natural convection in the large space of the transverse cylinder and the forced convection in the air-swept single tube. The Nu numbers of natural convection and forced convection are calculated by using the experimental correlation formula of natural convection in the large space of the transverse cylinder and the Churchill–Bernstein criterion formula of forced convection in the air swept out of a single round tube, and Equations (22) and (23) are shown [33] as the corresponding experimental correlation formula of natural convection and the criterion formula of forced convection:

$$\mathrm{Nu} = n_1(\mathrm{Gr}\cdot\mathrm{Pr})^{n_2} \tag{22}$$

$$\mathrm{Nu} = 0.3 + 0.62\mathrm{Re}^{m_1}\mathrm{Pr}^{m_2}\left(1+\left(\frac{\mathrm{Re}}{282000}\right)^{m_3}\right)^{m_4} \Big/ \left(1+\left(\frac{0.4}{\mathrm{Pr}}\right)^{m_5}\right)^{m_6} \tag{23}$$

where n_1, n_2, and m_1–m_6 denote the exponents of the equations, as specified in Table 1. Gr denotes the Grashof number, a measure that physically characterizes the ratio of the buoyancy force to the viscous force, and it is calculated using Equation (24), and Pr denotes the Prandtl number, a measure that physically characterizes the ratio of the momentum diffusion capacity to the heat diffusion capacity, and it is calculated using Equation (25).

$$\mathrm{Gr} = \frac{g\alpha_{\mathrm{vair}}\Delta T\rho_{\mathrm{air}}^2 d_{\mathrm{owo}}^3}{\nu_{\mathrm{air}}^2} \tag{24}$$

$$\mathrm{Pr} = \frac{\nu_{\mathrm{air}}c_{\mathrm{pair}}}{k_{\mathrm{air}}} \tag{25}$$

where α_{vair} is the coefficient of the body expansion of the air; g is the acceleration of gravity, 9.81 $\mathrm{m/s^2}$; ΔT is the temperature difference, K; ρ_{air} is the density of the air, $\mathrm{kg/m^3}$; ν_{air} is the kinetic viscosity of the air, Pa·s; k_{air} is the thermal conductivity of the air, W/(m·K); d_{owo} is the outer diameter of the external tank, m; c_{pair} is the constant-pressure specific heat of the air, J/(kg·K).

Table 1. Parameters in the Nu number correlation formula and criterion number.

Border Classification	Gr Range		n_1		n_2	
Natural convection	>4.65 × 10⁹ → $>4.65 \times 10^9$		0.11		1/3	
	5.76×10^8~4.65×10^9		0.0165		0.42	
	1.43×10^4~5.76×10^8		0.48		1/4	
Forced convection	m_1	m_2	m_3	m_4	m_5	m_6
	1/2	1/3	5/8	4/5	2/3	1/4

2.4. Solution Procedure

The heat transfer calculation model of the LHe-4 tank includes the heat transfer module of the vacuum multilayer adiabatic layer and the heat transfer module of the thermal bridge, and MATLAB 2021a is used for the programming and solving. Figure 2 shows the flow chart of the LHe-4 tank heat transfer model. The heat transfer model of the adiabatic structure mainly includes the heat transfer calculation of the adiabatic structure, the heat transfer calculation of the thermal bridge, the heat leakage calculation of the inner tank, and the calculation of the maximum and minimum critical liquid nitrogen flow rates of the liquid nitrogen screen. When carrying out the heat transfer calculation of the adiabatic structure, the temperature distribution of the multilayer adiabatic structure is updated by using Equation (26), and the calculation is iterated repeatedly until the residual difference between the heat flux on the wall surface of the outer tank and the heat flux of the outer adiabatic layer are less than 0.01 W.

$$T_i = T_1 + \frac{\sum_{0}^{i} R_i}{R_{\mathrm{tot}}}(T_n - T_1) \tag{26}$$

where T_i is the temperature of the current node, K; T_1 is the temperature of the first node, K; T_n is the temperature of the last node, K; $\sum R_i$ is the sum of the thermal resistances before the current node, K/W; R_{tot} is the total heat transfer thermal resistance, K/W.

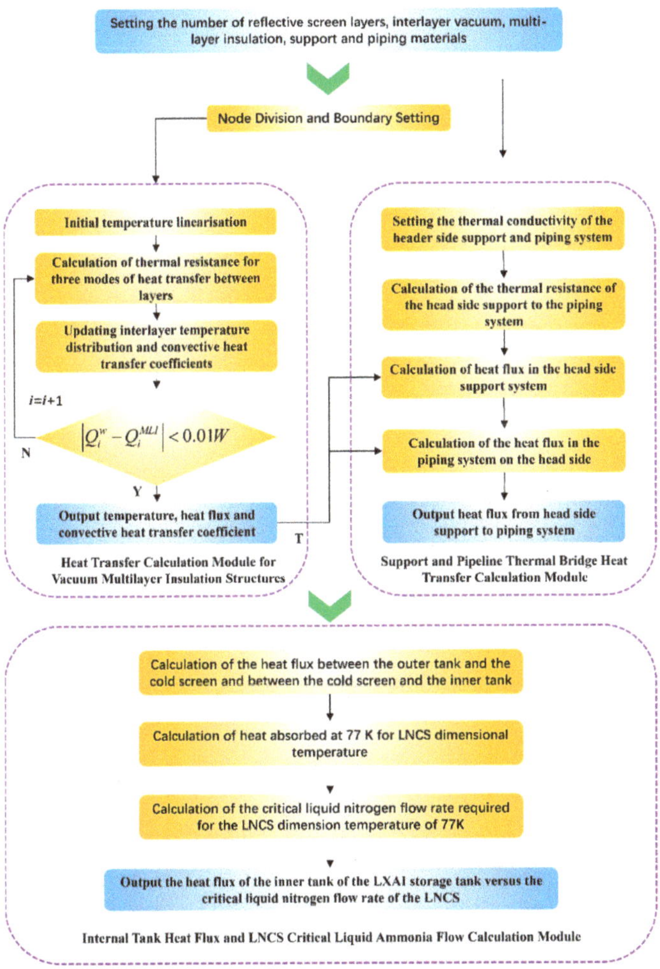

Figure 2. Methane–alcohol system liquid-phase component fraction prediction result.

In solving the thermal bridge heat transfer calculation module, Fourier's law is used to calculate the heat flux of the thermal bridge based on the temperatures of the inner wall surface of the inner tank and the outer wall surface of the outer tank obtained from the vacuum multilayer adiabatic heat transfer module [2]. To ensure the maximum performance of the LNCS, the temperature of the LNCS needs to always be maintained at around 77 K, and therefore a critical value of the liquid nitrogen flow in the LNCS coil exists. In addition, the literature states that the liquid nitrogen evaporation rate in the liquid nitrogen coil of the LHe-4 tank is between 10% and 15%; therefore, the maximum critical flow rate (m_{max}) (10% liquid nitrogen evaporation rate) and the minimum critical flow rate (m_{min}) (15% liquid nitrogen evaporation rate) of the LNCS coil can be calculated based on the liquid nitrogen evaporation rate.

3. Thermal Stratification–Self-Pressurization Modeling

The thermal response process in LHe-4 tanks includes key issues such as natural convection, thermal stratification, self-pressurization, and inter-phase heat and mass transfer. Firstly, a mathematical model of the thermal response of LHe-4 tanks is established based on the basic control equation describing the conservation of mass, momentum, and energy in the flow field, a turbulence model with a closed set of N-S equations, and a source term calculation model describing the inter-phase mass and energy transfer.

3.1. The VOF Model

We track the phase interface using the VOF model [24]. For the two-phase flow problem, the volume fraction equation for the *q*th phase is shown in Equation (27), and the volume fraction of the other phase is obtained using Equation (28):

$$\frac{1}{\rho_q}\left[\frac{\partial(\alpha_q\rho_q)}{\partial t} + \nabla\cdot(\alpha_q\rho_q v_q) = \sum_{p=1}^{n}\dot{m}_{pq} - \dot{m}_{qp}\right] \quad (27)$$

$$\sum_{q=1}^{n}\alpha_q = 1 \quad (28)$$

where ρ_q is the density of the *q*th phase, kg/m^3; α_q is the volume fraction of the *q*th phase. $\alpha_q = 0$ means that the cell is all liquid; $\alpha_q = 1$ means that the cell is all gas; $0 < \alpha_q < 1$ means that the cell contains the gas and liquid phases; v_q is the velocity of the *q*th phase, m/s; and the term on the right side of the equal sign in Equation (27) is the source term.

In the VOF model, the velocity field obtained from the solution is shared between the gas and liquid phases, and the momentum equation is shown in Equation (29):

$$\frac{\partial}{\partial t}(\rho v) + \nabla\cdot(\rho v v) = -\nabla p + \nabla\left(v\left(\nabla v + \nabla v^{T}\right)\right) + \rho g \quad (29)$$

where p denotes the pressure term.

The energy equation is also shared between the gas and liquid phases, and the energy (E) and temperature (T) are obtained by mass averaging. This is shown in Equation (30):

$$\frac{\partial}{\partial t}(\rho E) + \nabla\cdot(v(\rho E + p)) = \nabla\cdot(k_e\nabla T) + S_h \quad (30)$$

where k_e denotes the effective thermal conductivity; S_h denotes the energy source term (i.e., the heat exchange between the phases due to evaporation–condensation).

3.2. Turbulence Model

The *k-ε* model was selected among the commonly used turbulence models, such as the *k-ε* series model, the *k-w* series model, and the Reynolds pressure model [34]. The k and ε equations of the *k-ε* model are shown in Equations (31) and (32):

$$\frac{\partial}{\partial t}(\rho k) + \frac{\partial}{\partial x_j}(\rho k u_j) = \frac{\partial}{\partial x_j}\left[\left(\mu + \frac{\mu_t}{\sigma_k}\right)\frac{\partial k}{\partial x_j}\right] + G_k + G_b - \rho\varepsilon - Y_M \quad (31)$$

$$\frac{\partial}{\partial t}(\rho\varepsilon) + \frac{\partial}{\partial x_j}(\rho\varepsilon u_j) = \frac{\partial}{\partial x_j}\left[\left(\mu + \frac{\mu_t}{\sigma_\varepsilon}\right)\frac{\partial \varepsilon}{\partial x_j}\right] + \rho C_1\varepsilon S - \rho C_2\frac{\varepsilon^2}{k + \sqrt{v\varepsilon}} + \frac{\varepsilon}{k}C_{1\varepsilon}C_{3\varepsilon}G_b \quad (32)$$

where G_k denotes the turbulent kinetic energy due to the mean velocity gradient; Gb denotes the turbulent kinetic energy due to buoyancy; and Y_m denotes the effect of the pulsating expansion of the compressible turbulence on the total dissipation rate. $C_2 = 1.9$, $C_{1\varepsilon} = 1.44$, $C_{3\varepsilon}$ is an empirical constant, and σ_k and σ_ε denote the turbulent Prandtl numbers for the k and ε equations, $\sigma_k = 1.0$ and $\sigma_\varepsilon = 1.2$.

3.3. Phase-Change Model

The accurate calculation of the source terms of the continuity and energy equations due to the LHe-4 phase transition is the key to numerical calculations in the thermal response process of LHe-4 storage tanks. We adopt the previously modified Lee model, and its results are shown in Equations (33) and (34). The energy transfer accompanying the mass transfer process can be calculated using Equation (35), and the saturation temperature of the LHe-4 is fitted using data from NIST 1334:

$$T_l > T_{sat} \quad \dot{m}_{lv} = \mathrm{MCoe_e}\alpha_l\rho_l\left|\frac{T_l - T_{sat}}{T_{sat}}\right| = \varphi\Pi_e\mathrm{Coe_e}\alpha_l\rho_l\left|\frac{T_l - T_{sat}}{T_{sat}}\right| \quad (33)$$

$$T_v < T_{sat} \quad \dot{m}_{vl} = \mathrm{MCoe_c}\alpha_v\rho_v\left|\frac{T_v - T_{sat}}{T_{sat}}\right| = \varphi\Pi_c\mathrm{Coe_c}\alpha_v\rho_v\left|\frac{T_v - T_{sat}}{T_{sat}}\right| \quad (34)$$

$$S_h = S_m\Delta_l^v H \quad (35)$$

$$T_{sat} = 1.3495 \times p^{0.2486} \quad (36)$$

where T_{sat} denotes the saturation temperature, K; $\Delta_l^v H$ is the latent heat of the evaporation, J/kg; the subscripts l and v denote the liquid and gas phases; p denotes the pressure, kPa.

The evaporation frequency increases from 6.885×10^{-4} s^{-1} to 2.849×10^{-3} s^{-1} nonlinearly, while the condensation frequency increases from 1.609×10^{-4} s^{-1} to 6.519×10^{-4} s^{-1} nonlinearly. The evaporation frequency and condensation frequency temperature correlation equations shown in Equations (37) and (38) were developed using the fitting method. The relative deviations of the evaporation frequency–temperature correlation equations ranged from 0.27% to 9.94% with an AARD of 3.02%, while the relative deviations of the condensation frequency–temperature correlation equations ranged from 0.29% to 10.12% with an AARD of 3.06%. The parameters of the evaporation–condensation frequency–temperature correlation equation for the mLee model are shown in Table 2.

$$\mathrm{MCoe_e} = 0.1\left(\kappa_{e1} + \kappa_{e2}\left|\frac{T}{T_c} - \kappa_{e3}\right|^{\kappa_{e4}}\right) \quad (37)$$

$$\mathrm{MCoe_c} = \kappa_{c1} + \kappa_{c2}\left|\frac{T}{T_c} - \kappa_{c3}\right|^{\kappa_{c4}} \quad (38)$$

Table 2. Parameters of the evaporation–condensation frequency–temperature correlation equation for the mLee model.

MCoe$_e$		MCoe$_c$	
κ_{e1} (1/s)	6.231×10^{-3}	κ_{c1} (1/s)	1.441×10^{-4}
κ_{e2} (1/s)	1.10664	κ_{c2} (1/s)	0.02323
κ_{e3} (-)	0.76634	κ_{c3} (-)	0.76634
κ_{e4} (-)	0.77817	κ_{c4} (-)	0.76138

3.4. Physical Parameter Modeling

Parameters such as the saturated vapor pressure, enthalpy of the phase change, thermal conductivity, and kinetic viscosity are required by the numerical model. Therefore, we establish the computational equations for the above thermal and transport parameters based on NIST 1334 data.

3.4.1. Enthalpy of Phase Change

The accuracy of the LHe-4 phase-change enthalpy equation recommended by Ter Harnsel et al. [35] and Malkov et al. [36] is poor. For example, by comparing it with the phase-change enthalpy recommended by NIST Handbook 1334, it was found that the AARD of Malkov's equation [36] was 15.26% and the MRD was 17.87%. Therefore, based on the NIST data, the rational equation for the enthalpy of the phase change of LHe-4 was developed, as shown in Equation (39), with an AARD of 1.59% and an MRD of 5.27%, which is a significant improvement over the original calculation method:

$$\Delta_{\mathrm{l}}^{\mathrm{v}}H = \frac{26.14156 - 25.87419T_{\mathrm{r}}}{1 - 0.89635T_{\mathrm{r}} - 0.03869T_{\mathrm{r}}^2} \quad (39)$$

3.4.2. Thermal Conductivity

The effect of pressure on the thermal conductivity of He-4 was neglected in the previous study, and the binary equation for calculating the thermal conductivity of He-4 (EBC) applicable in the ranges of 3–50 K and 60–600 kPa was established based on the data from NIST 1334, as shown in Equation (40). The parameters of the EBC were regressed in three segments, and the first segment of the temperature (T^*) was calculated as a univariate function of the comparison pressure. The second segment temperature was 8 K, as shown in Equation (41). The values of the parameters for each temperature interval are shown in Table 3.

$$\lambda = \kappa_1\left(\frac{T}{T_c}\right) + \kappa_2\left(\frac{p}{p_c}\right) + \kappa_3\left(\frac{T}{T_c}\right)^2 + \kappa_4\left(\frac{p}{p_c}\right)^2 + \kappa_5\left(\frac{T}{T_c}\right)\left(\frac{p}{p_c}\right) + \kappa_6 \quad (40)$$

$$T^* = 0.0863 \times \left(\frac{p}{p_c}\right)^3 + 0.4488 \times \left(\frac{p}{p_c}\right)^2 + 0.8208 \times \left(\frac{p}{p_c}\right) + 0.5367 \quad (41)$$

Table 3. Regression parameters of EBC equation.

Parameters	κ_1	κ_2	κ_3	κ_4	κ_5	κ_6
$T \le T^*T_c$	3.4921	−0.0970	−2.1634	−0.01884	0.35744	0.3770
$T^*T_c < T < 8$	0.2785	1.4625	0.3566	0.05996	−0.9708	0.1996
$8 \le T \le 50$	0.5481	0.1082	−0.0133	0.0162	−0.0161	0.6491

The AARDs of the established EBCs were all within 2.28% and the MRDs were all within 6.81% for the temperature and pressure ranges discussed. The AARDs were all within 1.85% and the MRDs were all within 7.57% for the pressure ranges discussed.

3.4.3. Dynamic Viscosity

Adopting the effect of neglecting the pressure on the He-4 dynamic viscosity in the previous study, a binary equation for calculating the He-4 dynamic viscosity (EBV) was established based on the data from NIST Handbook 1334, which is applicable in the ranges of 3–50 K and 60–600 kPa, as shown in Equation (42). Similar to the EBC, the parameters of the EBV were regressed in three segments, with the first segment temperature (T^*) calculated as a unitary function of pressure, as shown in Equation (43), and the second segment temperature at 8 K. The values of the parameters for each temperature interval are shown in Table 4.

$$\nu = \kappa_1\left(\frac{T}{T_c}\right) + \kappa_2\left(\frac{p}{p_c}\right) + \kappa_3\left(\frac{T}{T_c}\right)^2 + \kappa_4\left(\frac{p}{p_c}\right)^2 + \kappa_5\left(\frac{T}{T_c}\right)\left(\frac{p}{p_c}\right) + \kappa_6 \quad (42)$$

$$T^* = 0.0863 \times \left(\frac{p}{p_c}\right)^3 + 0.4488 \times \left(\frac{p}{p_c}\right)^2 + 0.8208 \times \left(\frac{p}{p_c}\right) + 0.5447 \quad (43)$$

Table 4. Regression parameters of EBV equation.

Parameters	κ_1	κ_2	κ_3	κ_4	κ_5	κ_6
$T \leq T^*T_c$	−0.8622	−0.0158	−2.3249	−0.0732	0.8986	5.0565
$T^*T_c < T < 8$	0.7480	1.5709	0.2780	0.1015	−1.0457	0.1506
$8 \leq T \leq 50$	0.7918	0.2148	−0.0215	0.0058	−0.0244	0.7472

Within the thermodynamic ranges discussed, the AARDs are all within 1.63% and the MRDs are all within 7.56%. Within the thermodynamic ranges discussed, the AARDs are all within 1.31% and the MRDs are all within 8.45%.

3.5. CFD Modeling and Solution Methods

3.5.1. Numerical Calculation Method

Fluent commercial software was used to carry out the study of the thermal response characteristics, and a double-precision pressure-based solver was selected to solve the numerical model. Considering the long computation period of the thermal response process, the numerical model is solved by an implicit solver that allows a large time step of 0.1 s. The PISO (Pressure-Implicit with Splitting of Operators) algorithm is used to realize the pressure–velocity coupling, the second-order upwind format is used to discretize the control equations, the Body force weighted format is used to discretize the control equations, and the second-order windward format is used to discretize the control equations. The Body force weighted format is used to discretize the pressure term. Considering the presence of a distinct gas–liquid interface in the model, and that the focus of attention is on thermal stratification and self-pressurization rather than on the gas–liquid interface morphology, the volume fraction equations are discretized using the Compressive format. The convergence criterion for the continuity equation, momentum equation, volume fraction equation, turbulent kinetic energy, and dissipation rate equations is 1×10^{-3}, and that for the energy equation is 1×10^{-6}. For the heat flux under the natural convection boundary condition, 16.569 W is chosen as the boundary of the thermal response model for the storage tank. The initial temperature is 3.98 K, and the initial pressure is 80 kPa. Meanwhile, the local initialization of the gas-phase region is carried out. The mLee phase-change model is written by Visual Studio 12.0 software and loaded by the UDF compilation method of the Fluent platform, which is used for calculating the source terms of the mass conservation equation and the energy conservation equation.

3.5.2. Computational Domain Mesh and Irrelevance Validation

The LHe-4 tank studied in this thesis is a large horizontal tank, and the diameter of the cylindrical part of the tank is 3 m. Since the thermal response phenomenon mainly occurs in the cross section perpendicular to the axis of the tank, to save the cost of numerical computation, the central cross section of the cylindrical part of the tank was used as the computational domain to carry out the study of the thermal response characteristics. Based on the ICEM meshing platform, a structured mesh was established, and the mesh at the gas–liquid interface was encrypted with a determinant quality of 0.95. At the same time, considering that the natural convection near the wall surface of the tank was the main focus of the thermal response characteristics study, the mesh near the wall surface was encrypted. The computational domain and meshing results are shown in Figure 3.

ICEM was used to delineate five O-structured grids with different grid numbers, which were 83,286, 113,976, 135,186, 156,886, and 184,416. The grid masses were all above 0.95, and the grid aspect ratios were in the range of 1.0 to 2.2. The temperature distributions of different numbers of grid schemes were calculated with the temperature distribution of the plumbline in the liquid-phase region, and the results of the grid-independence validation are shown in Figure 3. An analysis of Figure 3 shows that when the number of grids was increased from 156,886 to 184,416, the difference in the temperature profiles along the plumbline in the liquid-phase region was very small, and the relative deviation of the mean temperature at the free-liquid surface was below 0.1%. Therefore, the grid with the number 156,886 was selected as the grid for the thermal response numerical model.

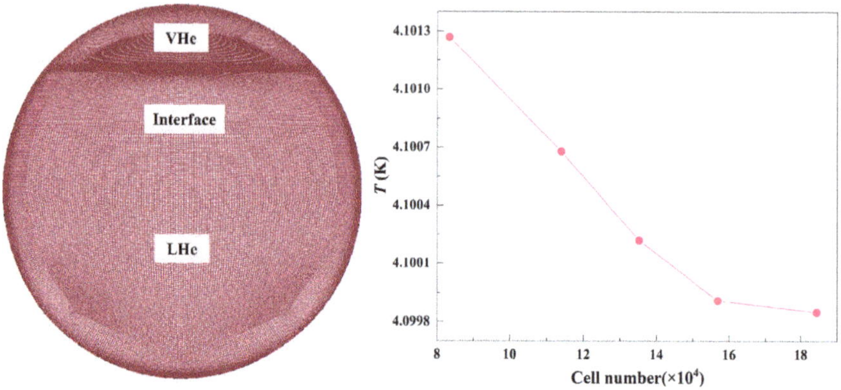

Figure 3. Computational domain meshing results and mesh-independence verification results.

4. Traditional Thermodynamic Modeling

4.1. Thermodynamic Model Establishment

Based on the isothermal saturation and homogeneous model [37], a thermodynamic model was established to calculate the warming and self-pressurization curves in the LHe-4 storage tank. The saturated homogeneous model considers that the heat enters the storage tank, its thermal state inside the tank changes continuously, and the temperatures of the two gas and liquid phases are equal. According to the law of the conservation of energy, the heat intruded into the tank is equal to the sum of the heat gained in the liquid and gas phases:

$$T^* = 0.0863 \times \left(\frac{p}{p_c}\right)^3 + 0.4488 \times \left(\frac{p}{p_c}\right)^2 + 0.8208 \times \left(\frac{p}{p_c}\right) + 0.5447 \tag{44}$$

where Q_{leak} denotes the heat intruded into the tank, W; ΔQ_1 and ΔQ_2 denote the heat gained in the liquid and gas phases, W.

$$Q_{leak} = Q_{MLI} + Q_s + Q_p \tag{45}$$

$$\Delta Q_l = c_{pl} m_l (T_{i+1} - T_i) \tag{46}$$

$$\Delta Q_v = c_{pv} m_v (T_{i+1} - T_i) \tag{47}$$

where Q_{MLI}, Q_s, and Q_p are the heat entering the tank through the adiabatic layer, the supporting thermal bridge, and the pipeline thermal bridge, W; c_{pl} and c_{pv} denote the constant-pressure specific heat of the liquid phase and the gas phase, J/(kg·K); m_l and m_v denote the masses of the liquid phase and the gas phase, kg; T_{i+1} and T_i denote the calculated endpoint temperature and the initial temperature, K.

4.2. Thermodynamic Model Solution

The initial temperature of the LHe-4 tank is taken as 3.98 K and the initial filling rate is taken as 85%. The thermodynamic model solution procedure is shown in Figure 4. Firstly, the endpoint temperature is calculated based on the initial temperature, and then the pressure of the tank is obtained with the minute evaporation rate (MER).

$$\text{MER} = \frac{Q_{leak} \times 60}{m_l H_{vap}} \tag{48}$$

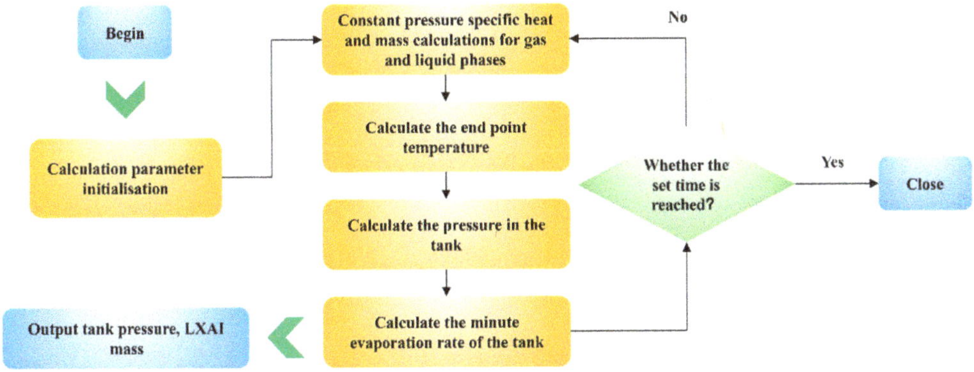

Figure 4. Methane–alcohol system liquid-phase component fraction prediction result.

5. Results and Discussion

5.1. Experimental Validation of the Numerical Model

The benchmark test of the numerical model was carried out using the results of the self-pressurization experiments of the K-Site liquid hydrogen storage tank at NASA Lewis Research Centre, tested by Hasan et al. [38] and Van Dresar et al. [39]. Further experimental validation of the developed numerical model was carried out to comprehensively demonstrate its correctness and reasonableness. The model validation was carried out using the engineering application data of the typical self-pressurization of LHe-4 tanks from the literature published by Weat [37]. A 40 m^3 LHe-4 tank developed by Linde for Kansas in 1966 was used for the validation. The inner tank diameter of the LHe-4 tank was 2.362 m, the length of the cylindrical section was 10.480 m, the depth of the head curved surface was 0.45 m, and the surface area was 3.71 m^2. The filling rate of the tank was 90%, and the level of LHe-4 was at 0.918 m. The heat leakage from the tank was 10 W (0.117 W/m^2). Since the natural convection in this tank mainly occurs near the wall of the cylindrical part of the tank, and the tank is symmetric about the Z = 0 plane, a two-dimensional model was built to save computational resources. A structured mesh was established using ICEM 18.1 software with a mesh quality of 0.9 or more. The results of the model validation are shown in Figure 5.

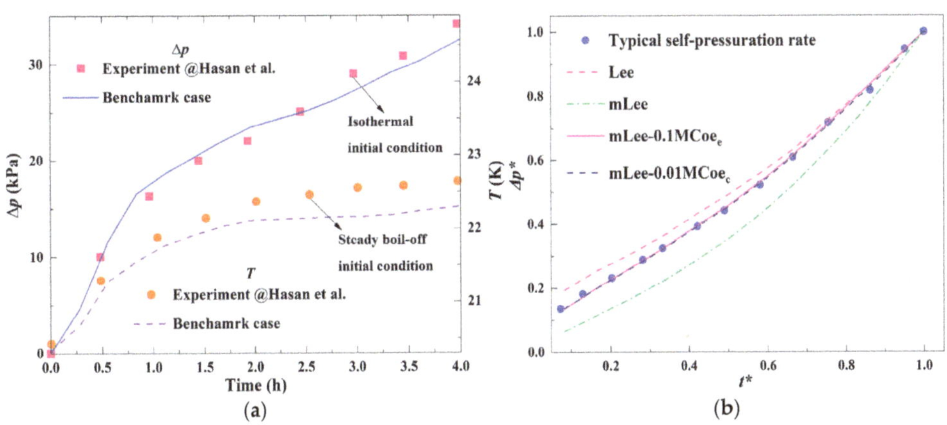

Figure 5. Numerical model calculation results. (**a**) Temperature and pressure variation with time; (**b**) Pressure rise curve in the gas-phase region.

Analyzing the variation patterns of the temperature and pressure with time shown in Figure 5a, it can be seen that under isothermal initial conditions, the variation pattern of the pressure with time in the storage tank is consistent with the experimental results of Hasan et al. However, the temperature prediction value is lower than the experimental value, which is because the steady-state evaporation initial condition of the experiment is different from the isothermal initial condition of the numerical calculation, but Hasan et al. did not provide the temperature data under the isothermal initial condition in their article [38]. Thus, it is demonstrated that the developed numerical model can obtain a valid numerical solution, which can be used to predict the thermal response characteristics of cryogenic storage tanks. The comparison of the self-boosting curve with the typical self-boosting engineering application data is shown in Figure 5b. The relative deviation of the average self-boosting rate from the benchmark value is 14.32%, and therefore the mLee model was used to carry out the study of the thermal response characteristics of the LHe-4 storage tank.

5.2. Analysis of Heat Transfer Calculation Results

5.2.1. Natural Convection Boundary Conditions

The heat transfer calculation results of the adiabatic structure of the LHe-4 tank under the natural convection boundary conditions are shown in Figure 6. Analyzing Figure 6a, the residual curve decreases sharply in the first three iterations, and the residual is less than 10^{-3} when the iteration is carried out to the 10th step. Figure 6b shows the number of iterations at each temperature. Analyzing Figure 6c, the existence of the LNCS leads to an interruption in the temperature distribution of the tank insulation structure, the temperature gradient of the outer insulation layer gradually decreases from the inside to the outside, and the value of its temperature increases with the ambient temperature. The higher the ambient temperature, the higher the temperature gradient at the outer side of the cold screen. At the same time, when the ambient temperatures are 300 K, 320 K, and 340 K, the temperature differences between the outer wall surface of the storage tank and the environment are 0.287 K, 0.349 K, and 0.425 K, which shows that the temperature difference rises with the ambient temperature, and when the ambient temperature rises by 1 K, the temperature difference increases by 0.003–0.004 K, and the gradient of the temperature difference increases with the increase in the ambient temperature.

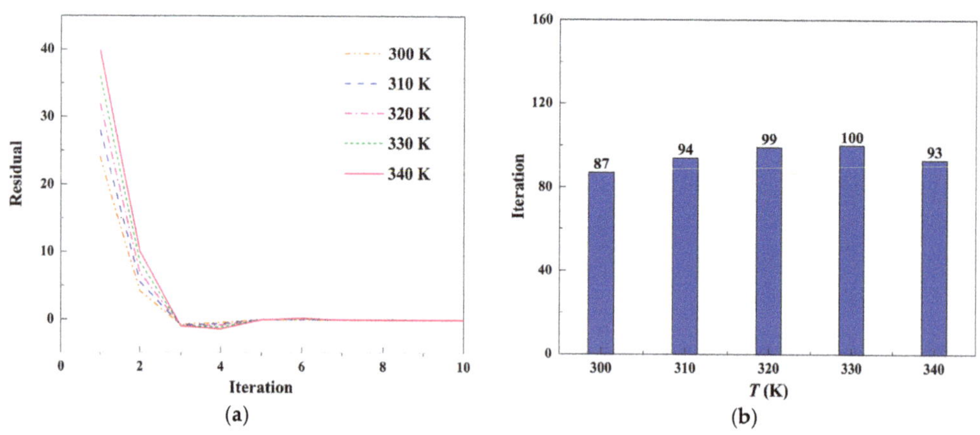

Figure 6. *Cont.*

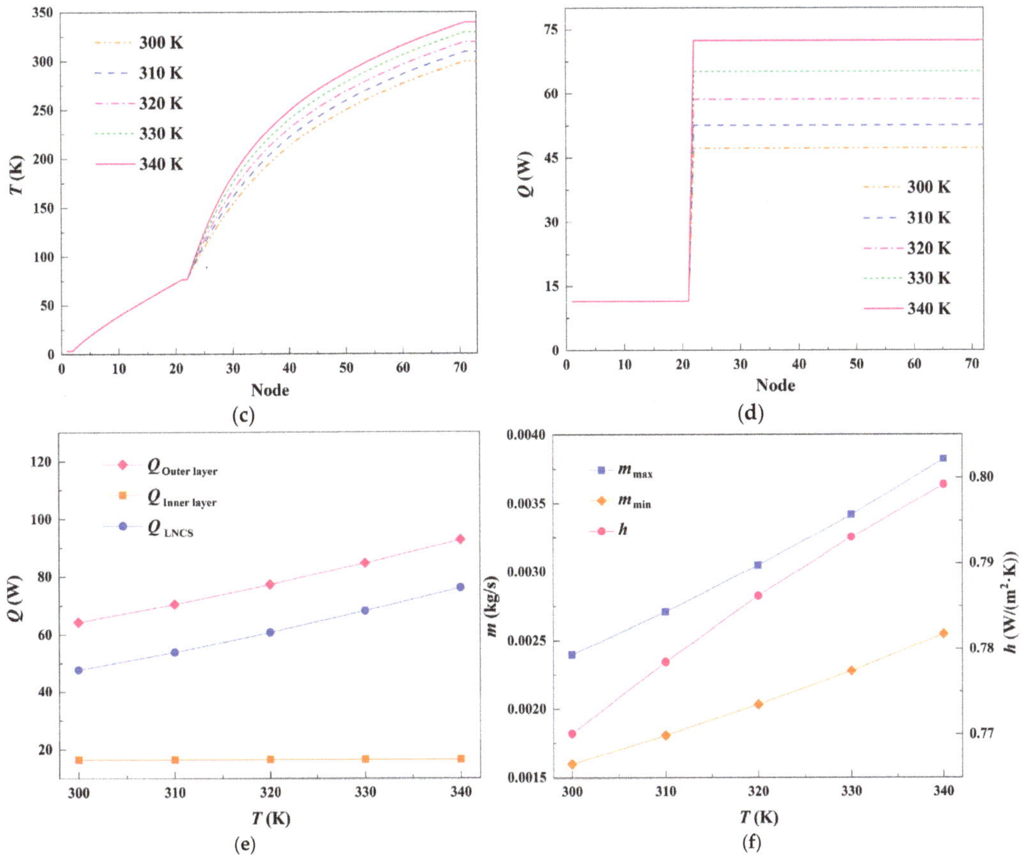

Figure 6. Heat leakage from the tank under the natural convection boundary. (**a**) Residual curve; (**b**) Iteration number; (**c**) Temperature distribution; (**d**) Heat flux distribution; (**e**) Heat flux versus temperature; (**f**) Convective heat transfer.

The temperature distribution curves of the inner adiabatic layer of the LHe-4 tank overlap, and the temperature gradient change is not obvious, which is because the LNCS temperature is always 77 K, so the heat flux of the inner adiabatic layer is 11.548 W, which can be analyzed from Figure 6d. The heat flux of the outer adiabatic layer increases when the ambient temperature increases, and when the ambient temperatures are 300 K, 320 K, and 340 K, the heat fluxes of the outer adiabatic layer are 47.212 W, 58.651 W, and 72.534 W, whereas the increased heat flux of the outer adiabatic layer is absorbed by the LNCS, and therefore the heat flux of the inner adiabatic layer does not increase with the ambient temperature increase. The heat flux of the outer adiabatic layer ($Q_{\text{Outer layer}}$) and the heat flux of the inner adiabatic layer ($Q_{\text{Inner layer}}$) are shown in Figure 6e. It can be seen that the increase in the ambient temperature led to an increase in the heat flux of the outer adiabatic layer according to the approximate linear law, and the heat fluxes of the outer adiabatic layer at ambient temperatures of 300 K, 320 K, and 340 K are 64.336 W, 77.308 W, and 92.724 W. However, the heat flux of the inner adiabatic layer did not show a positive correlation with the ambient temperature because the LNCS absorbed the heat from the increase in the heat flux of the outer adiabatic layer due to the increase in the ambient temperature.

The convective heat transfer coefficients and critical liquid nitrogen flow rates are shown in Figure 6f. The analysis shows that the maximum and minimum critical flow rates of the LNCS are positively correlated with the ambient temperature. For example, when the ambient temperatures are 300 K, 320 K, and 340 K, the maximum critical flow rates of the liquid nitrogen in the coil are 2.398×10^{-3} kg/s, 3.050×10^{-3} kg/s, and 3.823×10^{-3} kg/s, and the minimum critical flow rates are 1.599×10^{-3} kg/s, 2.033×10^{-3} kg/s, and 2.549×10^{-3} kg/s. In addition, the convective heat transfer coefficients between the outer tank and the environment of the LHe-4 tank were positively correlated with the temperature, but they were all less than 1 W/(m^2·K). The convective heat transfer coefficients are 0.770 W/(m^2·K), 0.786 W/(m^2·K), and 0.799 W/(m^2·K) for ambient temperatures of 300 K, 320 K, and 340 K.

5.2.2. Forced Convection Boundary Conditions

The temperature distribution of the multilayer adiabatic structure of the LHe-4 tank is shown in Figure 7. The analysis shows that the temperature distribution law of the adiabatic structure under the forced convection boundary condition is the same as that under the natural convection boundary condition. The temperature difference between the vacuum sandwich between the cold screen and the inner adiabatic layer and between the outer tank and the outer adiabatic layer is smaller, which is because the vacuum sandwich has only radiative heat transfer and its thermal resistance is smaller than that of the adiabatic structure. Under the boundary condition of forced convection, the influence of the ambient temperature on the temperature distribution of the adiabatic structure of the storage tank is more significant than that of the wind speed of forced convection. When the wind speed is 2 m/s and the ambient temperatures are 300 K, 320 K, and 340 K, the temperature differences between the outer wall surface of the tank and the outside environment is 0.047 K, 0.061 K, and 0.078 K, respectively, and when the wind speed is increased to 5 m/s, the temperature differences are 0.022 K, 0.029 K, and 0.038 K, respectively. It can be seen that the increase in the ambient temperature results in the temperature difference between the outer wall surface and the environment increasing gradually, while the increase in the wind speed leads to the temperature difference between the outer wall surface and the environment increasing gradually. It can be seen that an increase in the ambient temperature leads to a gradual increase in the temperature difference between the outer wall and the environment, while an increase in the wind speed leads to a gradual decrease in the temperature difference between the outer wall and the environment. The calculated residuals at different wind speeds are less than 10^{-3} after 10 steps of iterative calculations, and the residual curves at a wind speed of 3 m/s are shown in Figure 7c.

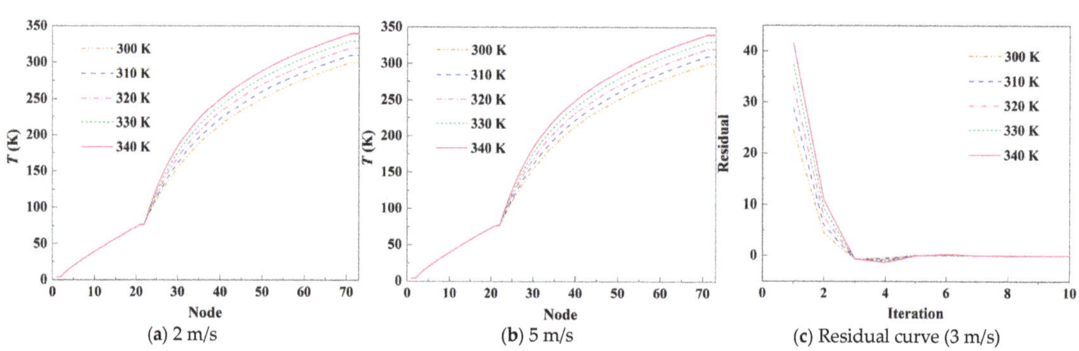

Figure 7. Temperature distribution of the adiabatic layer under forced convection boundary conditions.

The heat flux distribution of the adiabatic structure under the forced convection boundary condition is shown in Figure 8. It can be seen that the temperature distributions of the inner adiabatic layer are almost the same because the cold screen exerts the maximum performance (i.e., the temperature is maintained at about 77 K). The heat flux of the inner adiabatic layer is 11.548 W, which is the same as that under the natural convection boundary condition because the boundary of the outer tank wall does not affect the heat flux of the inner adiabatic layer, which is attributed to the cold screen. The heat flux of the outer adiabatic layer of the tank increases with the increase in the ambient temperature and forced convection wind speed. The heat fluxes of the outer adiabatic layer at ambient temperatures of 300 K, 320 K, and 340 K are 47.319 W, 58.807 W, and 72.760 W, respectively, for a wind speed of 1 m/s. The corresponding heat fluxes are 47.349 W, 58.853 W, and 72.831 W for a wind speed of 5 m/s. The ambient temperatures of the outer adiabatic layer at ambient temperatures of 300 K, 320 K, and 340 K are 47.349 W, 58.853 W, and 72.831 W, respectively. The influence of the ambient temperature on the heat flux of the outer insulation layer is much larger than that of the wind speed.

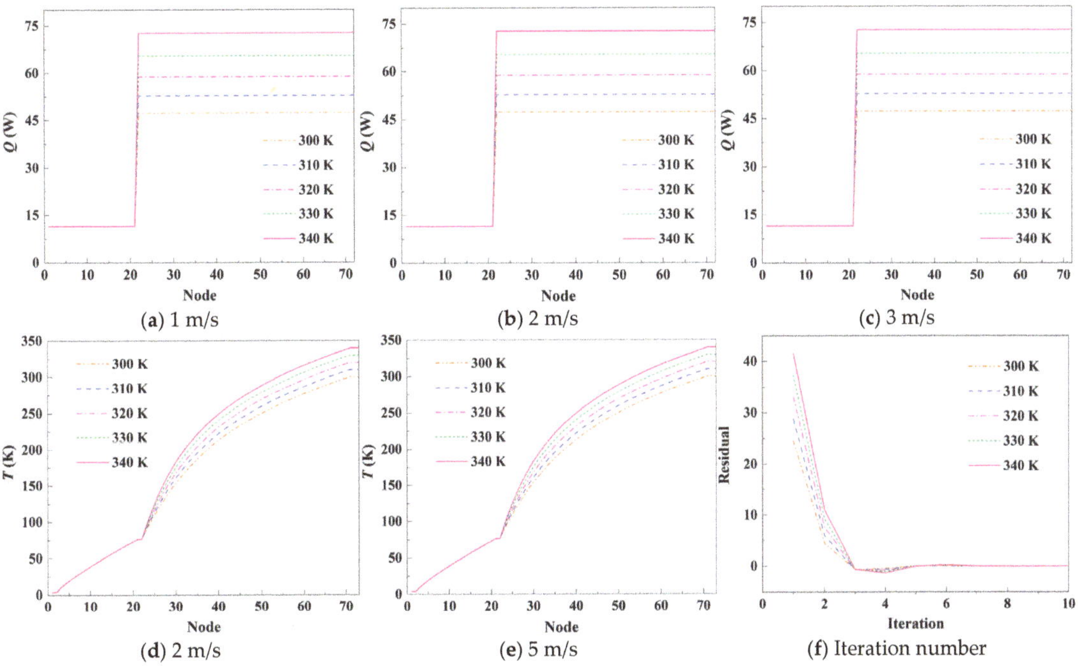

Figure 8. Heat flux distribution in the adiabatic layer under forced convection boundary.

Figure 9 shows the results of the heat transfer calculations for the storage tank under the forced convection boundary. Analyzing Figure 9a–e, it can be seen that the heat flux of the outer adiabatic layer increases in an approximately linear manner due to the increase in the ambient temperature, and the effect of the wind speed on the heat flux of the outer adiabatic layer is much smaller than that of the ambient temperature, which is because the vacuum degree of the adiabatic layer is 10^{-3} Pa, the daily evaporation rate is 0.39%, and the adiabatic performance is better. When the ambient temperature is 300 K and the wind speeds are 1 m/s, 3 m/s, and 5 m/s, the heat fluxes of the outer adiabatic layer are 64.460 W, 64.488 W, and 64.494 W, respectively; and the corresponding heat fluxes when the ambient temperature is 340 K are 72.760 W, 72.816 W, and 72.831 W. In addition, under the same ambient temperature, the rate of the heat flux increasing with the wind speed decreases gradually, and the higher the temperature, the higher the rate of the increase in the heat flux. However, due to the formation of a cold wall in the LNCS, the heat flux of the inner adiabatic layer remains constant. Analyzing the critical liquid nitrogen fluxes of the LNCS in Figure 9a–e, the maximum and minimum critical fluxes of the LNCS are positively correlated with the ambient temperature due to the significant increase in the heat flux of the outer adiabatic layer with the temperature. Taking the convection velocity of 5 m/s as an example, when the ambient temperatures are 300 K, 320 K, and 340 K, the maximum critical liquid nitrogen flow rates of the LNCS are 2.406×10^{-3} kg/s, 3.061×10^{-3} kg/s, and 3.840×10^{-3} kg/s, respectively, and the minimum critical flow rates are 1.604×10^{-3} kg/s, 2.041×10^{-3} kg/s, and 2.560×10^{-3} kg/s, respectively. It can be seen that, to absorb the heat flux increment of the outer adiabatic layer caused by the increase in the ambient temperature, the maximum and minimum critical liquid nitrogen flow rates increase with the increase in the ambient temperature, and the gradient of the critical flow rate increases gradually with the ambient temperature increase.

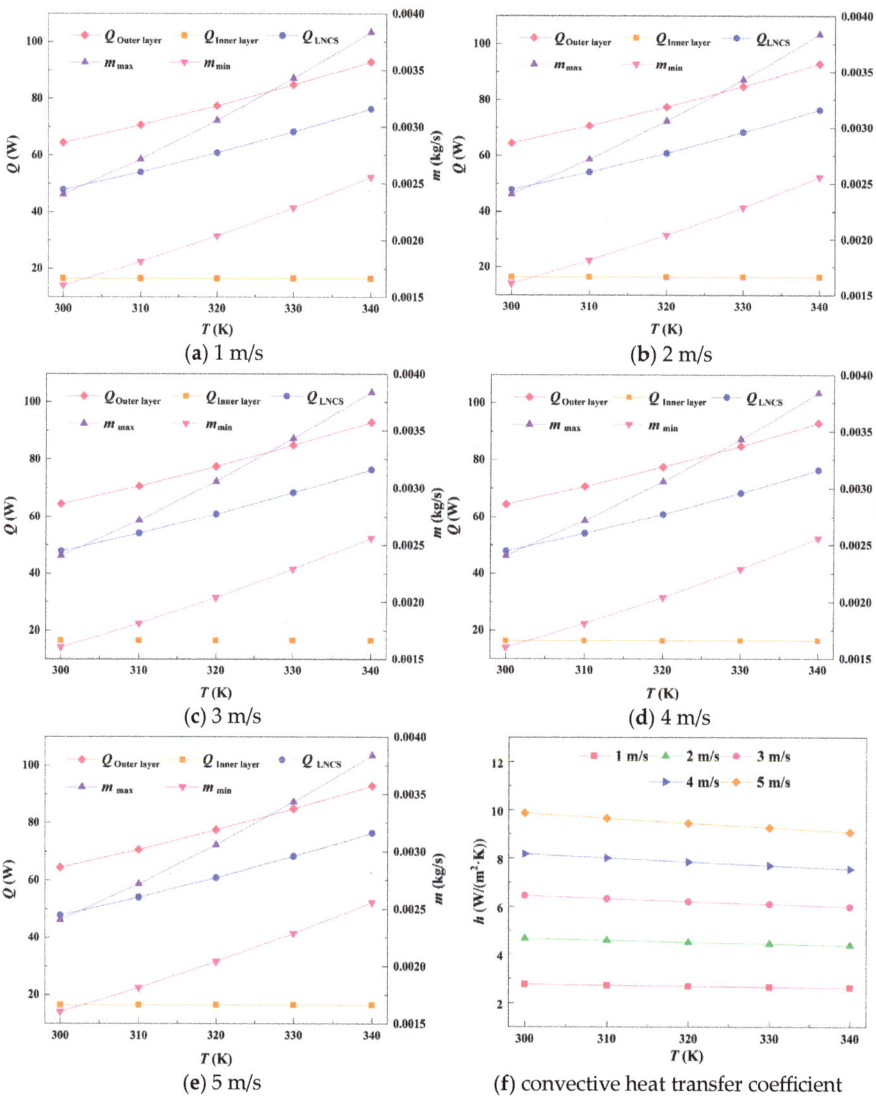

Figure 9. Calculated heat transfer from tanks with forced convection boundary.

The convective heat transfer coefficient of the outer wall surface of the storage tank is shown in Figure 9f. Analysis shows that the convective heat transfer coefficient of the outer container of the storage tank and the environment gradually decreases with the increase in the ambient temperature. The convective heat transfer coefficients at an ambient temperature of 300 K and wind speeds of 1 m/s, 3 m/s, and 5 m/s are 2.768 W/(m^2·K), 6.458 W/(m^2·K), and 9.875 W/(m^2·K), respectively, while the corresponding convective heat transfer coefficients at an ambient temperature of 340 K are 2.615 W/(m^2·K), 5.988 W/(m^2·K), and 9.079 W/(m^2·K), respectively. In addition, the higher the wind speed, the more significant the decrease in the convective heat transfer coefficient with the ambient temperature.

5.3. Thermal Stratification Characteristics Analysis

5.3.1. Temperature Distribution

The temperature distribution in the liquid-phase region at different moments is shown in Figure 10. It can be seen that with the numerical calculation, the high-temperature zone at the gas–liquid interface of the storage tank continuously extends to the inside of LHe-4. Taking the 3.9805 K isotherm as an example, the 3.9805 K isotherm is 0.02 m above the gas–liquid interface at 1 h, while the isotherm is 0.93 m below the gas–liquid interface at 3 h. This is because the heat at the wall surface of the tank is transferred to the free-liquid surface by natural convection, and the heat is gradually transferred to the LHe-4 below the free-liquid surface. At 3 h, a stable isotherm and temperature gradient are formed in the tank (i.e., thermal stratification is formed).

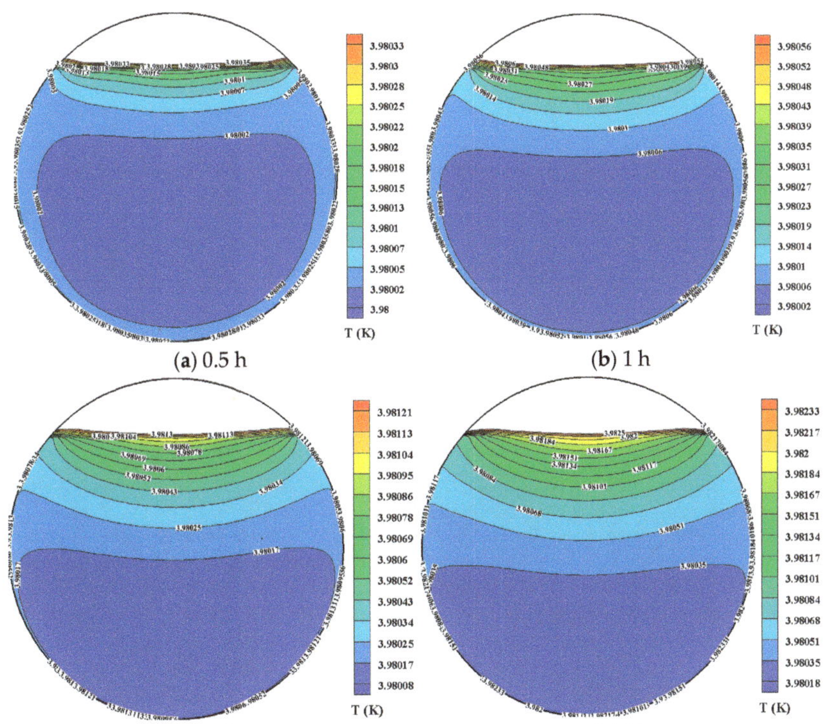

Figure 10. Temperature distribution in the liquid-phase region at different moments.

The temperature distribution of each path and monitoring point of the storage tank is shown in Figure 11. Analyzing Figure 11a, the temperature of LHe-4 increases slowly from the bottom of Line 1 upwards, and the region from y = 0.983 m to y = 1.5 m is the gas-phase region, where the temperature increases significantly. The temperature in the region from y = 0.983 m to y = 1.5 m is higher than that in the region from y < 0.983 m and the temperature gradient is larger than that in the region from y < 0.983 m. In the gas-phase region, the temperature of the VHe-4 increases sharply with the y-coordinate and reaches a maximum at the tank wall. At 1 h, 2 h, and 3 h, the temperatures at the tank wall are 3.983 K, 3.985 K, and 3.990 K, respectively.

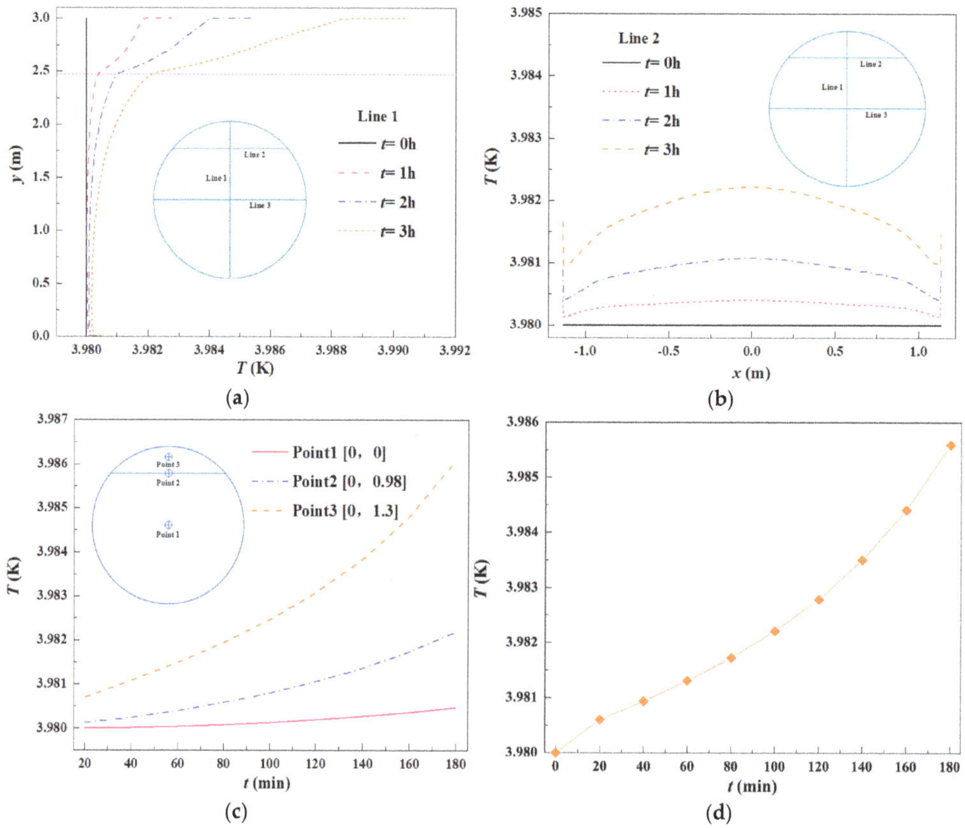

Figure 11. Temperature distribution of tank paths and monitoring points. (**a**) Line1 temperature distribution; (**b**) Line 2 temperature distribution; (**c**) Flow field monitoring point temperature; (**d**) Temperature rise curve in the gas-phase region.

Analyzing Figure 11b, due to the strong evaporation point at the wall in contact with the gas–liquid interface, the heat-absorbing effect of evaporation leads to significantly lower temperatures at the two ends of Line 2 compared to the middle of Line 2. In addition, the higher temperature at the wall causes the temperature at the two endpoints of Line 2 to be higher than the temperature at the strong evaporation point. Meanwhile, the temperature of Line 2 increases with time, and its distribution pattern changes significantly with time.

Analyzing Figure 11c, the temperature of Point 2 rises faster than that of Point 1 due to its proximity to the gas–liquid interface. The temperature at Monitoring Point 3 in the gas-phase zone is higher than that at Point 1 and Point 2, and the rate of the temperature rise is larger than that at Point 1 and Point 2. At 180 min, the temperatures at Point 1, Point 2, and Point 3 are 3.98048 K, 3.9822 K, and 3.9861 K, respectively. The temperature-rise curves in the gas-phase zone are shown in Figure 11d. The analysis shows that the temperature in the gas-phase region increases with time in a nonlinear pattern, and that the temperature rise rate is small at the initial stage and larger at the later stage of the calculation, which is caused by thermal stratification. The average temperature rise rate before 100 min is 2.21×10^{-5} K/min, and the average temperature rise rate after 100 min is 4.23×10^{-5} K/min.

5.3.2. Characterization of Thermal Stratification Degree

The temperature distribution, temperature gradient, and thermal stratification degrees in the thermal stratification zone are shown in Figure 12. The temperature distribution in the region of y = 0 m and y = 0.983 m is shown in Figure 12a. The analysis shows that the temperature in this region is increasing with time, and the temperature gradient is gradually increasing. For example, the temperatures at the gas–liquid interface at 1 h, 2 h, and 3 h are 3.9804 K, 3.9811 K, and 3.9822 K. Analyzing Figure 12b shows that the average temperature gradient in the region of y = 0 m and y = 0.983 m increases with time, and that the average temperature gradients in the region at 1 h, 2 h, and 3 h are 3.84×10^{-4} K/m, 9.12×10^{-4} K/m, and 18.00×10^{-4} K/m.

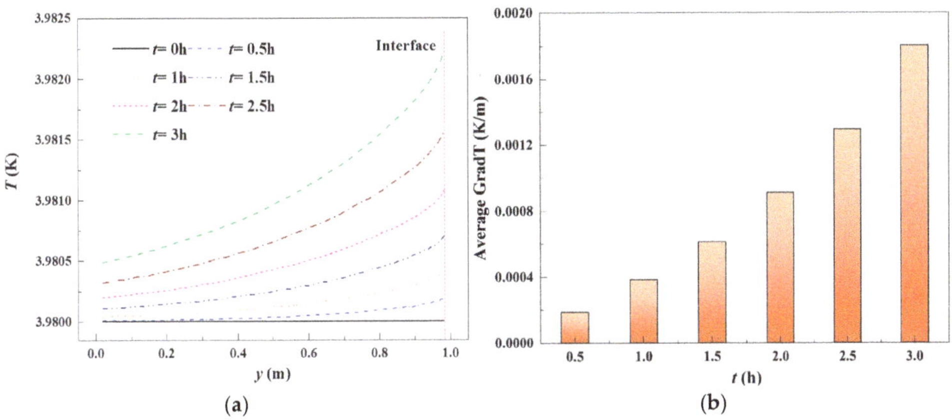

Figure 12. Temperature distribution, temperature gradient, and degree of thermal stratification in the thermally stratified zone. (**a**) Temperature distribution; (**b**) Average temperature gradient.

5.4. Flow Field Pressure Distribution Characteristics

Figure 13 shows the pressure distribution in each path and monitoring point of the storage tank. It can be seen that along Line 1, the pressure distribution in the gas-phase region of the storage tank is uniform, but there still exists a slight gradient, which is due to the low temperature of the VHe-4. The pressure in the liquid-phase zone is distributed in a linear pattern, and the increase in the gas-phase pressure leads to the corresponding increase in the liquid-phase pressure, as shown in Figure 13a. The pressures at the bottom of the tank at 1 h, 2 h, and 3 h are 83,451.7 Pa, 83,795.3 Pa, and 84,515.8 Pa, respectively, which are much higher than those in the gas-phase zone, and therefore the bottom of the tank is the most hazardous location inside the storage tank. Analyzing Figure 13b, the pressures at the three observed monitoring points increased with time according to the same pattern, with a small rate of increase in the early stage and a large rate of increase in the later stage. At 100 min, the pressures at Point 1, Point 2, and Point 3 were 81,758.80 Pa, 80,518.69 Pa, and 80,465.83 Pa, respectively.

The self-pressurization curve of the gas-phase region of the tank is shown in Figure 13c. The vertical pink dotted line in Figure 13a indicates the liquid level line. It can be seen that the self-pressurization curve of the gas-phase region obeys the nonlinear time evolution law, which coincides with the self-pressurization law in the literature [18]. The pressures in the gas-phase region of the storage tank are 80,268.66 Pa, 80,612.27 Pa, and 81,332.53 Pa, respectively, at 1 h, 2 h, and 3 h, and the average self-pressurization rate is 0.12 Pa/s. It is worth noting that the self-pressurization rate increases gradually with the calculation, and the average self-pressurization rates in the periods of 0–100 min and 100–180 min are 0.079 Pa/s and 0.179 Pa/s, respectively. The reason for this phenomenon is that thermal stratification intensifies the self-pressurization rate.

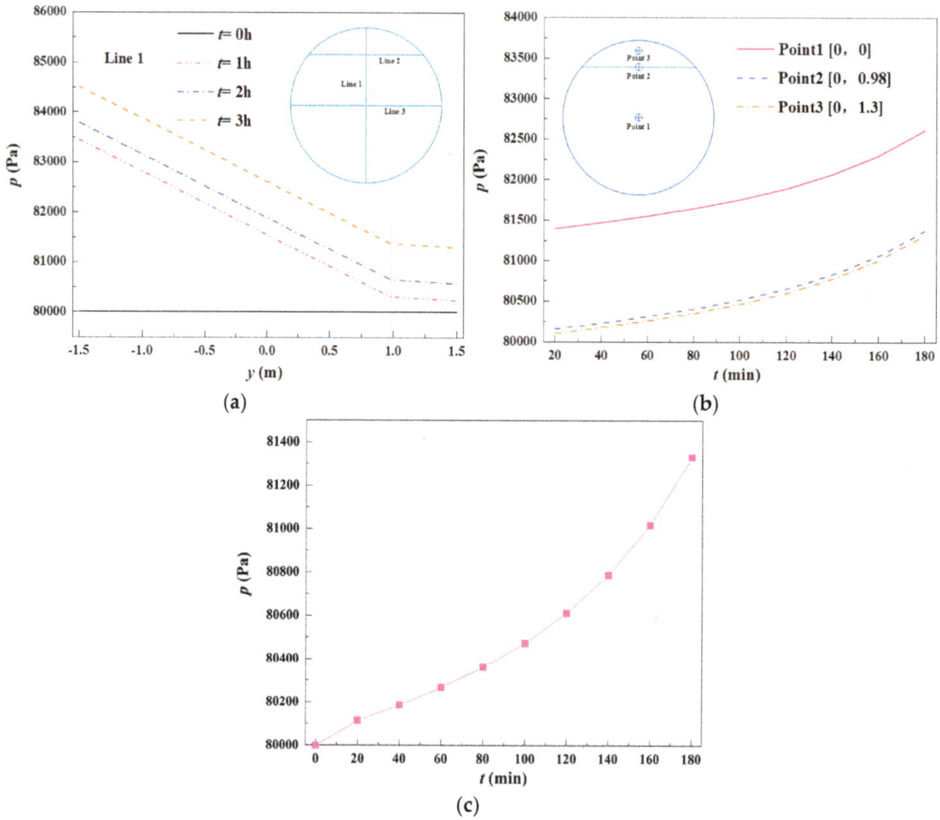

Figure 13. Pressure distribution by tank paths and monitoring points. (**a**) Line 1 pressure distribution; (**b**) Monitoring point pressure; (**c**) Self-pressurization curves in the gas-phase region of the tank.

5.5. Heat-Up and Self-Pressurization Calculations

Figure 14 shows the solution results of the thermodynamic model within 180 min. It can be seen that the tank pressure calculated by the thermodynamic model increased in an approximately linear pattern within 180 min, but the calculated value of the model is lower than that of the CFD calculation. This is because the thermodynamic model is based on the principle of phase equilibrium and the saturated homogeneous model; i.e., the temperatures of the VHe-4 and LHe-4 in the storage tank are always equal, and the supercooled state of the LHe-4 is neglected. Analyzing Figure 14b, the temperature inside the storage tank calculated by the thermodynamic model increases with an approximate linear law, and the trend of the temperature change is similar to that calculated by the CFD model. At 180 min, the temperature calculated by the thermodynamic model is 3.9847 K, with an absolute deviation of 0.0009 K from the CFD model. Analyzing Figure 14c as the calculation proceeds, the temperatures of the LHe-4 and VHe-4 masses in the tank are always equal, ignoring that the supercooled state of the LHe-4. 4 and VHe-4 decreased and increased with a linear law, respectively, and the mass of the LHe-4 decreased by 13.91 kg while the mass of the VHe-4 increased by 0.24 kg in 180 min. The MERs of the storage tank ranged from $5.57 \times 10^{-4}\%$ to $5.58 \times 10^{-4}\%$. The reason that the tank temperature, pressure, LHe-4 mass, VHe-4 mass, and MER calculated by the thermodynamic model varied in a linear pattern is that the heat flux at the tank wall is constant and the thermodynamic model does not consider the effect of thermal stratification on the self-boosting pressure.

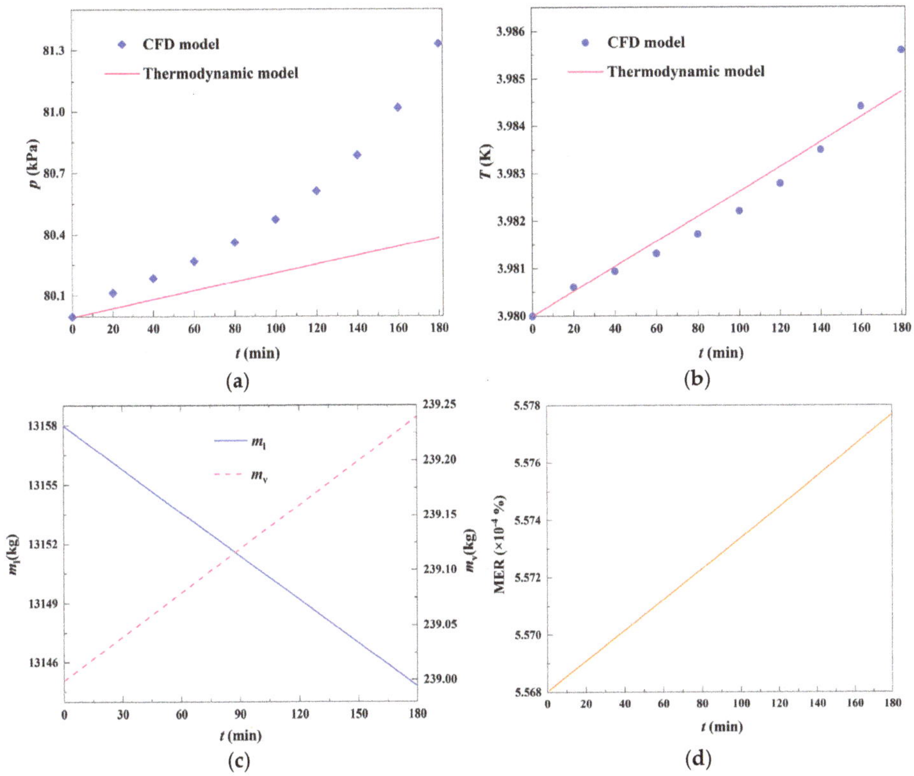

Figure 14. Thermodynamic model solution results in 180 min. (**a**) Self-boosting curve comparison; (**b**) Comparison of warming curves; (**c**) VHe-4 and LHe-4 masses; (**d**) Minute evaporation rate.

Figure 15 shows the changes in the temperature, pressure, and mass of the storage tank in different calculation periods. The analysis shows that when the calculation period is 1 d, 10 d, and 20 d, the pressure in the storage tank basically increases in a linear pattern, and the pressures in the storage tank at 1 d, 10 d, and 20 d are 83.09 kPa, 112.07 kPa, and 144.69 kPa, respectively. Considering that the pressure in the storage tank obtained by the thermodynamic model is on the low side, the estimation of the time of the storage tank storage is no longer carried out.

During the calculation periods of 1 d, 10 d, and 20 d, the temperature inside the tank increases with an approximately linear pattern, and the longer the calculation period, the more significant the nonlinearity of the temperature curve. The temperatures of the tank at 1 d, 10 d, and 20 d are 4.02 K, 4.33 K, and 4.62 K, respectively. As the calculation proceeds, the mass of LHe-4 and the mass of VHe-4 inside the tank decrease and increase, respectively, with an approximately linear pattern, and the longer the calculation period, the more significant the nonlinearity of the curve. The masses of the LHe-4 at 1 d, 10 d, and 20 d are 13,052.26 kg, 12,067.01 kg, and 10,875.31 kg, respectively. The longer the calculation period, the more significant the nonlinearity of the curves. At 1 d, 10 d, and

20 d, 13,052.26 kg, 12,067.01 kg, and 10,875.31 kg of LHe-4 and 240.94 kg, 260.61 kg, and 289.17 kg of VHe-4 were obtained, respectively.

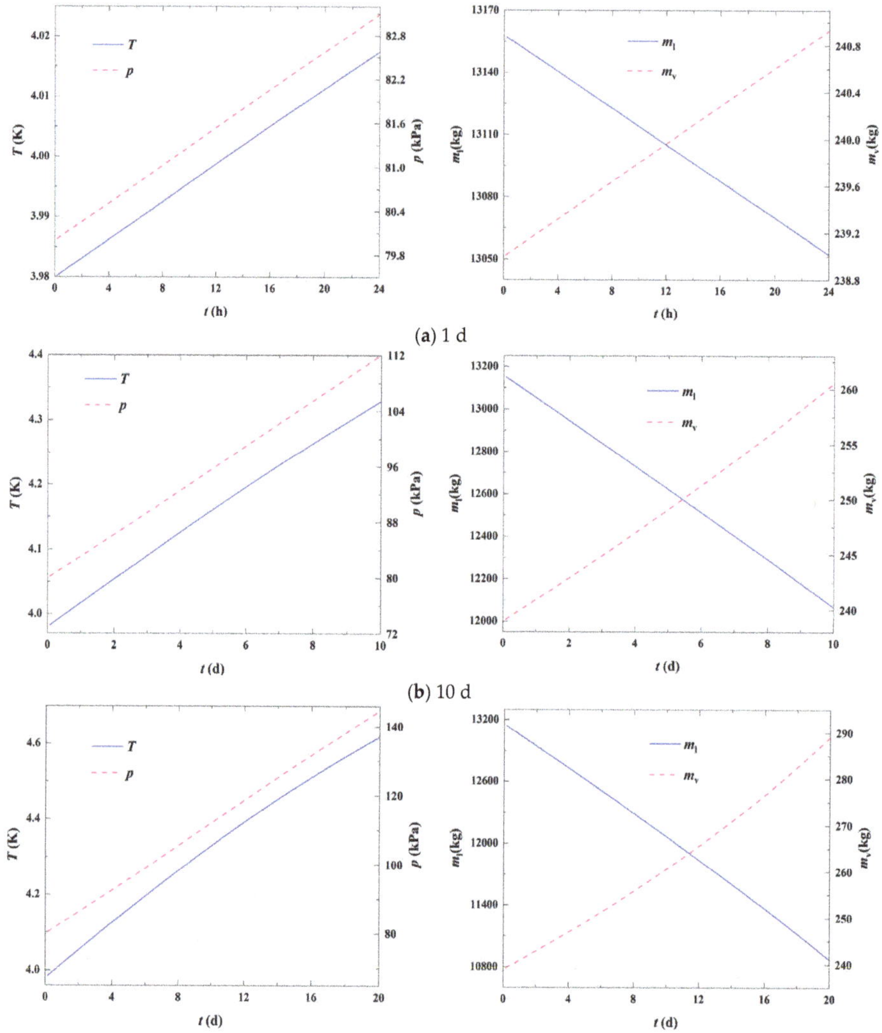

(a) 1 d

(b) 10 d

(c) 20 d

Figure 15. Variation in tank temperature, pressure, and mass for different calculation periods.

6. Conclusions

In this work, the heat transfer model of the composite adiabatic structure with a vacuum multilayer and an LNCS was established based on the Layer-by-Layer model to obtain the heat flux of the adiabatic layer of the storage tank. The mathematical model and numerical model of the thermal response of the LHe-4 storage tank were established based on the VOF model. The thermal stratification of the LHe-4 and the self-pressurization characteristics of the storage tank, as well as its temporal and spatial evolution laws, were analyzed. Finally, based on the saturated homogeneous model, the warming and self-boosting curves of the LHe-4 storage tank were calculated using MATLAB.

The degree of thermal stratification gradually increased with time, and the degrees of thermal stratification at 1 h, 2 h, and 3 h are 1.023, 1.028, and 1.036, respectively. The heat at the wall surface was transferred to the free-liquid surface by natural convection, resulting in the evaporation of LHe-4. At the same time, the heat was gradually transferred to the lower part of the liquid surface (the 3.9805 K isotherm extended to 0.93 m below the liquid surface at 3 h), resulting in thermal stratification in the upper part of the liquid-phase region. The degree of thermal stratification of the tank increased with time, and the degrees of thermal stratification are 1.023, 1.028, and 1.036 at 1 h, 2 h, and 3 h, respectively. The temperature of the gas-phase region increases with time with a nonlinear law, in which the wall in contact with the gas–liquid interface is a strong evaporation point, and the heat-absorbing effect of evaporation leads to lower temperatures at the two ends of the phase interface compared with that in the middle. The pressure in the gas-phase region obeys a nonlinear time evolution law, the thermal stratification at the late stage of the calculation leads to an increase in the self-pressurization rate, and the average self-pressurization rate for 0–180 min is 0.12 Pa/s.

The pressure inside the tank obtained by the thermodynamic model is lower than that of the CFD calculation. This is because the thermodynamic model ignores the supercooled state of the LHe-4. The trend of the temperature change inside the tank obtained by the thermodynamic model is similar to that calculated by the CFD model.

Author Contributions: Conceptualization, J.X.; methodology, F.L.; validation, J.Z. and C.L. (Chao Li); formal analysis, Q.L.; investigation, C.L. (Changjun Li); writing—original draft, W.J.; writing—review and editing, J.X., S.F. and L.L. All authors have read and agreed to the published version of the manuscript.

Funding: This research received no external funding.

Data Availability Statement: The original contributions presented in the study are included in the article, further inquiries can be directed to the corresponding author.

Conflicts of Interest: Authors Jing Xu, Fa'an Liu, Jianguo Zhang, Chao Li and Qinghua Liu were employed by the China Petroleum Engineering & Construction Corp. North China Company. The remaining authors declare that the research was conducted in the absence of any commercial or financial relationships that could be construed as a potential conflict of interest.

ıeferences

. Kerley, G.I. *An Equation of State for Helium*; Kerley Technical Services Research Report; Kerley Technical Services: Appomattox, VA, USA, 2004.
. Zhang, C.; Li, C.; Jia, W.; Pang, Y. Thermodynamic study on thermal insulation schemes for liquid helium storage tank. *Appl. Therm. Eng.* **2021**, *195*, 117185. [CrossRef]
. William, J.N.; Richard, H.C.; Bartek, A.G. Stop squandering helium: Establish a global agency to build a sustainable market for this precious commodity. *Nature* **2012**, *485*, 573–575.
. Ortiz Vega, D.O. A New Wide Range Equation of State for Helium-4. Ph.D. Thesis, Texas A&M University, College Station, TX, USA, 2013; pp. 1–155.
. Zuo, Z.; Zhu, W.; Huang, Y.; Wang, L.; Tong, L. A review of cryogenic quasi-steady liquid-vapor phase change: Theories, models, and state-of-the-art applications. *Int. J. Heat Mass Transf.* **2023**, *205*, 123916. [CrossRef]
. Partridge, J.K. Fractional consumption of liquid hydrogen and liquid oxygen during the space shuttle program. In *AIP Conference Proceedings*; American Institute of Physics: Ellipse College Park, MD, USA, 2012; Volume 1434, pp. 1765–1770.
. Hasan, M.F.; Zheng, A.M.; Karimi, I. Minimizing boil-off losses in liquefied natural gas transportation. *Ind. Eng. Chem. Res.* **2009**, *48*, 9571–9580. [CrossRef]
. Zakaria, Z.; Baslasl, M.S.O.; Samsuri, A.; Ismail, I.; Supee, A.; Haladin, N.B. Rollover phenomenon in liquefied natural gas storage tank. *J. Fail. Anal. Prev.* **2019**, *19*, 1439–1447. [CrossRef]
. Shuang, J.J.; Liu, Y.W. Efficiency analysis of depressurization process and pressure control strategies for liquid hydrogen storage system in microgravity. *Int. J. Hydrogen Energy* **2019**, *44*, 15949–15961. [CrossRef]
0. Ma, Y.; Zhu, K.; Li, Y.; Xie, F. Numerical investigation on chill-down and thermal stress characteristics of a LH2 tank during ground filling. *Int. J. Hydrogen Energy* **2020**, *45*, 25344–25356. [CrossRef]
1. Wang, Y.; Cui, L.; Zhang, C.; Kong, L.; Jia, W. CFD investigation of thermal and pressurization performation in LHe tank. *Int. J. Eng. Manuf.* **2020**, *10*, 12–28.

12. Ferrín, J.L.; Pérez-Pérez, L.J. Numerical simulation of natural convection and boil-off in a small size pressurized LNG storage tank. *Comput. Chem. Eng.* **2020**, *138*, 106840. [CrossRef]
13. Lee, W.H. A pressure iteration scheme for two-phase flow modeling. *Eng. Phys.* **1980**, *1*, 407–431.
14. Joseph, J.; Agrawal, G.; Agarwal, D.K.; Pisharady, J.C.; Sunil Kumar, S. Effect of insulation thickness on pressure evolution and thermal stratification in a cryogenic tank. *Appl. Therm. Eng.* **2017**, *111*, 1629–1639. [CrossRef]
15. Gursu, S.; Sherif, S.A.; Veziroglu, T.N.; Sheffield, J.W. Analysis and Optimization of Thermal Stratification and Self-Pressurization Effects in Liquid Hydrogen Storage Systems—Part 1: Model Development. *J. Energy Resour. Technol.* **1993**, *115*, 221–227. [CrossRef]
16. Kumar, S.P.; Prasad, B.V.S.S.S.; Venkatarathnam, G.; Ramamurthi, K.; Murthy, S.S. Influence of surface evaporation on stratification in liquid hydrogen tanks of different aspect ratios. *Int. J. Hydrogen Energy* **2007**, *32*, 1954–1960. [CrossRef]
17. Tatom, J.W.; Brown, W.H.; Knight, L.H.; Coxe, E.F. Analysis of Thermal Stratification of LH2 in Rocket Propellant Tank. *Adv. Cryog. Eng.* **1964**, *9*, 265–272.
18. Das, S.P.; Chakraborty, S.; Dutta, P. Studies on Thermal Stratification Phenomenon in LH2 Storage Vessel. *Heat Transf. Eng.* **2004**, *25*, 54–66. [CrossRef]
19. Zhang, W.; Cheng, J.; Song, Y.; Chen, Z.; Chen, W.; Gao, R. Experimental and numerical study of the liquid nitrogen accumulator based on two-phase thermal stratification. *Int. J. Therm. Sci.* **2023**, *184*, 107937. [CrossRef]
20. Stewart, M.; Moder, J.P. Self-Pressurization of a Flightweight, Liquid Hydrogen Tank: Simulation and Comparison with Experiments. In *52nd AIAA/SAE/ASEE Joint Propulsion Conference*; American Institute of Aeronautics and Astronautics: Reston, VA, USA, 2016.
21. Kang, M.; Kim, J.; You, H.; Chang, D. Experimental investigation of thermal stratification in cryogenic tanks. *Exp. Therm. Fluid Sci.* **2018**, *96*, 371–382. [CrossRef]
22. Gong, Y.W.; Lin, W.S.; Gu, A.Z.; Lu, X.S. A simplified model to predict the thermal response of PLG and its influence on BLEVE. *J. Hazard. Mater.* **2004**, *108*, 21–26. [CrossRef]
23. Liu, Z.; Li, Y.; Zhou, G. Study on thermal stratification in liquid hydrogen tank under different gravity levels. *Int. J. Hydrogen Energy* **2018**, *43*, 9369–9378. [CrossRef]
24. Saleem, A.; Farooq, S.; Karimi, I.A.; Banerjee, R. A CFD simulation study of boiling mechanism and BOG generation in a full-scale LNG storage tank. *Comput. Chem. Eng.* **2018**, *115*, 112–120. [CrossRef]
25. Fu, J.; Sunden, B.; Chen, X. Analysis of Self-Pressurization Phenomenon in a Cryogenic Fluid Storage Tank With VOF Method. In Proceedings of the ASME 2013 International Mechanical Engineering Congress and Exposition, San Diego, CA, USA, 15–21 November 2013.
26. Hands, B.A. Problems due to superheating of cryogenic liquids. *Cryogenics* **1988**, *28*, 823–829. [CrossRef]
27. Lu, J.; Xu, S.; Deng, J.; Wu, W.; Wu, H.; Yang, Z. Numerical prediction of temperature field for cargo containment system (CCS) of LNG carriers during pre-cooling operations. *J. Nat. Gas Sci. Eng.* **2016**, *29*, 382–391. [CrossRef]
28. Li, J.; Guo, Z.; Zhang, Y.; Zhao, J.; Li, K.; Hu, W. Thermal stratification and self-pressurization in a cryogenic propellant storage tank considering capillary effect in low-gravity. *Int. J. Therm. Sci.* **2023**, *194*, 108597. [CrossRef]
29. McIntosh, G.E. Layer by layer MLI calculation using a separated mode equation. In *Advances in Cryogenic Engineering*; Springer Berlin/Heidelberg, Germany, 1994; pp. 1683–1690.
30. Zheng, J.; Chen, L.; Wang, J.; Xi, X.; Zhu, H.; Zhou, Y.; Wang, J. Thermodynamic analysis and comparison of four insulation schemes for liquid hydrogen storage tank. *Energy Convers. Manag.* **2019**, *186*, 526–534. [CrossRef]
31. Cunnington, G.R.; Tien, G.L. A study of heat-transfer process in multiplayer insulation. In *Progress in Astronautics and Aeronautics*, Academic Press: New York, NY, USA, 1970; Volume 23, pp. 111–126.
32. Corruccini, R.J. Gaseous heat conduction at low pressures and temperatures. *Vacuum* **1959**, *7-8*, 19–29. [CrossRef]
33. Michaelis, N.; Welsch, F.; Kirsch, S.M.; Schmidt, M.; Seelecke, S.; Schütze, A. Experimental parameter identification for elastocaloric air cooling. *Int. J. Refrig.* **2019**, *100*, 167–174. [CrossRef]
34. Liu, Z.; Li, Y. Thermal physical performance in liquid hydrogen tank under constant wall temperature. *Renew. Energy* **2019**, *130*, 601–612. [CrossRef]
35. Ter Harmsel, H.; Van Dijk, H.; Durieux, M. The heat of vaporization of helium. *Physica* **1967**, *36*, 620–636. [CrossRef]
36. Malkov, M. *Physical and Technical Foundations of Deep Cooling Handbook*; State Energy Building: Moscow, Russia, 1963.
37. Weat, J. Ballancing liquid and gas sales at helium facilities with liquefiers. *GryGas Int.* **2004**, *42*, 35–38.
38. Hasan, M.; Lin, C.; Vandresar, N. Self-pressurization of a flightweight liquid hydrogen storage tank subjected to low heat flux. In Proceedings of the ASME/AIChE National Heat Transfer Conference, NASA Technical Memorandum, Minneapolis, MN, USA, 28–31 July 1991; Report Number: NAS 1.15:103804.
39. Van Dresar, N.; Hasan, M.; Lin, C. Self-pressurization of a flightweight liquid hydrogen tank-effects of fill level at low wall heat flux. In Proceedings of the 30th Aerospace Sciences Meeting and Exhibit, Reno, NV, USA, 6–9 January 1992.

energies

MDPI

Article

Use of Pressure Transient Analysis Method to Assess Fluid Soaking in Multi-Fractured Shale Gas Wells

Jun Zhang [1], Boyun Guo [2,*] and Majid Hussain [2]

[1] College of Petroleum Engineering, Northeast Petroleum University, Daqing 163319, China; zhangjun410410@nepu.edu.cn

[2] Department of Petroleum Engineering, University of Louisiana at Lafayette, Lafayette, LA 70504, USA; majid.hussain1@louisiana.edu

* Correspondence: boyun.guo@louisiana.edu

Abstract: Multi-stage hydraulic fracturing is a key technology adopted in the energy industry to make shale gas and shale oil fields profitable. Post-frac fluid soaking before putting wells into production has been found essential for enhancing well productivity. Finding the optimum time to terminate the fluid-soaking process is an open problem to solve. Post-frac shut-in pressure data from six wells in two shale gas fields were investigated in this study based on pressure transient analysis (PTA) to reveal fluid-soaking performance. It was found that pressure-derivative data become scattering after 1 day of well shut in. The overall trend of pressure-derivative data after the first day of well shut in should reflect the effectiveness of fluid soaking. Two wells exhibited flat (zero-slope) pressure derivatives within one week of fluid soaking, indicating adequate time of fluid soaking. Four wells exhibited increasing pressure derivatives within one week of fluid soaking, indicating inadequate time of fluid soaking. This observation is consistent with the reported well's Estimated Ultimate Recovery (EUR). This study presents a new approach to the assessment of post-frac fluid-soaking performance with real-time shut-in pressure data.

Keywords: shale gas/oil; post-fracturing; fluid soaking; well productivity; test analysis

1. Introduction

It is generally recognized that post-fracturing fluid soaking can improve productivity of multi-fractured shale gas and oil wells. The mechanism of fluid soaking for improving well productivity is believed to be the imbibition of hydraulic fracturing fluid into the shale, causing micro-cracks of the shale that increases formation permeability [1]. While some shale gas/oil wells have undergone month-long shut-in times to allow adequate water soaking and achieved significant productivity improvement, some other wells were killed by the long-time soaking process [2]. Among many investigators, Makhanov et al. [3], Gupta et al. [4], and Zolfaghari et al. [5] investigated the spontaneous water imbibition in shale formations at the laboratory scale. Their work concluded that water imbibition occurred due to the preferential adsorption of water molecules by clay particles. Factors affecting the rate of water imbibition include the type and concentration of salts and surfactants in the water and fluid viscosity. The numerical simulation work performed by Fakcharoenphol et al. [6] suggested that the spontaneous imbibition during fluid soaking can improve early gas production but will have negligible effects on long-term productivity. As the fluid imbibition progresses, the clayey minerals inside the shale and at the surfaces of cracks begin to swell and partially plug the shale pore and the induced cracks. The fluid imbibition, on one hand, can promote formation of cracks in shale; it can also reduce permeabilities of the shale matrix and the conductivity of cracks due to clay-swelling on the other hand. This effect should reduce the effectiveness of longtime fluid soaking for well-productivity improvement. Although much of the literature has demonstrated the negative effect of clay-swelling on fluid transfer in porous media, it is extremely difficult, if not impossible, to quantitatively describe the time-dependent flow resistance to fluid flow in real cracks. An in-depth review of the latest research on capillary behavior in shale gas/oil reservoirs was given by Cai et al. [7], which highlights the challenges in modeling spontaneous imbibition processes due to the intricate imbibition mechanisms.

Because the effectiveness of fluid soaking is soaking-time dependent, and there is a reverse trend beyond critical soaking time, prediction of the critical soaking time becomes essential. Guo et al. [8] presented a hypothetical method to predict the optimum soaking time based on the theory of fracture closure without validation. Guo et al. [9] proposed a simplified approach to solving the problem based on the theory of fluid imbibition in cracks, which is again not proven for its accuracy. Obviously, how to evaluate the effectiveness of fluid soaking becomes an open problem to solve. Due to the lack of valid means of predicting the optimum fluid-soaking time for achieving the maximal well productivity, it is worthwhile to evaluate the effectiveness of fluid soaking and identify the optimal fluid-soaking time using real-time data obtained during the fluid-soaking period. Recently, pressure fall-off data during fluid soaking have become available for some multi-fractured shale gas wells. The pressure transient analysis (PTA) method was proposed in this study to fulfill its purpose.

The PTA method is first reviewed in this work. Pressure fall-off data collected from six shale gas wells in the southwestern region of China were then analyzed. Data analysis allowed us to gain some insight into the effectiveness of fluid soaking on these wells. Findings are consistent with the reported well's Estimated Ultimate Recovery (EUR).

2. Method Description

Pressure transient analysis is employed in this investigation for assessment of fluid-soaking processes. Traditionally, the PTA examines pressure data collected from well tests to reveal unknown reservoir properties such as permeability, near-well formation damage, fracture properties, reservoir geometry, boundaries, etc. PTA reveals these property values based on flow regime identification and data interpretation. The log-log diagnostic plot approach is normally used for identification of flow regimes. This approach plots the pressure differential and its derivative in log-log scale. Depending on the test type (flow test, pressure buildup test, and fall-off test), the pressure differential is defined differently. For flow tests, the differential is defined as the initial reservoir pressure minus the time-dependent flowing bottom-hole pressure. For pressure buildup tests, the pressure differential is defined as the time-dependent shut-in bottom-hole pressure minus the last flowing bottom-hole pressure recorded right before well shutting in. For pressure fall-off tests, the pressure differential is defined as the initial shut-in pressure minus the time-dependent shut-in pressure, or pressure drop. The post-frac fluid-soaking periods are considered as pressure fall-off tests. The pressure-drop Δp used in this study is defined as

$$\Delta p = p_0 - p(t) \tag{1}$$

where p_0 is the initial well shut-in pressure, and $p(t)$ is the shut-in pressure at the shut-in time t, both in MPa. The pressure derivative used in this study is defined as

$$p' = \frac{d\Delta p}{d\ln(t)} = t\frac{d\Delta p}{dt} \tag{2}$$

In conventional oil and gas reservoirs, for non-fractured horizontal wells, the cross-plot of the pressure differential data and its numerical derivative allows for identification of wellbore storage, vertical radial flow, pseudo-linear flow, and pseudo-radial flow based on early-time unit slope, early-time zero slope, half slope, and late-time zero slope in the pressure-derivative curve. Late-time sharp increases in both pressure differential data and its derivative data indicate pseudo-steady flow in closed systems with no-flow boundaries. Late-time zero slope in pressure differential data and a sharp-dropping slope (approaching negative infinity) in its derivative data indicate steady flow in closed systems with constant pressure boundaries.

In shale gas formations/reservoirs, theoretically the cross-plots of pressure differential data and its derivative data should infer additional fluid transport mechanisms such as gas desorption. Lewis and Hughes [10] work showed a match of analytical models to the real pressure differential data by tuning some model parameters. Due to the ultralow permeability of shale, transient regimes have been observed across the entire life of a well [11,12]. PTA on a hydraulically fractured horizontal shale gas well considered characteristics of shale formations and the conductivity of hydraulic fractures, cluster spacing, and enhanced permeability of stimulated zones. Lee et al. [13] generated some type curves for multi-fractured shale gas reservoirs. Their mathematical model considered the effects of dual porosity, Klinkenberg effect, desorption, non-Darcy flow, and skin effect. Their type curves show fluid flow from natural fractures in the near-wellbore region, from the matrix in the near wellbore into the natural fracture, from the equilibrium pressure of the matrix and natural fracture in the near-wellbore region, and from a flow region comprising vertical radial flow, semi-radial flow, bilinear flow, linear flow, pseudo-radial flow, and boundary-dominated flow. It is clear from their type curves that long desorption is required for rich adsorbed gas to flow from the matrix surface to the natural fracture before the degree of fracture can be observed. However, the desorption period was not seen from the real data analyses presented by Lewis and Hughes [10] or Lee et al. [13]. The valley signature for the desorption period was not observed by Zeng et al. [14] either. Afagwu et al. [15] pointed out that the non-Darcy, fracture spacing, and skin effects cannot be analyzed with the Lee et al. [13] curve because the pressure-derivative plot shows the same valley signature irrespective of changes in these parameters.

Based on the experimental studies conducted by Zou et al. [16] and Shaibu [1], we believe that the net pressure, which is defined as the fracture fluid pressure minus the minimum formation stress, creates the same additional compressive stress inside the shale layers. Due to the different values of Young's modulus of shale layers, the same additional stress will create different strain values in different layers. As a result, the shale layers will deform according to different amounts, creating different displacements at the fracture face. This should create shear failure, or cracks, at the interfaces of shale layers.

It is logical to assume that the layer cracks grow with fluid-soaking time. Figure 1 depicts top views of two anticipated adjacent vertical fractures with cracks at early and late-soaking time. Figure 2 shows side views of the two adjacent vertical fractures with cracks. It is expected that the cracks at mid-height are longer than cracks at the fracture tips due to the high level of formation at mid-height. Without knowing the true profile of the crack front, an elliptic profile is assumed in this study. Fluid flowing to the cracked region is assumed to follow a radial pattern.

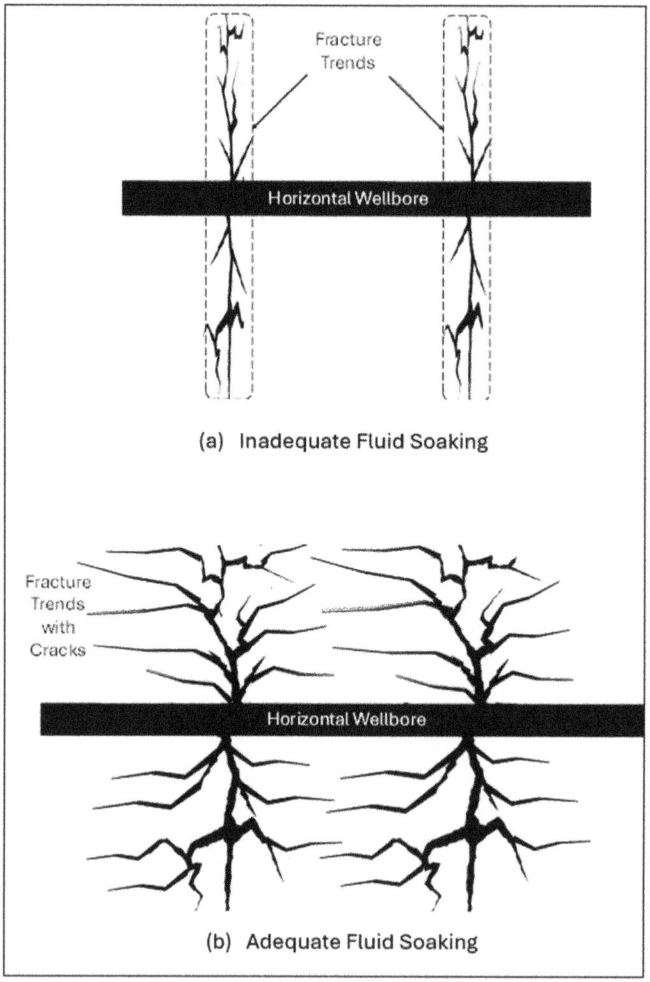

Figure 1. Top views of two anticipated fractures with lateral cracks.

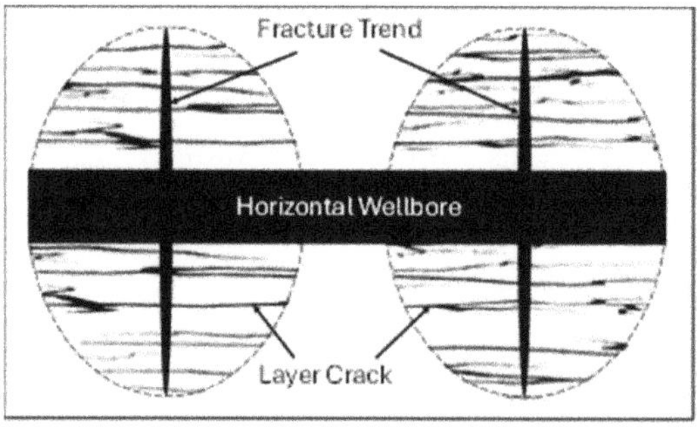

Figure 2. Plan view of two anticipated fracture trends with layer cracks.

The well shut-in time for post-frac fluid soaking usually lasts for only a few days or a few weeks at most. Therefore, the shut-in pressure data gathered from multi-fractured shale gas/oil wells are not adequate for obtaining properties of reservoir/fracture in the regions away from the wellbores. However, because PTA captures the fluid flow condition (transmissibility) at the front of pressure waive propagation, it is possible to make an assessment of fluid-soaking performance around the hydraulic fractures. Such a procedure is exercised in the case studies that follow.

3. Case Studies

Six shale gas wells completed between 2021 and 2023 with multi-fracturing technology in a shale gas field in a southwestern region of China were studied in this project. These wells are from the L2 and D2 areas. Wells in the L2 area have close geological conditions, while wells in the D2 area have similar geological conditions. All these wells have a close engineering background. Wells from the L2 area include L2-7-1, L2-9-1, L2-9-2, L2-9-3, and L2-9-4. Well D2-1-1 is from the D2 area.

3.1. Well L2-7-1

Well L2-7-1 is one of the horizontal wells drilled from platform L2-7. It was completed in August of 2021. Other horizontal wells drilled from the same platform are Well L2-7-2, Well L2-7-3, Well L2-7-4, and Well L2-7-5. However, the post-frac soaking pressure was measured for Well L2-7-1 only. Figure 3 presents pressure-drop data and their numerical derivative. The first peak in derivative might relate to fracture closure, not wellbore storage, because the after-flow process can quickly disappear for the water in the wellbore. The decline in the pressure derivative between 0.2 day and 2 days may be due to the flow from the matrix in the near-wellbore region into the fractures. The numerical derivative data trend reaches a low value of 1.2 MPa at 2 days. After 2 days, the derivative data becomes scattering in nature, not allowing for identification of any flow regimes, but it has a rough trend of going up, implying that the reservoir transmissibility is becoming lower, away from the wellbore region. This may be an indication of inadequate fluid-soaking time. The owner of the well reported the well's Estimated Ultimate Recovery (EUR) to be about 5.4×10^6 m^3, which is lower than the EUR of other wells drilled in the same platform (10.4×10^6 m^3 to 12.2×10^6 m^3).

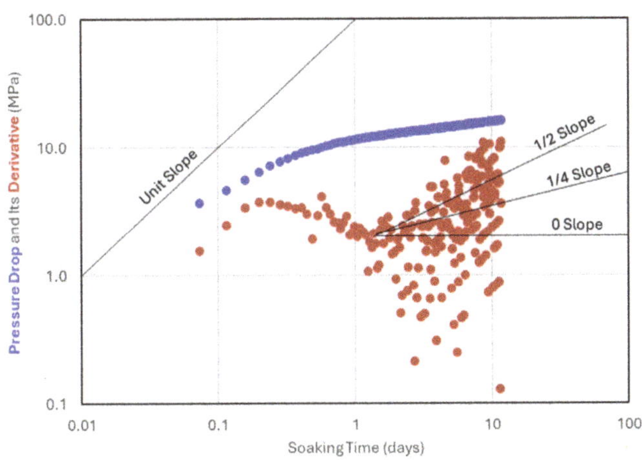

Figure 3. Pressure-drop data and their numerical derivative for Well L2-7-1.

3.2. Well L2-9-1

Well L2-9-1 is the first horizontal well drilled from platform L2-9. It was completed in September of 2021. Figure 4 presents pressure-drop data and their numerical derivative. The decline in the pressure derivative between 0.1 day and 1 day may be due to the flow from the matrix in the near-wellbore region into the fracture. The numerical pressure-derivative data trend reaches a low value of 0.9 MPa at the end of the first day. This low value of pressure derivative reflects a high value of reservoir transmissibility in the near-wellbore region. After the first day, the pressure-derivative data becomes scattered, but it has a rough trend with a zero slope, implying that the reservoir transmissibility is becoming stable once farther away from the wellbore region. This may be an indication of adequate fluid-soaking time. The owner of the well did not report the well's EUR. It is expected that this well should have a higher EUR compared to Well L2-7-1 because of the higher reservoir transmissibility near the wellbore and stable transmissibility away from the wellbore region.

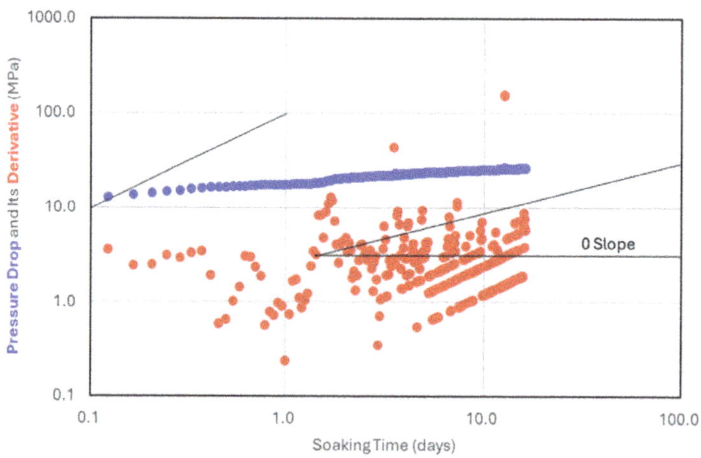

Figure 4. Pressure-drop data and their numerical derivative for Well L2-9-1.

3.3. Well L2-9-2

Well L2-9-2 is the second horizontal well drilled from platform L2-9. It was completed in October of 2023. Figure 5 presents pressure-drop data and its numerical derivative. The early-time data shows a peak in derivative. It might relate to fracture closure, not wellbore storage, because the after-flow period of water in the wellbore disappears quickly. The decline in the pressure derivative between 0.4 day and 1 day may be due to the flow from the matrix in the near-wellbore region into the fracture. The numerical pressure-derivative data trend reaches a value of 1.6 MPa at the end of the first day. This relatively high value of pressure derivative reflects a relatively low value of reservoir transmissibility in the near-wellbore region. After the first day, the pressure dropped five times due to likely frac-hits from offset wells during the shut-in period, which distorted the pressure-derivative data trend, but the derivative data still form an overall increasing trend, implying that the reservoir transmissibility is becoming lower away from the wellbore region. This may be an indication of inadequate fluid-soaking time. The owner of the well did not report the well's EUR. It is expected that this well should have a lower EUR compared to Well L2-9-1 because of the lower reservoir transmissibility near the wellbore and decreasing transmissibility away from the wellbore region.

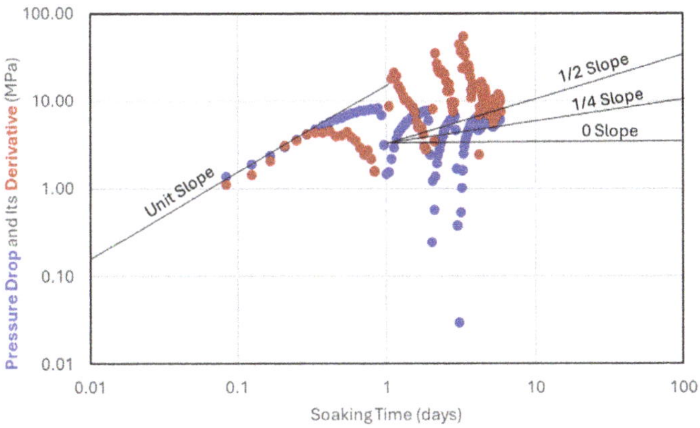

Figure 5. Pressure-drop data and their numerical derivative for Well L2-9-2.

3.4. Well L2-9-3

Well L2-9-3 is the third horizontal well drilled from platform L2-9. It was completed in October of 2023. Figure 6 presents pressure-drop data and its numerical derivative. The early-time data with the unit slope of the derivative should reflect either the fracture storage effect or fracture closure. The decline in the pressure derivative between 0.4 day and 1 day may be due to the flow from the matrix in the near-wellbore region into the fractures. The numerical pressure-derivative data trend reaches a value of 2 MPa at the end of the first day. This relatively high value of pressure derivative reflects a relatively low value of reservoir transmissibility in the near-wellbore region. After the first day, the pressure dropped two times due to likely frac-hits from offset wells during the shut-in period, which distorted the pressure-derivative data trend, but the derivative data still form an overall increasing trend, implying that the reservoir transmissibility is becoming lower away from the wellbore region. This may be an indication of inadequate fluid-soaking time. The owner of the well did not report the well's EUR. It is expected that this well should have a lower EUR compared to Well L2-9-1 because of the lower reservoir transmissibility near the wellbore and decreasing transmissibility away from the wellbore region.

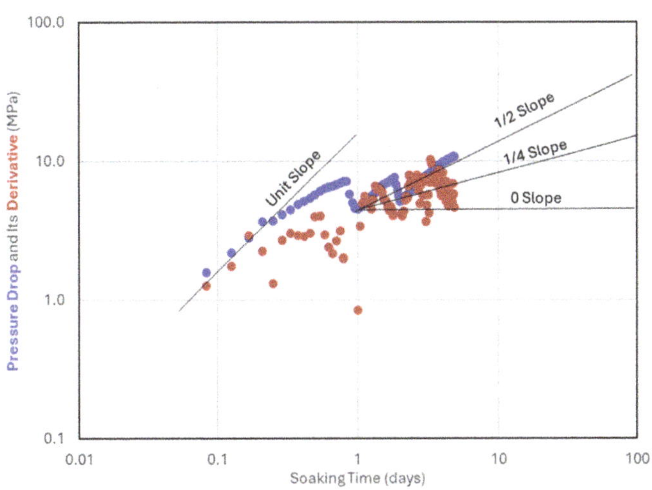

Figure 6. Pressure-drop data and their numerical derivative for Well L2-9-3.

3.5. Well L2-9-4

Well L2-9-4 is the fourth horizontal well drilled from platform L2-9. It was completed in October of 2023. Figure 7 presents pressure-drop data and its numerical derivative. Because a wellbore storage process can quickly disappear for the water in the wellbore, the first peak in derivative might relate to fracture closure. The decline in the pressure derivative between 0.5 day and 1 day may be due to the flow from the matrix in the near-wellbore region into the fractures. The numerical pressure-derivative data trend reaches a value of 4 MPa at the end of the first day. This high value of pressure derivative reflects a low value of reservoir transmissibility in the near-wellbore region due to ineffective fracturing. After the first day, the derivative data become scattered and form an overall stable trend with a zero slope, implying that the reservoir transmissibility is not becoming lower away from the wellbore region. This may be an indication of adequate fluid-soaking time. The owner of the well did not report the well's EUR. It is expected that this well should have a lower EUR compared to Well L2-9-1 because of the lower reservoir transmissibility near the wellbore despite the stable transmissibility away from the wellbore region.

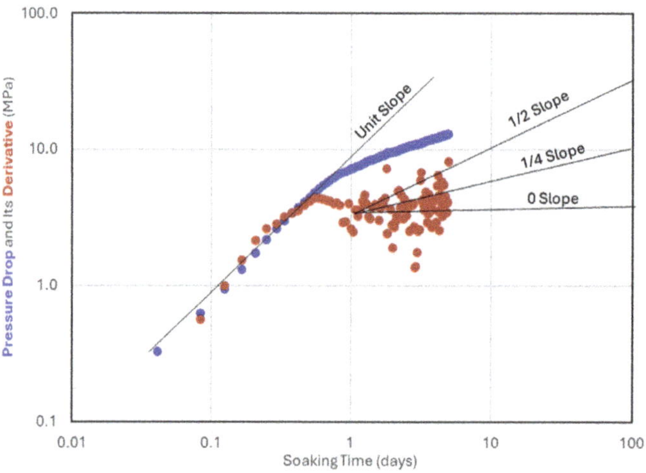

Figure 7. Pressure-drop data and their numerical derivative for Well L2-9-4.

3.6. Well D2-1-1

Well D2-1-1 is one of the horizontal wells drilled from platform D2-1. It was completed in June of 2023. Other horizontal wells drilled from the same platform are Well D2-1-2, Well D2-1-3, Well D2-1-4, and Well D2-1-5. However, the post-frac soaking pressure was measured for Well D2-1-1 only. Figure 8 presents pressure-drop data and their numerical derivative. The decline in the pressure derivative between 0.2 day and 0.3 day may be due to the flow from the matrix in the near-wellbore region into short fractures. The numerical derivative data trend reaches a low value of 1.2 MPa at 0.3 day, indicating high transmissibility near the wellbore. From 0.3 day to 1 day, the derivative increases from 1.2 to 4 MPa, indicating a decreasing trend of reservoir transmissibility in the equilibrium pressure of the matrix and natural fracture in the near-wellbore region. The derivative data becomes scattered after 1 day, not allowing us to draw a clear trend with a specific slope, but it has a rough trend of slowly going up, implying that the reservoir transmissibility is becoming even lower away from the wellbore region. This may be an indication of inadequate fluid-soaking time. The owner of the well reported the well's EUR of 10.9×10^6 m^3. Post-frac fluid soaking was not run on other wells in the platform. The owner reported these wells' EURs as ranging from 10.4 × 10^6 m^3 to 11.0 × 10^6 m^3.

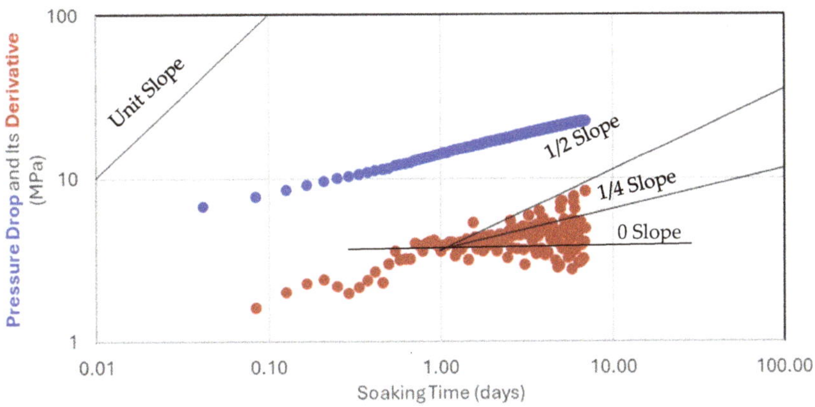

Figure 8. Pressure-drop data and their numerical derivative for Well D2-1-1.

Table 1 presents a summary of the PTAs for these wells. A discussion of these findings is given in the next section.

Table 1. Summary of findings from the PTAs for all wells.

Well No.	Soak Time (Days)	Early-Time Pressure Behavior	Late-Time Derivative Trends	Reservoir Transmissibility	Soaking Sufficiency	Remarks
L2-7-1	2	Sharp derivative decline, unit slope	Scattered after Day 2	Low	Insufficient	Indicates inadequate soaking time.
L2-9-1	7	Stable derivative, zero slope	Scattered after Day 7	High	Insufficient	Short soaking duration impacted results.
L2-9-2	5	Moderate decline, no unit slope	Increasing derivative after Day 5	Moderate	Insufficient	Pressure bleed-offs distorted trends.
L2-9-3	5	Decline with unit slope	Increasing derivative after Day 5	Low	Insufficient	Low transmissibility indicates inadequate soaking.
L2-9-4	10	Stable derivative, zero slope	Stable trend with zero slope	High	Sufficient	Demonstrates effective soaking.
D2-1-1	10	Decline, no unit slope	Scattered after Day 1, increasing	Low	Insufficient	Observed insufficient transmissibility improvement.

4. Discussion

Lee et al. [13] theory, the effects of dual porosity, Klinkenberg effect, desorption, non-Darcy flow, and skin effect are identifiable based on type-curve matching. Their theory is not fully supported by the shut-in pressure data obtained from the post-frac fluid-soaking periods of the six wells in this study. The early-time data from three of the six wells allow for identification of the fluid flow in the fractures in the near-wellbore region because unit slopes of data trends are found. The flows from the matrix in the near-wellbore region into fractures are identified in all the six wells. The regime of equilibrium pressure of the matrix and natural fracture in the near-wellbore region are identified from one well only. In general, the pressure-derivative data becomes scattered after 1 day of well shut in. The overall trend of pressure-derivative data after the 1 day of well shut in should reflect the effectiveness of fluid soaking.

Although the 1/2-, 1/4-, and late-unit-slope trend lines are marked in the derivative plot, the paper does not claim finding these signatures from the derivative plot. The effectiveness of soaking is evaluated based on a comparison between the reservoir transmissibility in the near-wellbore region and that in the away-from-wellbore region. Data smoothing will not make these derivative trends clearer, but it will cause some information loss regarding the real condition. Even though these trends could be clearly identified, they cannot be used for evaluation of soaking effectiveness.

Of the six wells studied, four had a soaking time of less than 10 days. Among these, Well L2-9-4 and Well D2-1-1 exhibited distinct differences in soaking performance. Well L2-9-4 showed sufficient soaking time, as evidenced by stable reservoir transmissibility and pressure-derivative trends, while Well D2-1-1 was deemed to have insufficient soaking time due to scattered pressure derivatives and declining transmissibility after the early-soaking period. Although Well L2-9-1 also exhibited scattered pressure derivatives on days 7-8, this was attributed to its unique reservoir characteristics (e.g., local natural fractures) and operational factors (e.g., pumping rate and fluid properties). The conclusion regarding D2-1-1's soaking time insufficiency stems from observed trends rather than a direct comparison to L2-9-1, as each well's performance is influenced by its specific geological and operational context. However, there was a steady upward trend observed in the case of Well D2-1-1 with no exact slope compared to Well L2-9-1 where the derivative data had a rough trend toward the horizontal direction with zero slope. Therefore, there is a slight indication that Well D2-1-1 has insufficient soaking time.

It is seen that the six wells analyzed in this work had varying fluid-immersion times. This variation reflects the local geology (e.g., natural fractures) and operational practices (e.g., pumping rates and fluid properties). Thanks to the technology with which to obtain real-time pressure data and the PTA, which is suitable for analyzing the real-time data, the unique immersion time of fluid soaking allows for timely termination of fluid soaking on individual wells to achieve the maximal well productivity.

5. Limitations

The PTA method with analytical solutions used in this work is limited to isotropic and homogeneous reservoirs in symmetrical shapes. It requires that the hydraulic fractures be uniform with identical properties (porosity and permeability) and dimensions (length, height, and width). When the method is applied to analyzing pressure fall-off data, it requires the multi-stage fracturing operation to be completed in a short time, e.g., less than a day. Otherwise, the fluid-immersion times in different fractures can be significantly different, affecting the results of the interpretation using PTA.

This study is based on data from six shale gas wells in the southwest region of China, which represents a limited dataset. While this sample size provides valuable preliminary insights, it also presents a challenge in terms of generalizing the findings to broader contexts. The wells were selected to encompass a range of reservoir characteristics and operational practices, thereby offering a diverse perspective within the constraints of available data. Nonetheless, to enhance the robustness and applicability of the conclusions, future research should incorporate data from a larger number of wells across multiple regions and varying geological conditions. This expansion would not only validate the current findings but also refine the methodology for broader adoption in unconventional reservoirs worldwide.

The pressure-derivative data becomes periodically fluctuating and scattered after the first day. The reasons can be attributed to possible well leaking/pressure releasing that were not identified/reported. The use of wellhead pressure data instead of measured downhole pressure data (not available) can be another reason for the scattering nature of the pressure-derivative data. Other causes might be error/accuracy of the measuring instruments. Noise in data acquisition systems or insufficient sampling frequency can contribute to scattering in pressure-derivative data. These issues should be addressed in future applications of the PTA.

6. Conclusions

Post-frac shut-in pressure data from six wells in three platforms in two shale gas fields in the southwestern region of China were investigated in this study based on pressure transient analysis (PTA). The analysis allows us to draw the following conclusions:

1. The early-time data from three wells identified fluid flow in fractures in the near-wellbore region. The flow from the matrix in the near-wellbore region into fractures are identified in all six wells, indicating that all these wells were indeed fractured;
2. The regime of equilibrium pressure of the matrix and fracture in the near-wellbore region is identified in Well D2-1-1 only. This can be attributed to the long fractures created in this well;
3. In general, the pressure-derivative data becomes scattered after 1 day of well shut in. The overall trend of pressure-derivative data after the first day of well shut in should reflect the effectiveness of fluid soaking. This observation seems to be consistent with the reported well's Estimated Ultimate Recovery (EUR);
4. Only two wells exhibited flat (zero-slope) pressure derivatives within one week of fluid soaking, indicating adequate time of soaking. The fluid-soaking time for the other four wells is considered inadequate.

Author Contributions: Conceptualization, B.G. and J.Z.; methodology, B.G.; software, B.G.; validation, J.Z., B.G. and M.H.; formal analysis, J.Z.; investigation, M.H.; resources, B.G.; data curation, J.Z.; writing—original draft preparation, B.G.; writing—review and editing, J.Z.; visualization, M.H.; supervision, B.G.; project administration, B.G.; funding acquisition, J.Z. All authors have read and agreed to the published version of the manuscript.

Funding: This research was funded by the Northeast Petroleum University and the University of Louisiana at Lafayette.

Informed Consent Statement: Not applicable.

Data Availability Statement: Data is unavailable due to ethical restrictions.

Conflicts of Interest: The authors declare no conflict of interest.

References

1. Shaibu, R. Fracture Conductivity, Hydro-Cracking, and Proppant Embedment of Tuscaloosa Marine Shale Formation Cores. Ph.D. Thesis, University of Louisiana at Lafayette, Lafayette, LA, USA, 2021.
2. Ghanbari, E.; Dehghanpour, H. The Fate of Fracturing Water: A Field and Simulation Study. *Fuel* **2016**, *163*, 282–294. [CrossRef]
3. Makhanov, K.; Dehghanpour, H.; Kuru, E. Measuring liquid uptake of organic shales: A workflow to estimate water loss during shut-in periods. In Proceedings of the SPE Unconventional Resources Conference Canada, Calgary, AB, Canada, 5–7 November 2013; Paper SPE 167157.
4. Gupta, A.; Xu, M.; Dehghanpour, H.; Bearinger, D. Experimental Investigation for Microscale Stimulation of Shales by Water Imbibition During the Shut-in Periods. In Proceedings of the SPE Unconventional Resources Conference, Calgary, AB, Canada, 15–16 February 2017. [CrossRef]
5. Zolfaghari, A.; Dehghanpour, H.; Holyk, J. Water Sorption Behaviour of Gas Shales: I. Role of Clays. *Int. J. Coal Geol.* **2017**, *179*, 130–138. [CrossRef]

6. Fakcharoenphol, P.; Torcuk, M.; Bertoncello, A.; Kazemi, H. Managing shut-in time to enhance gas flow rate in hydraulic fractured shale reservoirs: A simulation study. In Proceedings of the SPE Annual Technical Conference and Exhibition, New Orleans, LA USA, 30 September–2 October 2013; Paper SPE 166098. [CrossRef]
7. Cai, J.; Sun, S.; Wang, H. Current Advances in Capillarity: Theories and Applications. *Capillarity* **2023**, *7*, 25–31. [CrossRef]
8. Guo, B.; Shaibu, R.; Hou, X. Crack Propagation Hypothesis and a Model to Calculate the Optimum Water-Soaking Period in Shale Gas/Oil Wells for Maximizing Well Productivity. *SPE Drill. Complet.* **2020**, *35*, 655–667. [CrossRef]
9. Guo, B.; Mahmood, M.; Wortman, P.; Nguyen, V. Prediction of the Optimal Post-Fracturing Soaking Time in Multifractured Shale Gas/Oil Formations on the Basis of Modeling of Fluid Imbibition. *SPE J.* **2024**, *29*, 4902–4916, paper SPE 221459. [CrossRef]
10. Lewis, A.M.; Hughes, R.G. Production data analysis of shale gas reservoirs. In Proceedings of the SPE Annual Technical Conference and Exhibition, Denver, CO, USA, 21–24 September 2008.
11. Cheng, Y. Pressure Transient Characteristics of Hydraulically Fractured Horizontal Shale Gas Wells. In Proceedings of the SPE Eastern Regional Meeting, Columbus, OH, USA, 17–19 August 2011; Paper SPE 147603. [CrossRef]
12. Ozkan, E.; Raghavan, R.S.; Apaydin, O.G. Modeling of fluid transfer from shale matrix to fracture network. *SPE Annu. Tech. Conf Exhib.* **2010**, *4*, 3314–3331. [CrossRef]
13. Lee, S.J.; Kim, T.H.; Lee, K.S. Type curves for pressure transient analysis of horizontal wells in shale gas reservoirs. In Proceedings of the SPE Middle East Oil and Gas Show and Conference, Manama, Bahrain, 10–13 March 2013.
14. Zeng, Y.; Wang, Q.; Ning, Z.; Sun, H. A Mathematical Pressure Transient Analysis Model for Multiple Fractured Horizontal Wells in Shale Gas Reservoirs. *Geofluids* **2018**, *2018*, 8065949. [CrossRef]
15. Afagwu, C.; Abubakar, I.; Kalam, S.; Al-Afnan, S.F.; Awotunde, A.A. Pressure-transient analysis in shale gas reservoirs: A review *J. Nat. Gas Sci. Eng.* **2020**, *78*, 103319. Available online: https://www.sciencedirect.com/science/article/abs/pii/S1875510020301736?via=ihub (accessed on 20 January 2025). [CrossRef]
16. Zou, Y.; Li, N.; Ma, X.; Zhang, S.; Li, S. Experimental Study on the Growth Behavior of Supercritical CO_2-Induced Fractures in a Layered Tight Sandstone Formation. *J. Nat. Gas Sci. Eng.* **2018**, *49*, 145–156. [CrossRef]

energies

MDPI

Article

Experimental Study on Surfactant–Polymer Flooding After Viscosity Reduction for Heavy Oil in Matured Reservoir

Xiaoran Chen [1,2,3], Qingfeng Hou [3,*], Yifeng Liu [3], Gaohua Liu [4], Hao Zhang [5], Haojie Sun [5], Zhuoyan Zhu [3] and Weidong Liu [2,3]

1 School of Engineering Science, University of Chinese Academy of Sciences, Beijing 100049, China; chenxiaoran22@mails.ucas.ac.cn
2 Institute of Porous Flow and Fluid Mechanics, Chinese Academy of Sciences, Langfang 065007, China; lwd69@petrochina.com.cn
3 PetroChina Research Institute of Petroleum Exploration and Development (RIPED), Beijing 100083, China; liuyifeng01@petrochina.com.cn (Y.L.); zhzhy@petrochina.com.cn (Z.Z.)
4 PetroChina Liaohe Oilfield Company, Panjin 124010, China; liugaohua@petrochina.com.cn
5 College of Safety and Ocean Engineering, China University of Petroleum, Beijing 102249, China; 2022215868@student.cup.edu.cn (H.Z.); 2023211021@student.cup.edu.cn (H.S.)
* Correspondence: houqingfeng@petrochina.com.cn

Abstract: An advanced enhanced oil recovery (EOR) method was investigated, employing a surfactant–polymer (SP) system in combination with a viscosity reducer for application in a heavy oil reservoir within the Haiwaihe Block, Liaohe Oilfield, in China. Significant advantages were observed through the combination of LPS-3 (an anionic surfactant) and OAB (a betaine surfactant) in reducing interfacial tension and enhancing emulsion stability, with the optimal results achieved at the ratio of 9:1. The BRH-325 polymer was found to exhibit superior viscosity enhancement, temperature resistance, and long-term stability. Graphene nanowedges were utilized as a viscosity reducer, leading to a viscosity reduction in heavy oil of 97.43%, while stability was maintained over a two-hour period. The efficacy of the combined system was validated through core flooding experiments, resulting in a recovery efficiency improvement of up to 32.7%. It is suggested that the integration of viscosity reduction and SP flooding could serve as a promising approach for improving recovery in mature heavy oil reservoirs, supporting a transition toward environmentally sustainable, non-thermal recovery methods.

Keywords: heavy oil; viscosity reduction; enhanced oil recovery; polymer; surfactant

1. Introduction

In recent years, unconventional oil and gas resources have become the core pillars of domestic oil and gas development, driven by advancements in exploration, production, and research [1–3]. As a key component of unconventional oil and gas reservoirs, heavy oil extraction plays a crucial role in bridging the significant gap in oil demand. Against the backdrop of achieving "carbon peak and carbon neutrality", cold recovery technology for heavy oil has emerged as a promising approach due to its economic, environmentally friendly, and efficient characteristics. Among these technologies, chemical flooding has been proven essential for enhancing the cold recovery efficiency of heavy oil. Surfactants and polymers are the most critical chemical agents in chemical flooding, as their synergistic interaction not only increases the swept volume but also significantly reduces the interfacial tension (IFT) between oil and solutions [4]. Compared to alkaline–surfactant-polymer flooding, alkaline-free surfactant–polymer (SP) systems exhibit advantages such as the reduced scaling and corrosion of wellbores, as well as relatively simpler handling of produced fluids [5].

Numerous scholars have studied the binary SP system for heavy oil recovery, with widespread consensus that its oil displacement efficiency surpasses that of either polymer flooding or surfactant flooding alone. For instance, Hocine S. et al. [6] demonstrated that introducing surfactants during secondary polymer injection increased the original oil-in-place (OOIP) recovery by approximately 23%. They concluded that the alkali-free surfactant–polymer formulation developed under reservoir conditions could serve as a valuable enhancement to polymer flooding, particularly for recovering heavy oil with viscosities exceeding 1000 mPa·s. Further findings indicated that under low-temperature and low-mineralization conditions, a simple mineralization gradient injection strategy effectively reduced surfactant adsorption. Romero-Zerón, L. et al. [7] conducted oil displacement experiments and found the supramolecular SP system exhibited favorable propagation and low adsorption characteristics in loosely consolidated sand systems. The formation of a stable and viscous displacement front was facilitated by the high structural strength of the system, resulting in a significant increase in swept volume. The synergistic effects of the surfactant and polymer are characterized by increased mobility control and reduced IFT at the oil–water interface. SP flooding could provide an average incremental recovery rate improvement of 19% compared to conventional polymer flooding. Kaili, L. et al. [8] screened an SP binary system tailored for Gudong heavy oil (350 mPacs viscosity at 50 °C) in the Shengli Oilfield, China. The IFT was observed to remain relatively stable at 80 °C and a salinity of 13,000 mg·L^{-1}. Furthermore, recovery rate was improved by 44.2% with SP binary flooding compared to water flooding. And the final recovery rate of SP flooding was 3.6% higher than polymer flooding alone.

Heavy oil is characterized by its high density and viscosity, and often contains significant impurities and waxy substances [9]. Its primary components include hydrocarbons (saturated hydrocarbons, cycloalkanes, and aromatic hydrocarbons), asphaltenes, and resins. The resins and asphaltenes exhibit high polarity, a high heteroatom content, and complex molecular structures, which are the primary contributors to the elevated viscosity of heavy oil [10]. The poor fluidity due to the high viscosity of crude oil has been considered as the main challenge in enhancing heavy oil recovery. Therefore, viscosity reducers are proposed as an effective solution to this issue.

Zhang, X. et al. [11] synthesized a novel polymeric surfactant, PAMOs, which incorporated 2-acrylamido-2-methylpropanesulfonic acid (AMPS) and sodium alpha olefin sulfonate (AOS) into the main carbon chain of acrylamide (AM). This viscosity reducer could penetrate the asphaltene molecular layer to form a "sandwich" structure, thereby weakening the intermolecular forces between asphaltene molecules. PAMOs also established strong hydrogen bonds with asphaltenes, disrupting their supramolecular structure and converting them into lower-molecular-weight forms. The hydrophobic long-chain spatial structure further inhibited the reaggregation of asphaltene molecules, effectively reducing the viscosity of heavy oil. Si, Y. et al. [12] also synthesized a boron-containing anionic–nonionic surfactant (SYW) through esterification and sulfonation reactions. The raw materials were 1,3-propylene glycol polyether (PPG), boric acid, maleic anhydride (MA), and sodium metabisulfite. It was revealed that SYW achieved a viscosity reduction rate of 97.3%. Moreover, when SYW was combined with oleic acid and ethanolamine in a 3:1:1 ratio to produce SYG, the viscosity reduction rate could be increased to 98.6%. The emulsion containing SYG exhibited a transition in the dispersed phase from water droplets to oil droplets. The oil droplets had a uniform particle size of 3.68 μm, which enhanced fluidity.

It is widely believed that viscosity reducers are an effective enhanced oil recovery (EOR) technique in addressing the challenges of heavy oil recovery. The novel graphene nanowedge viscosity reducer was employed in this study. It was a two-dimensional nanomaterial, and formed a wedge-shaped front at the oil–water–rock interface. The structural separation pressure was generated by the formation of this interface. This pressure could detach the oil film from the rock surface, thereby allowing it to enter the mobile phase. The modified graphene nanowedge with surfactant molecules was found to exhibit large hydrophilic groups and multiple lipophilic groups. These characteristics enabled significant interaction with heavy oil. Under the shear forces induced by the flow of displacement fluids, the modified graphene nanowedges were able to fragment dispersed oil clusters into smaller droplets. These droplets subsequently entered the aqueous phase. Zwitterionic surfactants in the aqueous phase enhanced this process by forming a molecular layer on the microdroplet surfaces, which prevented readhesion to the rock surface. Then, a stable oil-in-water emulsion with low viscosity and high fluidity was formed. As a result, the mobility and recovery of heavy oil could be improved.

The Haiwaihe Block of Liaohe Oilfield in China is a multi-layer sandstone reservoir, with an oil well section of 200–400 m. The matured reservoir is primarily composed of sedimentary rocks, with significant contributions from sandstone and siltstone. These rocks are characterized by their fine to medium grain size and relatively high porosity and permeability, which are conducive to oil accumulation. Overall, the matured reservoir exhibits medium porosity and medium-high permeability, with an average porosity of 28.7% and an average permeability of 858 mD. The reservoir also exhibits strong heterogeneity. After more than 30 years of water flooding, the water cut of the reservoir is 92.8% and the recovery rate is 44.8%. The temperature of the reservoir is 60 °C. The crude oil viscosity under reservoir conditions is 100 mPa·s, and the total dissolved solids (TDSs) are 1814.3 mg/L.

Currently, the block primarily employs thermal recovery techniques for conventional heavy oil extraction, which lead to substantial carbon emissions. The balancing between thermal and non-thermal recovery techniques is essential to align with carbon reduction targets and to fulfill development goals for enhanced oil production. In this study, a viscosity reducer was utilized to lower the viscosity of heavy oil, followed by the application of an optimized SP flooding system to increase oil displacement efficiency. This approach aims to enhance oil recovery by improving the flowability of heavy oil and facilitate the transition from thermal to water flooding techniques. This research could provide a technical basis for the application of viscosity-reduced SP flooding in heavy oil reservoirs.

2. Materials and Methods

2.1. Experimental Materials and Equipment

2.1.1. Surfactant Samples

The surfactants used in this study include petroleum sulfonate, betaine, and polyether. Five types of petroleum sulfonates, including LPS-1, LPS-2, LPS-3, LPS-4, and LPS-5, were provided by RIPED (Beijing, China). Betaine OAB (oleic acid amide propyl betaine), HSB1214 (alkyl C12–14 hydroxypropyl sulfobetaine), and HSB1618 (alkyl C16–18 hydroxypropyl sulfobetaine) were provided by Shandong Yousuo Chemical Technology Co., Ltd. (Linyi, China). The polyethers HH-B1, HH-B2, and HH-B3 were provided by Changzhou Haohua Chemical Co., Ltd. (Changzhou, China).

2.1.2. Polymer Samples

Three linear Hydrolyzed Polyacrylamides—HPAMs—(M_w = 2000 × 10^4, 2500 × 10^4 and 3000 × 10^4) and one high-branched HPAM (BRH-325) were employed as the polymers in this study. All of these polymers were provided by RIPED.

2.1.3. Viscosity Reducer Samples

Two types of traditional viscosity reducers, VR-SF and CA601S-HNS, were provided by Ningbo Fengcheng Advanced Energy Materials Research Institute Co., Ltd. (Ningbo, China). And a novel graphene viscosity reducer was provided by RIPED.

2.1.4. Oil Samples

The degassed and dehydrated heavy oil used in this study was sourced from the target block. The oil from the target block contained a large number of saturated hydrocarbons and aromatic hydrocarbons, with relatively low asphaltene content, and it was classified as an intermediate base oil. The water content of the heavy oil sample after dehydration was 0.8%. The density and viscosity of the crude oil were measured to be 0.965 g/cm^3 and 605 mPa·s at 60 °C, respectively. The kerosene used for diluting the heavy oil was provided by RIPED.

2.1.5. Core Samples

The core samples were saturated with brine to simulate reservoir conditions, after which permeability and porosity measurements were performed in the laboratory. The properties of the cores are collected in Table 1.

Table 1. Characteristics of cores.

Core	Length (cm)	Diameter (cm)	Porosity (%)	Water Permeability (mD)	Original Oil Saturation (%)
1st	9.979	2.501	33.02	1573	71.34%
2nd	9.990	2.497	33.22	1564	72.16%

2.1.6. Water

The water used in this study was brine, formulated to simulate the TDS of the produced water from the Haiwaihe Block. The properties of the brine are shown in Table 2.

Table 2. Parameters of brine from Haiwaihe Block.

Category	pH	Ion Concentration/(mg·L^{-1})							TDS (mg·L^{-1})
		Na^+	Ca^{2+}	Mg^{2+}	Cl^-	$SO_4{}^{2-}$	$CO_3{}^{2-}$	$HCO_3{}^-$	
Brine	7.86	523.77	23.36	10.73	300.18	11.95	33.18	911.13	1814.3

2.1.7. Experimental Equipment

The equipment utilized in this study included a Brookfield DV2T Viscometer (Brookfield Engineering, Middleborough, MA, USA) for viscosity measurements, an IFT Meter CNG-701 (CNG, Nantong, China) for interfacial tension (IFT) analysis, and a SARTORIUS BT4202S Electronic Balance (Sartorius, Göttingen, Germany) for precise weight determination. An IKA C-MAG HS 7 Magnetic Stirrer (IKA, Staufen, Germany) and an IKA EUROSTAR 20 Paddle Stirrer (IKA, Staufen, Germany) were employed for sample mixing, while a BINDER FED115 Drying Oven (BINDER, Tuttlingen, Germany) was used for sample preparation and drying. Additionally, a HAAKE MARS 60 Rotational Rheometer (Thermo Fisher Scientific, Waltham, MA, USA) was utilized to evaluate the rheological properties of the fluids.

2.2. Experimental Methods

2.2.1. Experiment for Screening Surfactant

The experimental procedures for screening surfactants under reservoir conditions were as follos: (1) Prepare individual surfactant solutions at varying concentrations (0.2%, 0.3%, 0.4%) using simulated brine. (2) Observe the dissolution behavior of surfactants at different concentrations and evaluate their solubility. (3) Measure the IFT (60 °C, 5000 revolutions per minute (rpm)) of each single surfactant solution across the prepared concentration range. (4) Identify surfactants with a high IFT reduction efficiency for further compounding. (5) Measure the IFT of the compounded surfactant solutions. (6) Use a paddle stirrer to blend the oil and compounded surfactant solution at a speed of 13,000 rpm for 30 s. (7) Observe and record the oil separation rate of the resulting emulsion to assess emulsion stability. (8) Screen the compounded surfactant solution that exhibits optimal emulsion stability and the strongest IFT reduction capability.

2.2.2. Experiment for Screening Polymer

Polymers were analyzed for key physicochemical properties, including solid content, water-insoluble content, filtration factor, and dissolution time. Additionally, the viscosity-average molecular weight of the polymers was determined under dilute solution conditions. The detailed parameters are provided in Table 3.

Table 3. Physicochemical parameters of polymers.

Category	BRH-325	HPAM (2000×10^4)	HPAM (2500×10^4)	HPAM (3000×10^4)
Appearance	White Powder	White Powder	White Powder	White Powder
Solid content (%)	89.38	89.55	89.42	89.27
Dissolution rate (min)	≤120	≤120	≤120	≤120
Water-insoluble content (%)	0.088	0.116	0.104	0.140
Filtration factor	1.007	1.006	1.047	1.035
Viscosity average molecular weight ($\times 10^6$)	2410	2080	2600	3090

The polymer screening process for reservoir applicability involved the following experimental procedures: (1) Prepare polymer solutions of varying concentrations using brine. Measure the viscosities of the solutions at 60 °C using a Brookfield DV2T viscometer operated at a rotational speed of 6 rpm. (2) Evaluate the rheological behavior of two polymer solutions by measuring viscosities at varying shear rates (0.01–1000 s^{-1}) using a rotational rheometer at 60 °C, and use power-law function (1) to fit the measured rheological curve. (3) Measure the viscosities of the polymer solutions using the viscometer at different temperatures (40 °C, 45 °C, 50 °C, 55 °C, 60 °C, 65 °C, 70 °C) to assess their temperature resistance. (4) Transfer 1500 mg·L^{-1} polymer solution into ampoules, connect them to the vacuum manifold system as shown in Figure 1, and evacuate to a pressure below 13.3 Pa for 1 h. (5) Inject nitrogen gas to restore atmospheric pressure, and seal the ampoules after repeating the process three times. (6) Place the ampoule in a drying oven set to 60 °C. At the designated aging intervals (1, 15, 30, 45, 60, 75, and 90 days), remove the ampoules, measure the viscosity, and calculate the viscosity retention rate. (7) Screen the polymer exhibiting strong viscosity-increasing properties, along with superior temperature resistance and long-term stability.

$$\eta_v = k \cdot \gamma^{n-1} \tag{1}$$

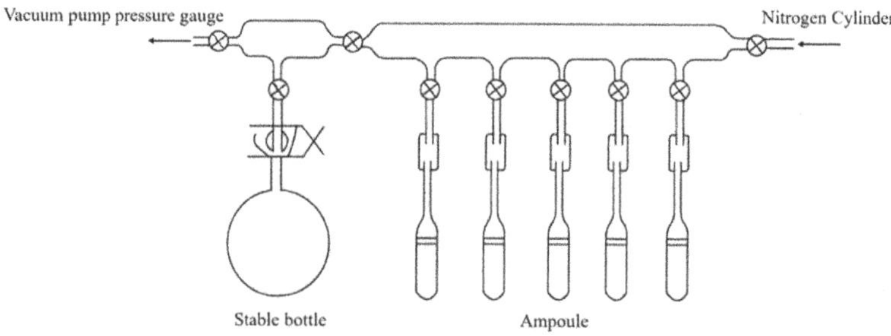

Figure 1. Vacuum manifold system.

2.2.3. Experiment for Surfactant and Polymer Compatibility

The compatibility of polymer and surfactant is evaluated under reservoir conditions following these experimental procedures, which are in accordance with the standard (SY/T 6424-2014) of China [13]: (1) Measure the IFT of the SP compound solution using the previously described methods. (2) Measure the viscosity of the SP compound solution using the methods outlined above. (3) Compare the IFT reduction performance of the SP compound solution to that of a single surfactant. (4) Evaluate the viscosity-increasing properties of the SP compound solution against a single polymer under identical conditions.

2.2.4. Experiment for Screening Viscosity Reducer

Screening a viscosity reducer suitable for reservoir conditions is conducted through the following experimental procedures: (1) Measure the viscosity of heavy oil from the target block under 60 °C. (2) Prepare an emulsion by mixing heavy oil with a viscosity reducer solution (0.4%) at an oil-to-water ratio of 4:1. (3) Measure and record the viscosity of the emulsion at regular intervals over a 2 h period. (4) Evaluate and screen the viscosity reducer based on the observed viscosity reduction rate within 2 h.

2.2.5. Experiment for Core Flooding

Core displacement experiments are conducted to evaluate the oil displacement performance of the SP compound solution and to investigate the effect of viscosity reducers on heavy oil SP binary flooding. The core flooding experimental equipment is shown in Figure 2. The detailed experimental procedures are as follows: (1) Compound the heavy oil from the target block with kerosene in an appropriate ratio to achieve a viscosity of 100 (±5) mPa·s under 60 °C; (2) place the core in a drying oven for 12 h and weigh the core column; (3) saturate the rock core with simulated brine for 12 h, vacuum it for 12 h, and then weigh it; (4) calculate pore size and porosity; (5) according to the permeability measurement method, the permeability of the rock core is measured using simulated brine. (6) Saturate the core column with oil under 60 °C until the water cut in the measuring tube no longer increases within 3 h, and record the volume of saturated oil; (7) place the core under 60 °C for 12 h of aging. (8) Conduct water injection experiments at a rate of 0.05 $mL \cdot L^{-1}$, monitor the produced water and oil volume, and calculate the water cut and the recovery efficiency at different times after the liquid flows out; (9) stop water injection when the water cut reaches 98%; (10) inject 0.3 PV viscosity reducer into the first core at a rate of 0.05 $mL \cdot L^{-1}$. Follow with water injection until the water cut again reaches 98%.

The second core was not injected with viscosity reducer. (11) Inject the SP compound solution into the core at a rate of 0.05 mL·L^{-1}. When the injection volume reaches 0.8 PV or more, stop injection and calculate the recovery efficiency (12). Conduct water injection experiments at a rate of 0.05 mL·L^{-1} to monitor the water and oil production. Stop injection when the effluent no longer contains oil. (13) Evaluate the oil displacement performance of SP compound solution based on recovery efficiency, and compare the results of two core experiments to research the effect of viscosity reducer on heavy oil SP binary flooding. All experiments during the injection phase were conducted in the incubator at a temperature of 60 °C.

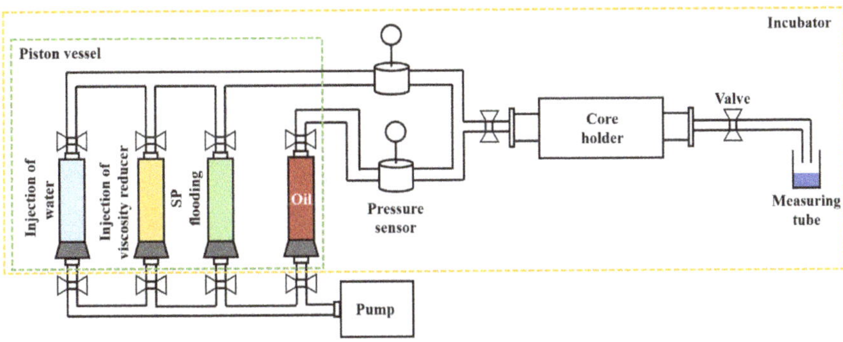

Figure 2. Schematic experimental installation of core flooding experiment.

3. Results

3.1. Screening of Surfactants

3.1.1. Solubility of Surfactants

As shown in Table 4, most petroleum sulfonates demonstrated favorable solubility in the simulated brine and showed no precipitation at concentrations up to 0.3%. However, at an LPS-4 concentration of 0.4%, slight precipitation was observed, whereas other petroleum sulfonates remained completely soluble. In terms of betaine surfactants, most exhibited a clear liquid state, except for HSB1618, as it displayed slight precipitation at a concentration of 0.4%. Among polyether surfactants, HH-B2 exhibited low solubility. Significant precipitate occurred at a concentration of 0.2%. It became completely insoluble when the concentration exceeded 0.3%.

Table 4. Evaluation of surfactant solubility.

Surfactants	Brine		
	0.2%	0.3%	0.4%
LPS-1	White turbid liquid	White turbid liquid	White turbid liquid
LPS-2	Clear liquid	Clear liquid	Clear liquid
LPS-3	Clear liquid	Clear liquid	Clear liquid
LPS-4	Clear liquid	Clear liquid	Slight precipitate
LPS-5	Clear liquid	Clear liquid	Clear liquid
OAB	Clear liquid	Clear liquid	Clear liquid
HSB1214	Clear liquid	Clear liquid	Clear liquid
HSB1618	Clear liquid	Clear liquid	Slight precipitate
HH-B1	Clear liquid	Clear liquid	Clear liquid
HH-B2	Significant precipitate	Completely insoluble	Completely insoluble
HH-B3	Clear liquid	Clear liquid	Clear liquid

These findings offered valuable insights into the solubility behavior of various surfactants in brine and provided a basis for screening optimal surfactant formulations in EOR applications within the target reservoir. Based on the solubility test results, ten out of the eleven surfactants (excluding HH-B2) were selected for subsequent experiments.

3.1.2. Interfacial Tension Reduction of Surfactants

Previous studies had demonstrated that surfactants exhibited effective oil displacement performance when the IFT was reduced to a magnitude of 10^{-3} mN/m or lower [14–16]. As listed in Table 5, three surfactants including LPS-3, OAB, and HSB1618 exhibited better IFT reduction properties than the others. The anionic surfactant LPS-3 exhibited superior performance, attributed to the presence of sulfonic acid groups (-SO_3H) in its molecular structure. These functional groups impart strong hydrophilicity, significantly reducing the IFT between oil and solution and enhancing oil-washing efficiency. OAB and HSB1618, as amphoteric surfactants, demonstrated excellent surface activity and emulsification properties.

Table 5. IFT between oil and solution with different concentrations of surfactants.

Surfactant	Concentration (%)	IFT ($mN \cdot m^{-1}$)
LPS-1	0.2	1.09
	0.3	1.60×10^{-1}
	0.4	1.17×10^{-1}
LPS-2	0.2	2.11
	0.3	2.18×10^{-1}
	0.4	8.08×10^{-2}
LPS-3	0.2	1.26×10^{-1}
	0.3	1.92×10^{-3}
	0.4	5.36×10^{-4}
LPS-4	0.2	5.06
	0.3	1.06
	0.4	9.83×10^{-1}
LPS-5	0.2	4.39
	0.3	4.40×10^{-2}
	0.4	1.15×10^{-2}
OAB	0.2	5.92×10^{-3}
	0.3	7.88×10^{-3}
	0.4	5.91×10^{-3}
HSB-1214	0.2	1.67×10^{-1}
	0.3	1.04×10^{-1}
	0.4	1.22×10^{-1}
HSB-1618	0.2	1.23×10^{-2}
	0.3	2.25×10^{-3}
	0.4	2.76×10^{-3}
HH-B1	0.2	1.25
	0.3	2.66
	0.4	5.44
HH-B3	0.2	1.55
	0.3	1.61
	0.4	5.44

Figures 3–5 illustrate that increasing the concentration from 0.3% to 0.4% resulted in minimal change in IFT across all surfactants. Notably, for LPS-3 and HSB1618, a minimum concentration of 0.3% was required to achieve IFT values at the magnitude of 10^{-3} mN/m. Based on these findings, combinations of LPS-3 with OAB and HSB1618 at a concentration of 0.3% were prepared and tested in varying ratios (9:1, 7:3, and 6:4) to investigate and optimize the synergistic effects on IFT reduction.

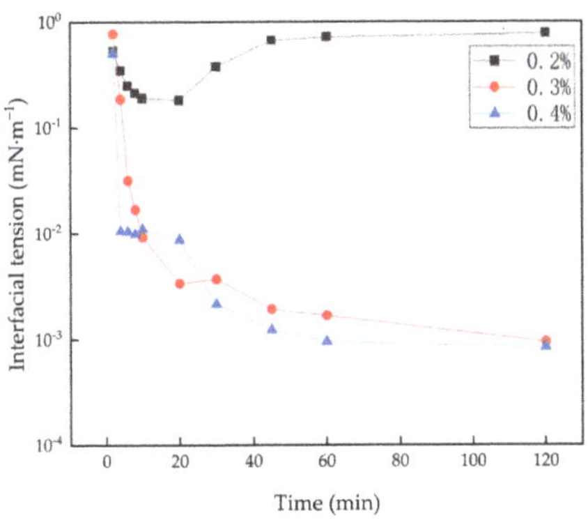

Figure 3. IFT between oil and solution over time for LPS-3 surfactant.

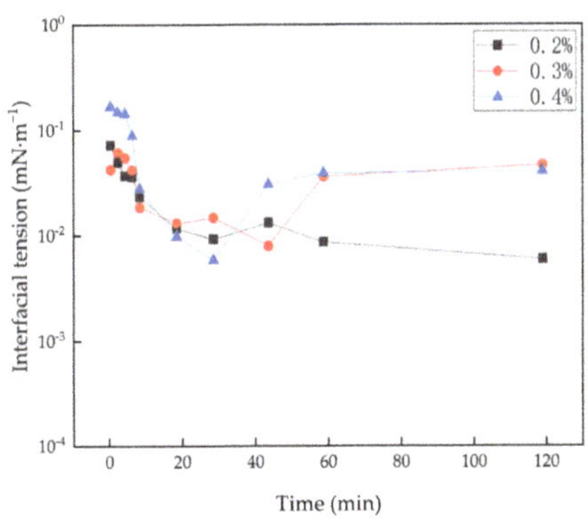

Figure 4. IFT between oil and solution over time for OAB surfactant.

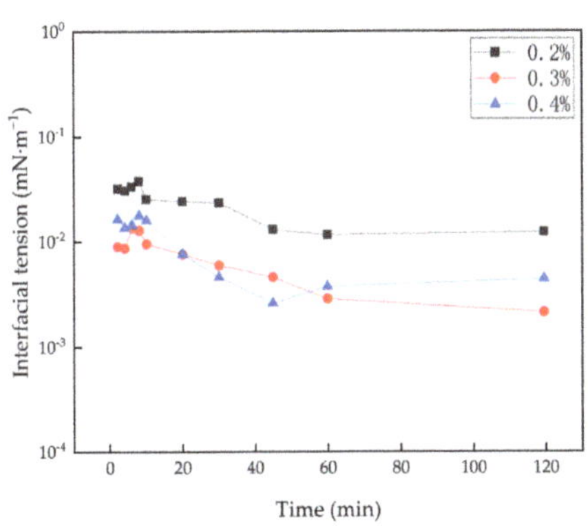

Figure 5. IFT between oil and solution over time for HSB1618 surfactant.

The experimental results for the compound solution of LPS-3 and HSB1618 were suboptimal. Among the tested ratios, the lowest IFT was 4.90×10^{-2} mN/m at the ratio of 6:4. And none of the three ratios achieved an IFT at a magnitude of 10^{-3} at a concentration of 0.3%. The compatibility between LPS-3 and HSB1618 was poor, and experimental results indicated that the antagonistic effect weakened the ability to reduce IFT.

In contrast, a significantly higher capacity to reduce IFT was observed for the compound solution of LPS-3 and OAB compared to either surfactant alone. As shown in Figure 6, an instantaneous IFT at a magnitude of 10^{-3} was achieved by the compound solution in a shorter time than that of the individual surfactants. Additionally, the IFT was further reduced to the magnitude of 10^{-4} within 20 min. Furthermore, the equilibrium IFT of all three ratios remained stable at 10^{-3} for 120 min, which indicated the enhanced stability of these solutions.

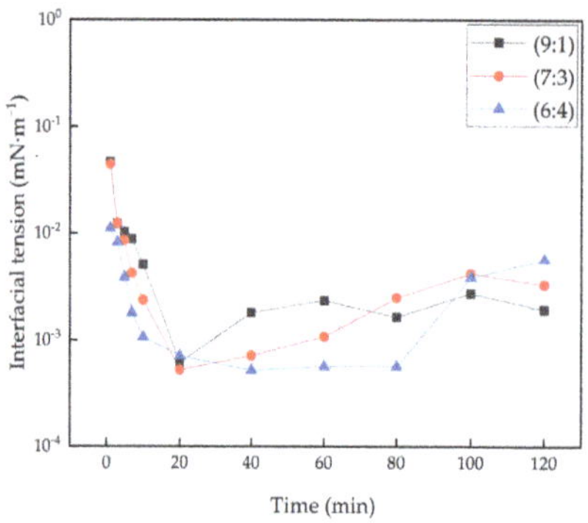

Figure 6. IFT between oil and solution over time for LPS-3/OAB compound solution.

The reduction in IFT through the synergistic effect of OAB and LPS-3 was demonstrated in the experiments. This phenomenon was hypothesized to result from the ability of OAB molecules to form a more compact molecular film at the oil–water interface. In contrast, when OAB was used individually, intermolecular repulsion tended to occur, potentially compromising the stability of the film. The addition of LPS-3 mitigated this repulsion through electrostatic interactions, which enhanced the compactness of the interfacial film. This improvement led to a significant reduction in IFT, as confirmed by the experimental data. Based on these results, the surfactant solution composed of LPS-3 and OAB was identified as the most effective formulation to reduce the IFT.

3.1.3. Emulsification Stability of Surfactants

Given the widely recognized importance of emulsifying ability in influencing oil displacement [17], the performance of the surfactant compound solutions was further assessed through emulsification stability experiments in this study. The time-dependent variation in oil separation rates for each emulsion is illustrated in Figure 7.

Emulsion stability was evaluated based on the oil separation rate, defined as the ratio of separated oil volume to the initial oil volume. As shown in Figure 7, OAB achieved an oil separation rate of 93.3% within the first hour, and complete oil separation occurred within one day. In comparison, LPS-3 reached 100% oil separation within the same period. The compound solutions of LPS-3 and OAB exhibited distinct emulsion stability characteristics depending on the mixing ratios. At a 6:4 ratio, the oil separation rate remained stable during the first hour. It rapidly increased and reached 100% within one day. In contrast, emulsions prepared with mixing ratios of 7:3 and 9:1 demonstrated markedly enhanced stability. Over the initial three days, the oil separation rates for the 7:3 and 9:1 were only 2.5% and 1.8%, respectively. These rates gradually increased over time and reached 45.3% and 24.7% on the 30th day.

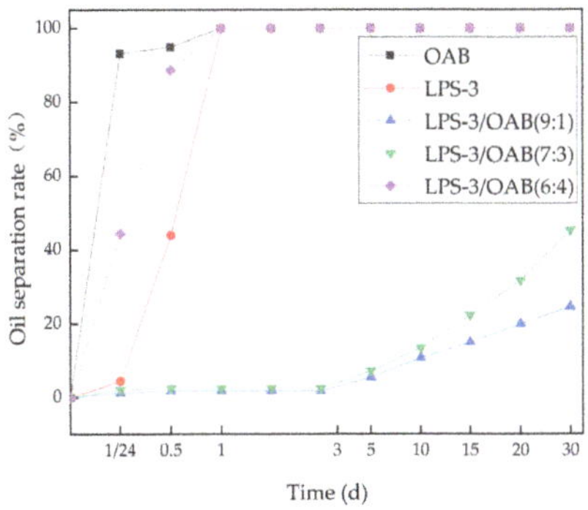

Figure 7. The oil separation rate of emulsions over time (Note: The x-axis scale is non-uniform due to the focus on the variation in oil separation rate on the first few days).

In summary, the compound solution with an LPS-3 to OAB ratio of 9:1 exhibited the highest emulsion stability. Its oil separation rate was only 24.7% after 30 days. Among all tested solutions, emulsions prepared from compound solutions demonstrated significantly greater stability compared to those formed by individual components. This highlighted the compatibility and synergistic effects between LPS-3 and OAB in stabilizing emulsions. This increased stability was attributed to the zwitterionic structure of betaine, where the incorporation of petroleum sulfonate enhanced charge complementarity. The resulting stable charge barrier effectively suppressed aggregation at the oil–water interface and further reinforced emulsion stability.

Based on these findings, the surfactant formula was defined as a 9:1 mixture of LPS-3 and OAB. The optimal concentration of compound solution was determined to be 0.3%.

3.2. *Screening of Polymer*

3.2.1. Viscosity-Increasing Properties of Polymers

Viscosity measurements were conducted to assess the viscosity-increasing properties of polymers under reservoir conditions. The viscosities of polymer solutions at different concentrations are shown in Figure 8.

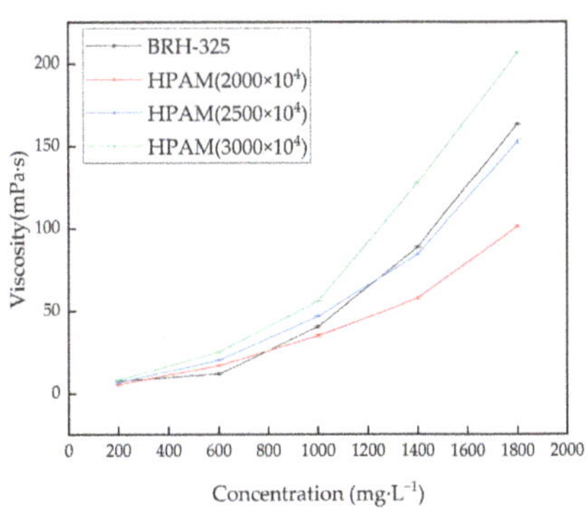

Figure 8. The viscosity of polymer solutions with concentration under 60 °C.

The HPAM (3000 × 10^4) demonstrated the best viscosity-increasing ability, followed by BRH-325, HPAM (2500 × 10^4), and HPAM (2000 × 10^4). According to the permeability and porosity characteristics of examined reservoirs, both BRH-325 and HPAM (2500 × 10^4) at a concentration of 1500 mg·L^{-1} were identified for further experiments.

Rheological experiments were conducted to further examine the viscosity-increasing properties of these polymers under varying shear rates. The apparent viscosities of polymer solutions were compared across shear rates from 0.01 s^{-1} to 1000 s^{-1}. As demonstrated in Figure 9, there was a decrease in apparent viscosity as the shear rate increased for both polymer solutions. It indicated pronounced shear-thinning behavior, a characteristic of pseudoplastic fluids. At high shear rates exceeding 100 s^{-1}, the apparent viscosities of both polymer solutions were comparable. In contrast, at low shear rates below 10 s^{-1}, the BRH-325 solution demonstrated a significantly higher apparent viscosity than the HPAM (2500 × 10^4) solution. These findings highlighted the superior performance of BRH-325 under low-shear conditions, which made it a promising candidate for viscosity increasing under reservoir conditions.

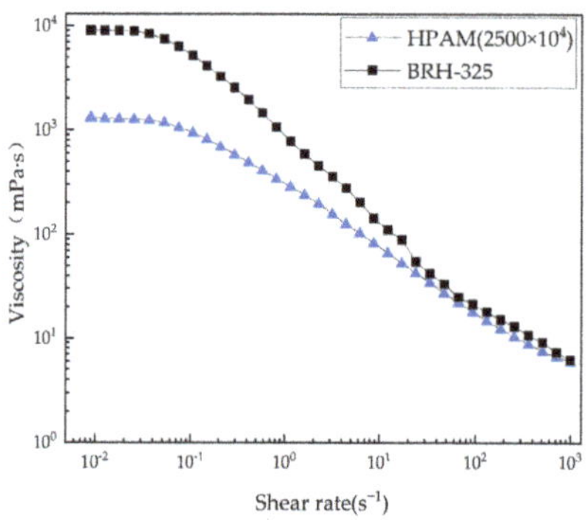

Figure 9. The viscosity of polymer solutions with shear rate under 60 °C.

Table 6 presents the power-law model parameters that describe the relationship between apparent viscosity and shear rate for BRH-325 and 2500 × 10^4 HPAM solutions under 60 °C. The power-law model effectively captured the pseudoplastic behavior of both polymer solutions within the shear rate range of 0.01 to 1000 s^{-1} ($R^2 > 0.99$). This model has been widely validated and employed to characterize the rheological properties for polymer solutions [18–22]. As shown in Table 6, the power-law exponent n of both solutions was less than 1, which confirmed the pseudoplastic fluid behavior. Additionally, the viscosity index k for BRH-325 was higher than that for the 2500 × 10^4 HPAM solution, consistent with the data presented in Figure 9. These results indicate that the BRH-325 solution demonstrated a better ability to increase viscosity compared to the 2500 × 10^4 HPAM solution.

Table 6. Power-law model parameters of polymer solutions at 60 °C.

Polymer	k (mPa·s^n)	n	R^2	Shear Rate (s^{-1})
BRH-325	820.98	0.243	0.9938	0.01–1000
HPAM (2500 × 10^4)	261.62	0.439	0.9932	0.01–1000

3.2.2. Temperature Resistance of Polymers

The viscosity of polymer solutions was measured at various temperatures to construct viscosity–temperature curves and analyze viscosity variation trends. The resulting curves are shown in Figure 10. The viscosity of both polymer solutions decreased as the temperature increased, which aligned with the typical viscosity–temperature behavior of polymers. It indicated that higher temperatures accelerated polymer molecule degradation, which resulted in reduced viscosity and a weakened thickening effect. A sharp decrease in viscosity was observed under temperatures exceeding 55 °C. Additionally, the viscosity reduction was more pronounced for the HPAM (2500 × 10^4) solution compared to BRH-325. Accordingly, BRH-325 not only exhibited superior viscosity-increasing properties at the same temperatures but also demonstrated stronger temperature resistance.

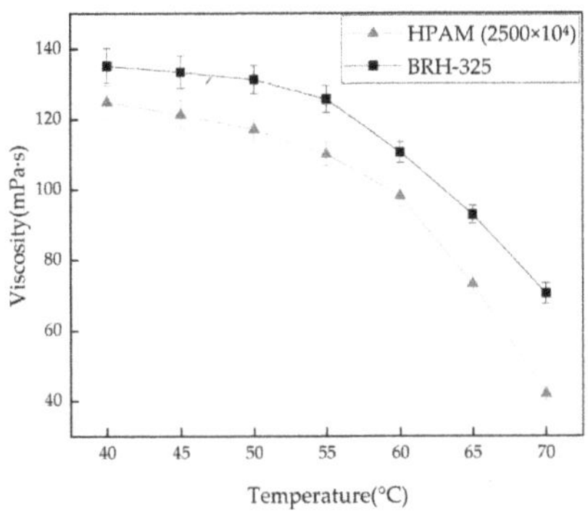

Figure 10. The viscosity of polymer solutions with temperature.

3.2.3. Long-Term Stability of Polymers

Polymer hydrolysis, particularly in polyacrylamide and its derivatives, alters charge distribution and affects viscosity enhancement. In the presence of divalent cations such as Ca^{2+} and Mg^{2+}, this process may lead to precipitation. Therefore, evaluating the aging performance of polymers was essential to assess their long-term stability and effectiveness [23]. The long-term stability was evaluated by monitoring viscosity changes over a 90-day period. As shown in Figure 11, both polymers exhibited a gradual decrease in viscosity over time. A gradual and consistent decrease in the viscosity of the BRH-325 solution was observed, whereas a significant decline in the viscosity of the HPAM (2500 × 10^4) solution was noted after 45 days. Overall, BRH-325 was found to exhibit greater long-term stability.

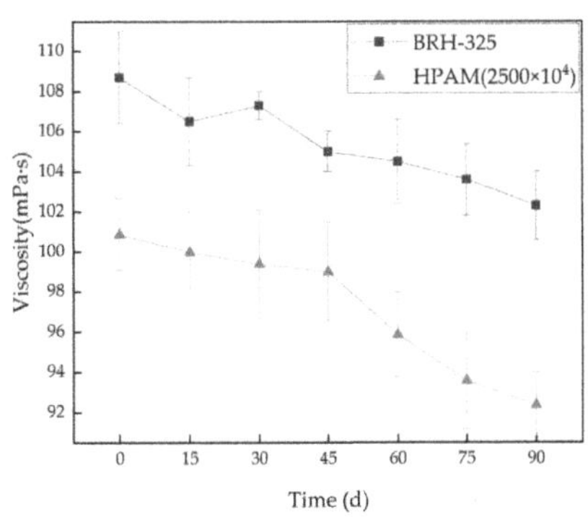

Figure 11. The viscosity of polymer solutions over time under 60 °C.

Figure 12 demonstrates that the viscosity retention rates of both polymers remain above the minimum standard requirement at 30 days and 90 days. Based on these findings, BRH-325 was selected for subsequent experiments at a concentration of 1500 mg·L^{-1}.

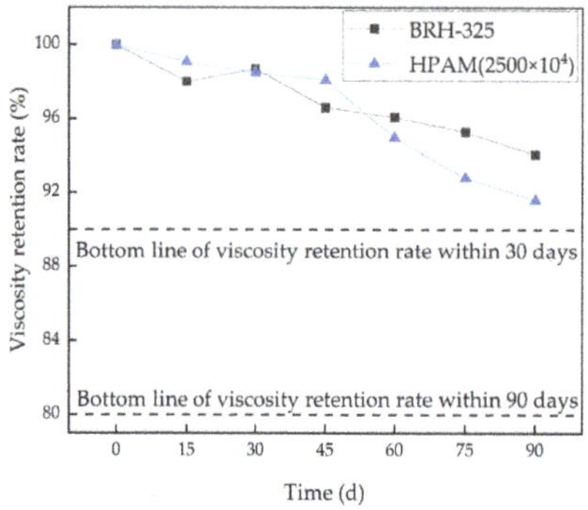

Figure 12. The viscosity retention rate of polymer solutions with aging time.

3.3. *Compatibility of Surfactants and Polymer*

The compatibility of the SP compound solution was evaluated through IFT and viscosity tests. The SP compound solution consisted of a 0.3% LPS-3/OAB (9:1) surfactant solution and a 1500 mg·L^{-1} BRH-325 solution. The IFT test was conducted according to the procedures outlined previously.

As observed in Figure 13, the instantaneous IFT of the SP compound solution was 2.03×10^{-4} mN/m, significantly below the 10^{-3} mN/m threshold. The IFT remained stable at the magnitude of 10^{-3} for 120 min. Furthermore, the SP compound solution achieved the magnitude of 10^{-3} within 5 min, outperforming the surfactant solution, which required 7 min. These results indicated that the addition of polymers did not negatively impact the IFT reduction performance of the surfactant system. Instead, the polymers exhibited a synergistic effect by modifying the arrangement of surfactant molecules at the oil–water interface. This interaction reduced the critical micelle concentration (CMC) and improved micelle stability [24,25]. The time required for the IFT to reach the magnitude of 10^{-3} was shorter, resulting in reduced time costs. This improvement presented significant implications for cost reduction and enhanced efficiency in oilfield operations.

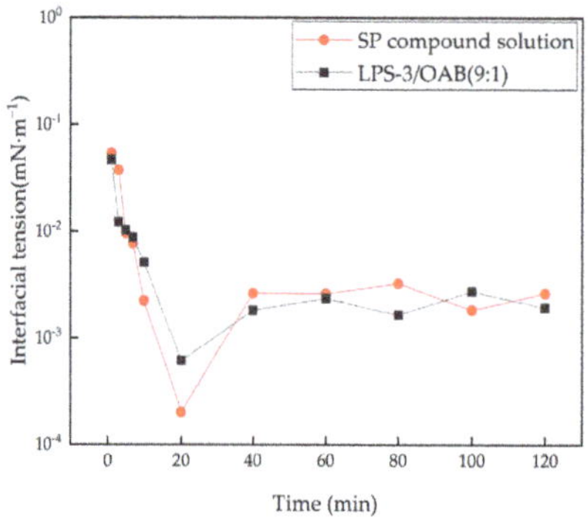

Figure 13. Comparison of IFT between SP compound solution and LPS-3/OAB (9:1) solution.

The viscosity test results for both the SP compound solution and the BRH-325 solution are shown in Figure 14. Both solutions exhibited a decrease in viscosity with increasing temperature. At lower temperatures (40–55 °C), the viscosities of the two solutions were comparable; however, a significant reduction in viscosity was observed, starting under 55 °C. Notably, the viscosity of the SP compound solution decreased less than that of the BRH-325 solution, which indicated better temperature resistance and superior viscosity-maintaining performance at elevated temperatures. These findings suggested that the incorporation of surfactants slightly improved both the viscosity and temperature resistance of the polymer. This further confirmed the compatibility of the components.

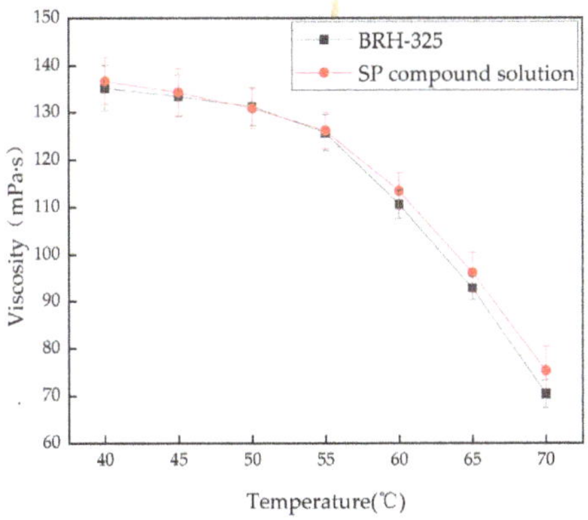

Figure 14. Comparison of viscosity between SP compound solution and BRH-325 solution.

3.4. Screening of Viscosity Reducer

The performance of viscosity reducers was evaluated through viscosity reduction and stability experiments. The initial viscosity of heavy oil was measured as 605.9 mPa·s under 60 °C using a viscometer. Three viscosity reducers were separately mixed with the heavy oil, and their viscosity changes over time were recorded, as shown in Figure 15. As depicted, the viscosity of heavy oil was reduced from 605.9 mPa·s to 48 mPa·s by VR-SF, which achieved a viscosity reduction rate of 92.08%. However, the viscosity gradually increased over time, reaching 485.9 mPa·s after 2 h, with the viscosity reduction rate declining to 19.81%. The CA601S-HNS demonstrated better initial performance. The viscosity of heavy oil was reduced from 605.9 mPa·s to 28.8 mPa·s, which corresponded to a viscosity reduction rate of 95.25%. After 2 h, the viscosity increased to 389.5 mPa·s, and the viscosity reduction rate decreased to 35.71%. Both VR-SF and CA601S-HNS exhibited a significant reduction in viscosity reduction rates over time. This indicated limited stability in maintaining viscosity reduction. In contrast, superior performance was demonstrated by the graphene viscosity reducer, with the viscosity of the heavy oil reduced to a minimum of 15.6 mPa·s, which achieved a viscosity reduction rate of 97.43%. Notably, the viscosity remained stable within 1 h, and slightly rebounded to 26.6 mPa·s after 2 h. This corresponded to a viscosity reduction rate of 95.61%. The graphene viscosity reducer not only achieved the highest viscosity reduction rate but also exhibited significantly better stability compared to VR-SF and CA601S-HNS.Therefore, the graphene viscosity reducer was proposed as an effective solution for reducing heavy oil viscosity.

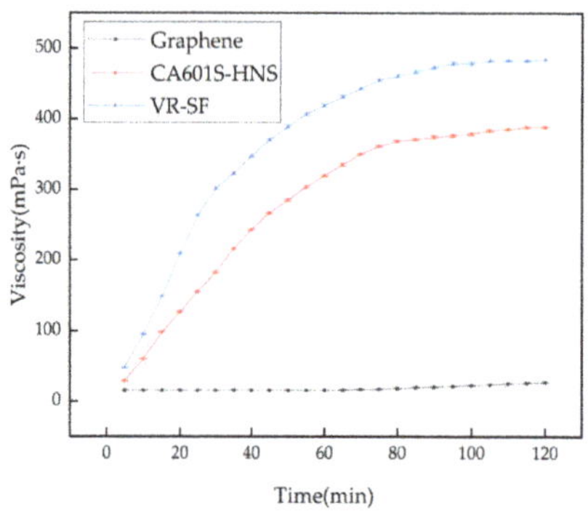

Figure 15. The viscosity of heavy oil with viscosity reducers over time.

3.5. Core Flooding Experiments

This study evaluated the effectiveness of the SP compound solution in core displacement experiments by recording produced fluid and pressure changes. The core sample parameters are listed in Table 1, while Figures 16 and 17 illustrate the variations in recovery efficiency, water cut, and pressure with injection volume for both cores.

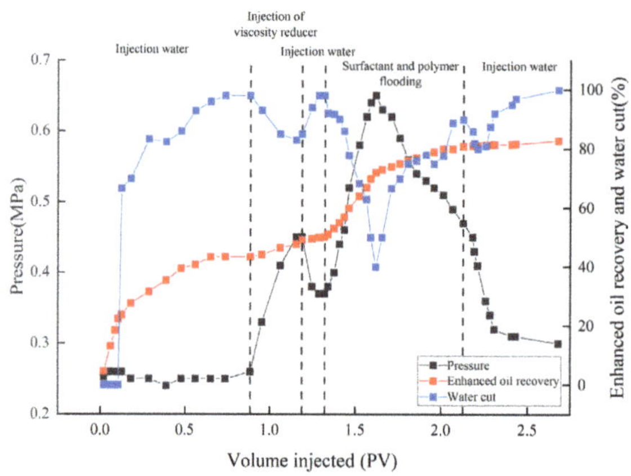

Figure 16. Profiles of the 1st core flooding experiment.

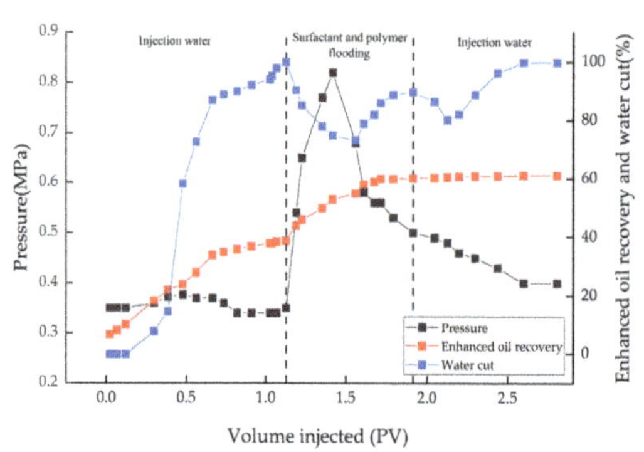

Figure 17. Profiles of the 2nd core flooding experiment.

As shown in Figure 16, during the initial water flooding, the pressure increased to 0.25 MPa. After injecting 0.12 PV, the water cut began to rise rapidly. By 0.88 PV, the water cut reached 98%, the pressure was 0.26 MPa, and the cumulative recovery efficiency was 43.4%. Subsequently, 0.3 PV of a graphene viscosity reducer was injected. This led to an increase in pressure to 0.45 MPa and a decrease in the water cut to 83%, accompanied by significant emulsification in the produced liquid. This indicated the formation of a low-viscosity oil-in-water emulsion at the displacement front, enhancing displacement efficiency by mobilizing high-viscosity crude oil. Following this phase, continued water injection raised the recovery efficiency to 50.1%, an improvement of 6.7%. Next, SP compound solution was injected. The displacement pressure rose, and the water cut decreased to a minimum of 40% at 0.3 PV injection, before increasing again. By the end of SP flooding, the water cut reached 90%, and the recovery efficiency was 80.9%. After subsequent water injection, recovery efficiency further increased to 82.8%, representing a 32.7% improvement due to SP flooding and subsequent water injection.

As shown in Figure 17, The initial water injection pressure for the 2nd core was 0.35 MPa. After injecting 0.29 PV, the water cut increased rapidly. At 1.1 PV injection, water flooding stopped with a water cut of 100%, pressure of 0.35 MPa, and a cumulative recovery efficiency of 38.8%. Injecting the SP compound solution raised displacement pressure and reduced water cut. The water cut reached a minimum of 73.3% at 0.44 PV injection before rising again. By the end of SP flooding (0.8 PV), the water cut reached 90%, and the recovery efficiency was 60.2%. Subsequent water flooding increased recovery efficiency to 61.2%, resulting in an overall enhancement of 22.4% due to SP flooding and water injection.

The experimental results showed that SP flooding could significantly enhance recovery efficiency on the basis of water flooding. The injection of the SP compound solution improved the viscosity of the displacement phase and expanded the swept volume. It also started the residual oil at the blind end, increased the number of capillaries, reduced the IFT, and stripped the heavy oil attached to the rock surface. These effects further improved the oil displacement efficiency. The addition of viscosity reducers could effectively improve the fluidity of heavy oil and reduce its viscosity. It proved that mobility control contributed greatly to the recovery efficiency [26–30]. The injection of graphene viscosity reducer effectively reduced the viscosity of heavy oil and increased the mobility of the displaced phase. However, the dynamic viscosity-increasing effect of the SP slug raised the viscosity of the displacement phase. It also increased the flow resistance, improved the resistance coefficient, and reduced the mobility of the displacement phase. When the mobility ratio was lower than 1, the affected volume increased and the recovery efficiency could be significantly increased.

4. Conclusions

This study investigated the potential of SP systems for enhanced oil recovery in heavy oil reservoirs, focusing on the Haiwaihe Block in Liaohe Oilfield. Key findings include the following:

(1) (LPS-3/OAB (9:1) demonstrated the best ability to reduce interfacial tension, achieving an IFT reduction of 10^{-4} mN/m and maintaining equilibrium at 10^{-3} mN/m. The system exhibited excellent emulsification stability, with an oil separation rate of only 24.7% after 30 days.

(2) BRH-325 showed stronger viscosity enhancement, temperature resistance, and long-term stability compared to HPAM (2500×10^4), with a viscosity retention rate of 94.1% after 90 days under 60 °C.

(3) The graphene viscosity reducer achieved a maximum viscosity reduction rate of 97.43% and maintained a 95.61% reduction after 2 h, outperforming other reducers.
(4) In the 1st Core, PS flooding increased oil recovery by 32.7%, reaching a total efficiency of 82.8%. In the 2nd Core, SP flooding increased oil recovery by 22.4%, achieving a final efficiency of 61.2%.

The SP system synergistically reduced IFT, improved emulsion stability, increased the viscosity of the displacement phase, and enhanced mobility control. These effects collectively expanded the swept volume, stripped residual oil, and significantly improved recovery efficiency. This research provides a practical basis for advancing chemical EOR technologies in heavy oil reservoirs.

Author Contributions: Conceptualization, Q.H. and W.L.; writing—original draft preparation, X.C.; writing—review and editing, X.C., Y.L., G.L., H.Z., Z.Z. and H.S. All authors have read and agreed to the published version of the manuscript.

Funding: This work was supported by the National Natural Science Foundation of China (Grant No. 42172050).

Data Availability Statement: The raw/processed data required to reproduce these findings cannot be shared at this time, as the data also form part of an ongoing study.

Acknowledgments: We would like to thank all the participants.

Conflicts of Interest: Author Gaohua Liu was employed by the company PetroChina Liaohe Oilfield Company. The remaining authors declare that the research was conducted in the absence of any commercial or financial relationships that could be construed as a potential conflict of interest.

References

1. Jia, C.-Z.; Pang, X.-Q.; Song, Y. Whole petroleum system and ordered distribution pattern of conventional and unconventional oil and gas reservoirs. *Pet. Sci.* **2023**, *20*, 1–19. [CrossRef]
2. Li, Q.; Li, Q.; Cao, H.; Wu, J.; Wang, F.; Wang, Y. The Crack Propagation Behaviour of CO_2 Fracturing Fluid in Unconventional Low Permeability Reservoirs: Factor Analysis and Mechanism Revelation. *Processes* **2025**, *13*, 159. [CrossRef]
3. Li, Q.; Li, Q.; Wu, J.; Li, X.; Li, H.; Cheng, Y. Wellhead Stability during Development Process of Hydrate Reservoir in the Northern South China Sea: Evolution and Mechanism. *Processes* **2024**, *13*, 40. [CrossRef]
4. Weidong, L.; Litao, L.; Guangzhi, L.; Luo, Z.; Yunyun, W.; Jiang, W. Experimental study on the mechanism of enhancing oil recovery by polymer–surfactant binary flooding. *Pet. Explor. Dev.* **2017**, *44*, 636–643.
5. Shangguan, Y.-N. *Parameter Design Optimization and Practice of Alkaline—Free Compound Flooding in Ordos Low Permeability Reservoir*; Springer: Singapore, 2024; pp. 1409–1418.
6. Hocine, S.; Magnan, A.; Degre, G.; Rousseau, N.; Rousseau, D. Alkaline-free surfactant polymer process for heavy oil. In Proceedings of the SPE EOR Conference at Oil and Gas West Asia, Muscat, Oman, 31 March–2 April 2014; p. SPE-169697-MS.
7. Romero-Zerón, L.; Jiang, X. Supramolecular Polymer-Surfactant System for Heavy Oil Recovery. In *Cyclodextrin-A Versatile Ingredient*; IntechOpen: London, UK, 2018.
8. Kaili, L.; Jijiang, G.; Lipei, F.; Haihua, P.; Ping, J.; Guicai, Z. Enhanced residual oil recovery by surfactant/polymer binary combination flooding after water flooding of heary oil. *Petrochem. Technol.* **2016**, *45*, 1519–1525.
9. Santos, R.; Loh, W.; Bannwart, A.; Trevisan, O. An overview of heavy oil properties and its recovery and transportation methods. *Braz. J. Chem. Eng.* **2014**, *31*, 571–590. [CrossRef]
10. Wang, Q.; Zhang, W.; Wang, C.; Han, X.; Wang, H.; Zhang, H. Microstructure of heavy oil components and mechanism of influence on viscosity of heavy oil. *ACS Omega* **2023**, *8*, 10980–10990. [CrossRef]
11. Zhang, X.; Guo, J.; Fei, D.; Wang, L.; Peng, Z.; Li, J.; Dong, J. Polymer surfactants as viscosity reducers for ultra-heavy oil: Synthesis and viscosity reduction mechanism. *Fuel* **2024**, *357*, 129871. [CrossRef]
12. Si, Y.; Zhu, Y.; Liu, T.; Xu, X.; Yang, J. Synthesis of a novel borate ester Anion-Nonionic surfactant and its application in viscosity reduction and emulsification of heavy crude oil. *Fuel* **2023**, *333*, 126453. [CrossRef]
13. *SY/T 6424-2014*; Chinese Oil and Gas Industry Standards: Analytical Method for Alkaline/Surfactant/Polymer Flooding System. National Energy Board: Beijing, China, 2014.

4. Bui, T.; Frampton, H.; Huang, S.; Collins, I.R.; Striolo, A.; Michaelides, A. Water/oil interfacial tension reduction—An interfacial entropy driven process. *Phys. Chem. Chem. Phys.* **2021**, *23*, 25075–25085. [CrossRef]
5. Pu, W.; Yuan, C.; Hu, W.; Tan, T.; Hui, J.; Zhao, S.; Wang, S.; Tang, Y. Effects of interfacial tension and emulsification on displacement efficiency in dilute surfactant flooding. *RSC Adv.* **2016**, *6*, 50640–50649. [CrossRef]
6. Li, H.; Chen, Z. Characteristics of interfacial tension and oil displacement efficiency with alkaline-surfactant-polymer flooding technology. *Acta Pet. Sin.* **2006**, *27*, 96.
7. Chen, X.; Gao, Y.; Wang, Y.; Zhang, T.; Peng, Y. A study on the emulsification ability and oil displacement efficiency of polymeric surfactant in porous media. *Colloids Surf. A Physicochem. Eng. Asp.* **2024**, *700*, 134690. [CrossRef]
8. Likavčan, L.; Košík, M.; Bílik, J.; Martinkovič, M. Determination of apparent viscosity as function of shear rate and fibres fraction in polypropylene. *IJEIT* **2014**, *4*, 23–26.
9. Azad, M.S. Influence of Polymer Concentration on the Viscous and (Linear and Non-Linear) Viscoelastic Properties of Hydrolyzed Polyacrylamide Systems in Bulk Shear Field and Porous Media. *Polymers* **2024**, *16*, 2617. [CrossRef]
:0. Gao, X.; Huang, L.; Xiu, J.; Yi, L.; Zhao, Y. A study on viscosity increasing and rheological properties of novel biopolymers in extreme reservoirs. *Polymers* **2023**, *15*, 4338. [CrossRef]
:1. Hashmet, M.R.; Onur, M.; Tan, I.M. Empirical correlations for viscosity of polyacrylamide solutions with the effects of concentration, molecular weight and degree of hydrolysis of polymer. *J. Appl. Sci.* **2014**, *14*, 1000–1007. [CrossRef]
:2. Dwier, S.; Garrouch, A.; Lababidi, H. Delineation of the effective viscosity controls of diluted polymer solutions at various flow regimes. *Fluid Phase Equilibria* **2024**, *584*, 114143. [CrossRef]
:3. Scott, A.J.; Romero-Zerón, L.; Penlidis, A. Evaluation of Polymeric Materials for Chemical Enhanced Oil Recovery. *Processes* **2020**, *8*, 361. [CrossRef]
:4. Bhardwaj, P.; Kamil, M.; Panda, M. Surfactant-polymer interaction: Effect of hydroxypropylmethyl cellulose on the surface and solution properties of gemini surfactants. *Colloid Polym. Sci.* **2018**, *296*, 1879–1889. [CrossRef]
:5. Yang, J.; Pal, R. Investigation of surfactant-polymer interactions using rheology and surface tension measurements. *Polymers* **2020**, *12*, 2302. [CrossRef]
:6. Fani, M.; Pourafshary, P.; Mostaghimi, P.; Mosavat, N. Application of microfluidics in chemical enhanced oil recovery: A review. *Fuel* **2022**, *315*, 123225. [CrossRef]
:7. Gbadamosi, A.O.; Junin, R.; Manan, M.A.; Agi, A.; Yusuff, A.S. An overview of chemical enhanced oil recovery: Recent advances and prospects. *Int. Nano Lett.* **2019**, *9*, 171–202. [CrossRef]
:8. Farajzadeh, R.; Wassing, B.; Lake, L. Insights into design of mobility control for chemical enhanced oil recovery. *Energy Rep.* **2019**, *5*, 570–578. [CrossRef]
:9. Ivanova, A.; Cheremisin, A. *Chemical Enhanced Oil Recovery: Where Do We Stand? Recent Advances and Applications*; IntechOpen: London, UK, 2022.
30. Hosny, R.; Zahran, A.; Abotaleb, A.; Ramzi, M.; Mubarak, M.F.; Zayed, M.A.; Shahawy, A.E.; Hussein, M.F. Nanotechnology Impact on Chemical-Enhanced Oil Recovery: A Review and Bibliometric Analysis of Recent Developments. *ACS Omega* **2023**, *8*, 46325–46345. [CrossRef] [PubMed]

energies

MDPI

Article

Two-Dimensional Physical Simulation of the Seepage Law of Microbial Flooding

Yongheng Zhao [1,2,*], Jianlong Xiu [2,3,*], Lixin Huang [3], Lina Yi [3] and Yuandong Ma [3]

1 University of Chinese Academy of Sciences, Beijing 100049, China
2 Institute of Porous Flow and Fluid Mechanics, Langfang 065007, China
3 Research Institute of Petroleum Exploration and Development, Beijing 100083, China; huanglixin69@petrochina.com.cn (L.H.); yilina69@petrochina.com.cn (L.Y.); mayuandong69@petrochina.com.cn (Y.M.)
* Correspondence: zhaoyongheng21@mails.ucas.ac.cn (Y.Z.); xiujianlong69@petrochina.com.cn (J.X.)

Abstract: The study of seepage laws during microbial enhanced oil recovery helps to elucidate the mechanisms behind microbial flooding, and the use of large-scale physical simulation experimental devices can more objectively and accurately investigate the seepage laws of microbes in porous media, and evaluate the oil displacement efficiency of microbial systems. In this study, physical simulation experiments of microbial flooding were conducted via a slab outcrop core, and the biochemical parameters such as the concentration of *Bacillus subtilis*, nutrient concentration, surface tension, and displacement pressure data were tracked and evaluated. The analysis revealed that the characteristics of the pressure field change during microbial flooding and elucidates the migration rules of microbes and nutrients, as well as the change rule of surface tension. The results show that after the microbial system is injected, cells and nutrients are preferentially distributed near the injection well and along the main flow paths, with the bacterial adsorption and retention capacity being greater than those of the nutrient agents. Owing to the action of microorganisms and their metabolites, the overall pressure within the model increased, From the injection well to the production well, the pressure in the model decreases stepwise, and the high-pressure gradient zone is mainly concentrated near the injection well. The fermentation mixture of *Bacillus subtilis* increased the injection pressure by 0.73 MPa, reduced the surface tension by up to 49.8%, and increased the oil recovery rate by 6.5%.

Keywords: microbial field; pressure field; large-scale physical simulation; microbial enhanced oil recovery

1. Introduction

Oil is a key energy source driving the economic development of human society and is also known as the "blood of industry", classified as a strategic resource [1]. Ensuring a long-term stable supply of oil resources has always been a significant strategic issue that requires special attention [2]. The purpose of improving oil recovery is to extract the original oil in place (OOIP) in the most economical and efficient manner possible. Based on the development process, oil extraction can be divided into three stages: primary oil recovery, secondary oil recovery, and tertiary oil recovery [3]. Primary oil recovery mainly relies on the natural pressure of the reservoir to extract oil and gas from the reservoir, with a recovery rate typically between 5 and 10% of the OOIP [4]. Secondary oil recovery primarily involves maintaining or enhancing reservoir pressure through water or gas injection to facilitate the production of additional crude oil, which can increase the recovery rate to between 10% and 40% of the original oil in place [5].

At present, many oil fields have entered a high water content period, increasing the difficulty of extraction, raising costs, reducing recovery rates, and correspondingly increasing pressure [6]. Approximately 70% of the world's conventional oil (one-third of the total recoverable oil) remains trapped in depleted and marginal reservoirs, where traditional primary and secondary oil recovery techniques are no longer effective in extracting the remaining oil from the rock pores [7]. At this point, tertiary oil recovery technology becomes particularly important. Tertiary oil recovery technology uses various physical, chemical, and biological methods to improve the seepage characteristics of the reservoir and the mobility of residual oil, thereby further increasing the recovery rate. The main techniques include microbial enhanced oil recovery (MEOR), chemical flooding, polymer flooding, miscible flooding, gas injection, and thermal methods [8].

MEOR technology refers to a tertiary oil recovery technology that utilizes the metabolic activities of microorganisms and their metabolites to effectively improve the physical properties of crude oil and reservoir formation conditions to increase oil recovery [9–11]. Compared to other tertiary recovery methods such as chemical flooding, thermal recovery, and gas injection [12], MEOR has the advantages of a simple process, low cost, rapid effectiveness, and environmental friendliness. It is a tertiary oil recovery technology with good development prospects and has been extensively applied in recent years [13,14]. However, MEOR requires a longer implementation period and has certain requirements for reservoir conditions. Table 1 lists the MEOR screening criteria parameters recommended by the U.S. Department of Energy (DOE), China National Petroleum Corporation (CNPC), and the Institute of Reservoir Research (IRS), among others [15].

Table 1. Screening standard parameters of microbial oil production reservoirs.

Parameters	DOE	CNPC	IRS
Temperature/°C	<70	30~60	<90
Viscosity/cp	……	30~150	<20
Permeability/mD	>100	≥50	>50
Salinity/g/L	<10	≤100	<10
Water cut/%	-	60~85	30~90
pH value	-	4~9	6~9
Pressure/MPa	-	10.5~20	<30
Oil saturation/%	>25	>25	>25

At present, the main mechanisms of MEOR proposed worldwide are as follows: (1) improving the oil washing efficiency and sweep rate of water flooding to increase oil recovery [16,17]; (2) improving the properties of crude oil by producing acids, gases, biosurfactants, biopolymers, and organic solvents to improve the recovery rate of crude oil [18,19]; (3) exopolysaccharides produced by microorganisms can form biofilms, which can flow deep into the reservoir and improve the fluidity of the remaining oil [20,21]; (4) biopolymers produced by microorganisms can increase the formation pressure and facilitate the extraction of crude oil [22–24].

In the field application of MEOR technology, four main methods are employed depending on the construction techniques and objectives: microbial single well huff and puff, microbial wax prevention and removal, microbial profile modification, and microbial flooding. The core of these methods involves the migration of microorganisms in porous media. Therefore, laboratory research is essential for providing theoretical guidance for the field application of MEOR [25].

The physical simulation of microbial flooding is a bridge between laboratory research and field tests, it can simulate the oil displacement process of microorganisms under reservoir conditions, and it can be used to study the growth, propagation, and migration laws of microorganisms in porous media [26–29]. In laboratory studies on MEOR, O'Bryan et al. [30] discovered that microbes can cause blockages within the rock core. This blockage restricts the microbes' migration to deeper parts of the reservoir, thereby affecting their ability to act on crude oil. The blockage is primarily caused by the accumulation of microbes and their metabolic by-products in the pores of the rock core, which reduces the permeability of the pores and throats. To address this issue, Jack et al. [31] through calculations and experimental research, pointed out that the size of microbes used for MEOR should be significantly less than 20% of the pore-throat size of the porous medium. This conclusion provides an important theoretical basis for the selection and optimization of microbes. In order to study the transport mechanisms and influencing factors of microorganisms in porous media, Lei et al. [32] found that through core displacement experiments the microorganisms exhibit convective diffusion characteristics when migrating in porous media, with strong adsorption and retention. The larger the permeability, the faster the diffusion speed and the less adsorption and retention. Bi et al. [33] studied the adsorption and retention patterns of microorganisms in porous media through isothermal adsorption experiments and flow experiments, discovering that the larger the specific surface area of the porous media, the more adsorption sites are available, and the greater the adsorption of bacterial bodies. Guo et al. [34] investigated the migration patterns of microorganisms in rock cores with different permeabilities and their oil displacement effects. They found that a permeability range of 100×10^{-3}~500×10^{-3} μm^2 is favorable for microbial growth, adsorption, and retention, resulting in the best oil recovery enhancement. This permeability range has been proposed as a criterion for selecting reservoirs suitable for microbial oil displacement. Sun et al. [35] utilized a 4.6×560 cm artificial long core to investigate the dynamic changes in nutrients during water flooding, revealing that nutrient loss during migration in the core was substantial and that a chromatographic separation phenomenon occurred between nutrients and microorganisms. Ke et al. [36] conducted indoor physical simulation experiments to study the migration and adsorption patterns of glucose, ammonium chloride, and potassium dihydrogen phosphate in porous media. They found that glucose has strong adsorption in the formation and is metabolized rapidly, making it difficult to penetrate deep into the reservoir. Ammonium chloride and potassium dihydrogen phosphate exhibit relatively weaker adsorption in the formation, but the migration speed of phosphate ions is slower than that of ammonium ions. However, these studies are all based on one-dimensional physical simulation experiments, using one-dimensional cylindrical cores. This approach can only consider the fluid flow in porous media as a linear motion, and it is unable to account for the changes in injection and production well patterns. As a result, there is a discrepancy between these studies and the actual conditions of oilfields [37]. It is necessary to use large-scale physical models to study the distribution characteristics of microorganisms and nutrients in two and three dimensions during the oil displacement process. Hu et al. [38] from the Shengli Oilfield Branch of Sinopec used a sand-filling model with a size of 80 cm × 80 cm × 5 cm to study the growth and metabolism of microbial communities in the process of indigenous microorganism flooding. At present, no large-scale physical model research focusing on the migration patterns of microorganisms has been conducted. To further elucidate the variations in two-dimensional biochemical parameters during MEOR, the presented study employed a planar outcrop core to conduct physical simulation experiments. By arranging 25 pressure measurements and sampling points on the core, key biochemical parameters during the microbial oil recovery process were monitored in real time. Subsequently, the acquired data were processed using Surfer15 software to generate distribution and variation maps of biochemical parameters such as pressure, microbial field, nutrient concentration, and interfacial tension. Based on these maps, an in-depth analysis was conducted to investigate the dynamic changes and interactions of these parameters during the microbial oil recovery process. Finally, a systematic evaluation of the oil displacement efficiency of the microbial system was performed, providing an important theoretical basis for the field application of microbial oil recovery technology.

2. Materials and Methods

2.1. Equipment

The large MEOR physical simulation experimental device is shown in Figure 1, which is composed of a drive system, confining pressure system, temperature control system, data acquisition system, and liquid production measurement system. The drive system is composed of a Quizix Q5000 pump (model Q5210-SS-A-AS-S, manufactured by Chandler Engineering, Tulsa, OK, USA) and a computer. It is capable of achieving a maximum pressure of 70 MPa and a maximum flow rate of 15 mL/min. The system can perform a constant flow rate or constant pressure injection while automatically measuring the displaced fluid volume. The confining pressure system consists of a confining pressure tracking pump (model HBWY1000/40, manufactured by Yangzhou Huabao Petroleum Instrument Co., Ltd., Yangzhou, China) and a Large-Scale Core Holder (model HBKC35/800, manufactured by Yangzhou Huabao Petroleum Instrument Co., Ltd.), with a maximum pressure tolerance of 35 MPa. The data acquisition system is composed of a pressure measuring system (model HBKZ-YC, manufactured by Yangzhou Huabao Petroleum Instrument Co., Ltd., Yangzhou, China) with a measuring accuracy of 0.01 MPa, temperature sensors, and a computer, which can realize real-time monitoring of both pressure and temperature. The produced liquid metering system is composed of a Computer-Controlled Multi-Functional Fully Automatic Fraction Collector (model CBS-C, manufactured by Shanghai HuXi Analysis Instrument Factory Co., Ltd., Shanghai, China), which enables the automatic collection and measurement of the produced liquid and reduces the error associated with human operation. The other instruments and equipment needed for the experiment, including their models and manufacturers, are shown in Table 2.

Figure 1. Large-scale physical simulation experimental equipment for MEOR.

Table 2. Experimental instruments and equipment.

Instrument/Equipment Name	Model Specifications	Manufacturer
Thermostatic Chamber	HX-3SP	Guangdong Aipei Testing Equipment Co., Ltd., Dongguan, China
Vacuum Pump	CDWARDSRV8	Edwards Vacuum Trading (Shanghai) Co., Ltd., Shanghai, China
Rotational Viscometer	DV-III	Brookfield, Toronto, ON, Canada
Multifunctional Thermostatic Shaker	ZQLY-300N	Shanghai Zhichu Instrument Co., Ltd., Shanghai, China
Fluorescence Microscope	Axio Imager	Carl Zeiss AG, Oberkochen, Germany
Crude Oil Electrical Dehydration Instrument	HB-S6C	Hai'an Fada Petroleum Instrument Technology Co., Ltd., Haian, China
Laminar Flow Cabinet	SJ-CJ-2FD	Suzhou Su Jie Medical Equipment Co., Ltd., Suzhou, China
Total Organic Carbon/Total Nitrogen Analyzer	multi N/C 3100	Analytik Jena AG, Jena, Germany
Manual Drop Shape Analyzer	FTA1000B	First Ten Angstroms, Inc., Portsmouth, VA, USA
Centrifuge	AvantI J-E	Beckman Coulter, Brea, CA, USA
Autoclave	HVE-50	HIRAYAMA Manufacturing Corp., Saitama, Japan

2.2. Design and Encapsulation of the Model

The core sample used in the experiment is a natural outcrop core, sourced from the Sichuan Basin and composed of sandstone. In the experiment, the reverse five-point injection-production pattern was adopted, with one injection and four production methods (Figure 2a). The central well is the injection well and the four corner wells are the production wells. A total of 25 pressure measuring points are arranged on the model, and then the core is encapsulated with epoxy resin. The encapsulated model is shown in Figure 2b.

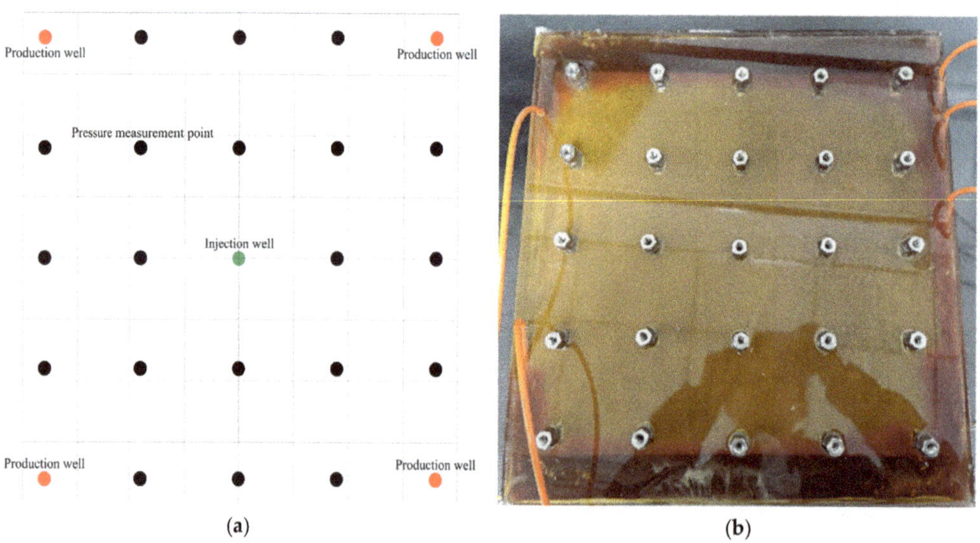

Figure 2. (**a**) Distribution of the injection and production well network, pressure measurement points, and sampling points; (**b**) Encapsulated core.

2.3. Materials

In this experiment, the crude oil and formation water used were both sourced from the Keshang Formation, Qizhong area, Karamay Oilfield. The formation temperature of this block is 39 °C, the viscosity of the formation crude oil is 5.6 mPa·s, and the salinity of the formation water is 15,726 mg/L.

The bacterial strain employed in the experiment was *Bacillus subtilis*, which was obtained from the laboratory strain bank of the Langfang Branch of the Research Institute of Petroleum Exploration and Development, China National Petroleum Corporation. The strain was preserved in glycerol tubes and stored at −80 °C for cryogenic conservation. The optimum growth temperature range of this strain was 25~40 °C, and the optimum growth pH range was 6~8. A total of 2 mL of bacterial suspension was inoculated into a 250 mL Erlenmeyer flask containing 100 mL of fermentation medium and then incubated in a shaking incubator at 39 °C for 72 h for subsequent experimental use. The fermentation medium used in the experiment consists of the following: tryptone 10.0 g/L, glucose 10.0 g/L, sodium nitrate ($NaNO_3$) 2.0 g/L, dipotassium hydrogen phosphate (K_2HPO_4) 0.2 g/L, and potassium dihydrogen phosphate (KH_2PO_4) 0.2 g/L, with a pH of 7.0. The nutrient solution used consists of the following: sucrose 30 g/L, sodium nitrate ($NaNO_3$) 3 g/L, dodecahydrate sodium phosphate ($Na_2HPO_4 \cdot 12H_2O$) 30 g/L, potassium dihydrogen phosphate (KH_2PO_4) 2.5 g/L, and heptahydrate magnesium sulfate ($MgSO_4 \cdot 7H_2O$) 0.8 g/L, with a pH of 7.2. The chemicals used in the experiment, their purity level, and manufacturers are shown in Table 3.

Table 3. Experimental chemicals.

Chemical Name	Purity Level	Manufacturer
NaCl	AR	Xilong Scientific Co., Ltd., Shantou, China
TRYPTONE	BR	Oxoid Ltd., Basingstoke, UK
Glucose	AR	Sinopharm Chemical Reagent Co., Ltd., Shanghai, China
$NaNO_3$	AR	Shandong Hosea Chemical Co., Ltd., Weifang, China
K_2HPO_4	AR	Sinopharm Chemical Reagent Co., Ltd., Shanghai, China
KH_2PO_4	PT	Tianjin Guangfu Technology Development Co., Ltd., Tianjin, China
Sucrose	AR	Xilong Scientific Co., Ltd., Shantou, China
$Na_2HPO_4 \cdot 12\,H_2O$	AR	Shaanxi Xihua Chemical Industry Co., Ltd., Baoji, China
$MgSO_4 \cdot 7\,H_2O$	CP	Jiangxi Xinhui Chemical Co., Ltd., Ji'an, China

2.4. Methods

The experimental parameters are shown in Table 4 and the experimental process is shown in Figure 3. The experiment was carried out according to the following steps:

1. The model was vacuumed and saturated with water. The encapsulated core was vacuumed and saturated with water for 24 h. Multiple-point vacuum extraction and water saturation were used to ensure high vacuum levels at different locations within the model, and the core pore volume was calculated on the basis of the volume of water saturated in the model.
2. The core was loaded into the chamber, and confining pressure was applied. The model was placed into the large-scale core holder and subjected to a confining pressure of 10 MPa, with the environmental pressure maintained at atmospheric pressure.
3. The model was saturated with oil. Water in the model was displaced with oil to form bound water within the model at an injection rate of 2 mL/min, and the initial oil saturation was calculated based on the volume of water displaced out of the model.

4. Primary water flooding was conducted. Waterflooding was performed at an injection rate of 2 mL/min until the water cut of the produced fluid reached 98%, followed by continued displacement for a certain period. The recovery efficiency of the waterflooding and the pressure changes during the displacement process were recorded.
5. The microbial system was injected. A volume of 1 PV (pore volume) of 5% *Bacillus subtilis* fermentation solution was injected into the model at an injection rate of 2 mL/min. Samples were collected at the sampling point to measure the concentrations of microorganisms, nutrients, and surface tension when the injection volume reached 0.3 PV, 0.6 PV, and 1 PV. Following these measurements, the well was sealed and shut in to culture the microbes for 7 days.
6. Secondary water flooding was conducted. Waterflooding was conducted at an injection rate of 2 mL/min until the water cut of the produced fluid reached 98%, after which the displacement was continued for an additional period of time. When the injection amounts reached 0.3 PV, 0.6 PV, and 1 PV, samples were taken from the sampling points to detect microbial and nutrient concentrations and surface tension, and the recovery efficiency and pressure changes during secondary water flooding were recorded.

Table 4. Experimental parameter.

Core Size	Permeability	Pore Volume	Porosity	Temperature	Initial Oil Saturation
40 × 40 × 3 cm	209.9 × 10^{-3} μm^2	823.3 mL	17.15%	39 °C	69.4%

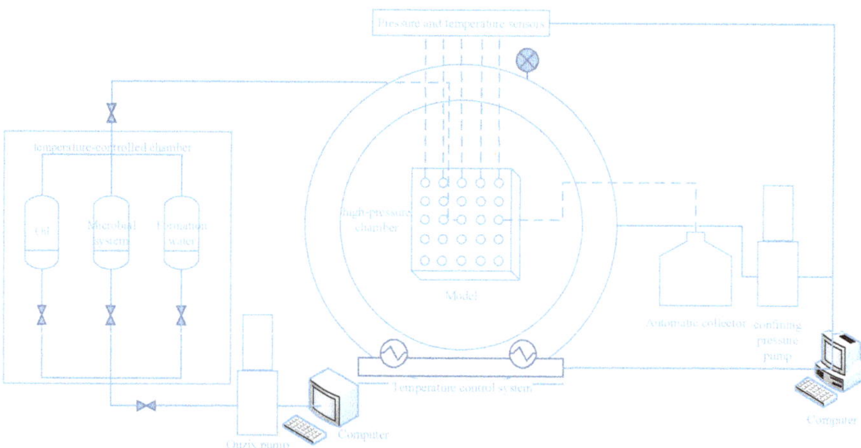

Figure 3. Flow chart of the large-scale physical simulation experiment for MEOR.

The microbial concentration was detected via hemocytometry [39], and the concentrations of total organic carbon and total nitrogen were detected via a multi-N/C3100 total organic carbon/total nitrogen analyzer.

3. Results and Discussion

3.1. Distribution and Variation in the Pressure Field

According to Darcy's law and the distribution law of the flow field potential, the changes in pressure reflect the dynamic flow of reservoir fluids. The analysis of pressure changes can reveal the flow states of reservoir fluids and study the seepage laws during the oil displacement process. The distribution and variation in the pressure field within the model during the displacement process are illustrated in Figure 4.

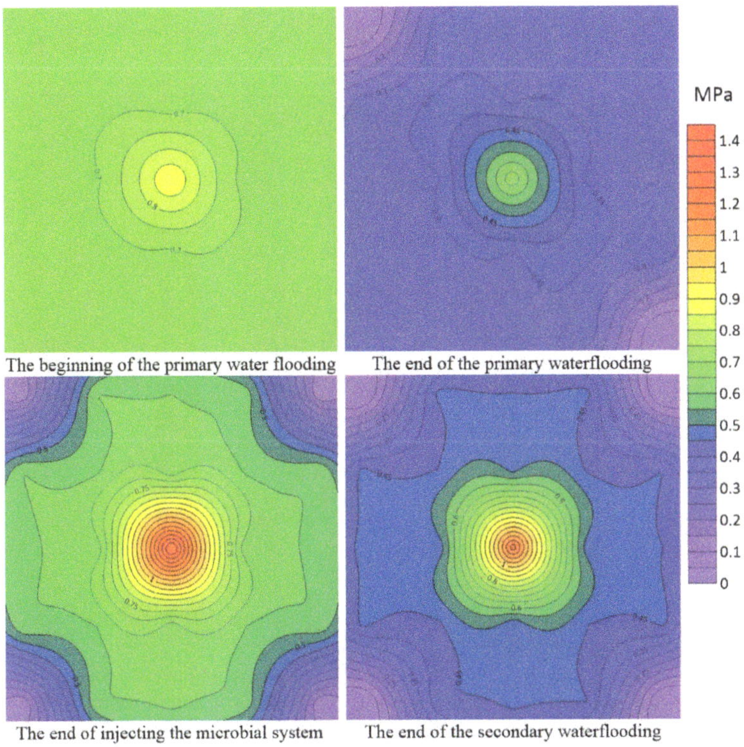

Figure 4. Distribution and variation in the pressure field during the displacement process.

A pressure drop funnel forms near the injection well during the primary water flooding process. That is, from the injection well to the production well, the pressure in the model decreases step by step, and the high-pressure gradient zone is mainly concentrated near the injection well. Due to the heterogeneity of the porous media, the path with high permeability and large pressure difference is preferred during the flow of fluid in porous media, thus forming an advantageous channel. At the end of primary water flooding, the pressure difference in the injection-production well in the main diagonal direction is the largest, which indicates that two main channels along the main flow paths are formed in the model. After the microbial system is injected, the overall pressure within the model increases due to the adsorption and retention effects of the microorganisms. The pressure near the injection well increases significantly, reaching 1.4 MPa. Additionally, the pressure differential between the injection well and production wells increases, and the high-pressure belt along the main flow paths gradually extends to both sides, enhancing the range of pressure influence. After secondary water flooding, the overall pressure within the model decreases, but it remains higher than the pressure at the end of the primary water flooding, indicating that the residual resistance generated by the adsorption and retention effects of the microorganisms within the model still existed.

3.2. Distribution and Variation in the Microbial Field

MEOR is a technology that utilizes the interactions between microorganisms and their metabolites with crude oil to improve the properties of the oil, thereby increasing the recovery rate. Studying the migration and distribution patterns of microorganisms in reservoirs during the microbial flooding process can help determine the effective range of microorganisms and provide theoretical support for the field application of MEOR. The distribution and variation in the microbial concentration within the model during the displacement process are illustrated in Figure 5.

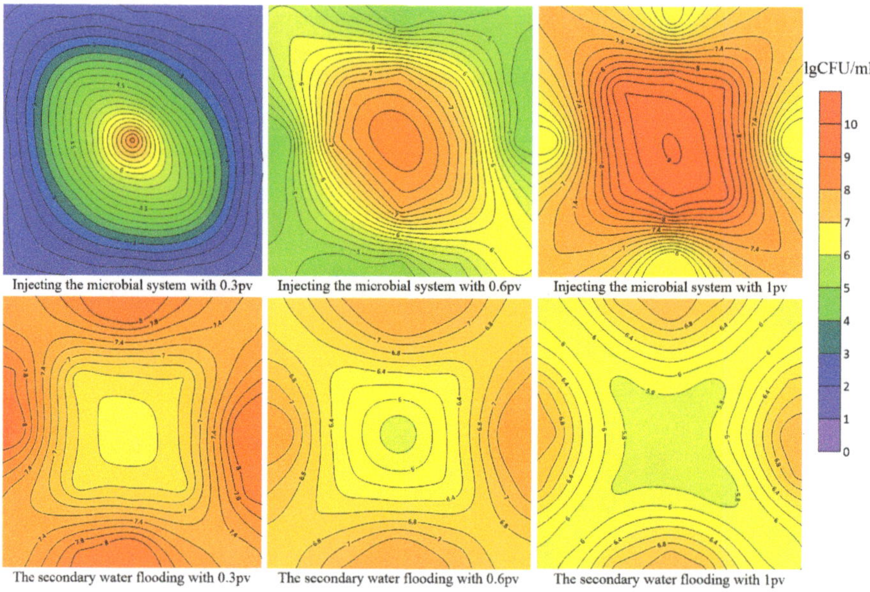

Figure 5. Distribution and variation in the microbial field during the displacement process.

Due to the presence of heterogeneity in the core, bacteria preferentially enter the high-permeability zones with the fluid. After being screened, they are retained within the core, forming a blockage in the high-permeability zones, thereby exerting a certain profile-adjusting effect. Figure 5 shows that in the early stages of microbial system injection, microorganisms are concentrated near the injection well and along the main flow paths. As the volume of the injected microbial system increases, the main flow paths become blocked because of microbial adsorption, retention, and exopolysaccharide production. The injected microbial system subsequently flows along nonmain pathways, effectively expanding the swept volume. By the end of the injection process, the microbial concentration near the injection well reached 10^8 colony-forming units (CFU)/mL. From the injection well to the production well, the microbial concentration in the model decreased progressively. The concentration of microorganisms along the main flow paths was greater than the concentrations on either side of these paths. This is because the high concentration of microbes in the near injection well area, due to their hydrophobicity and the sieving effect of the formation pores on the bacteria, cannot move forward rapidly with the water flow. As a result, the microbial concentration decreases with increasing distance from the injection well. This finding is consistent with the conclusion drawn by Li et al. [40], who found that the microbial concentration in the produced fluids of oil wells is typically one to two orders of magnitude lower than that in the injection well of injection wells.

In the second stage of water flooding, the bacteria in the core's main flow paths were largely flushed out due to detachment, unclogging, and dispersive diffusion caused by the water flow. As formation water continued to be injected, the bacterial concentration on both sides of the main flow lines also decreased by an order of magnitude. By the end of secondary water flooding, the concentration near the injection well and along the main flow paths decreases to below 10^6 CFU/mL. In the areas on both sides of the main flow paths, where water contact is less frequent, relatively high microbial concentration zones form, allowing microorganisms to continue their modifying effects on crude oil in these regions.

3.3. Distribution and Variation in Nutrients

By monitoring and regulating nutrient concentrations, the activities of microbial communities can be more effectively controlled to improve oil recovery. Therefore, studying the dynamic changes in nutrient concentrations during displacement is highly important for understanding the distribution and activity of microorganisms in reservoirs, and provides an important theoretical basis for optimizing the MEOR process. The ratio of the nutrient concentration in the model to the injection concentration in the displacement process is shown in Figure 6.

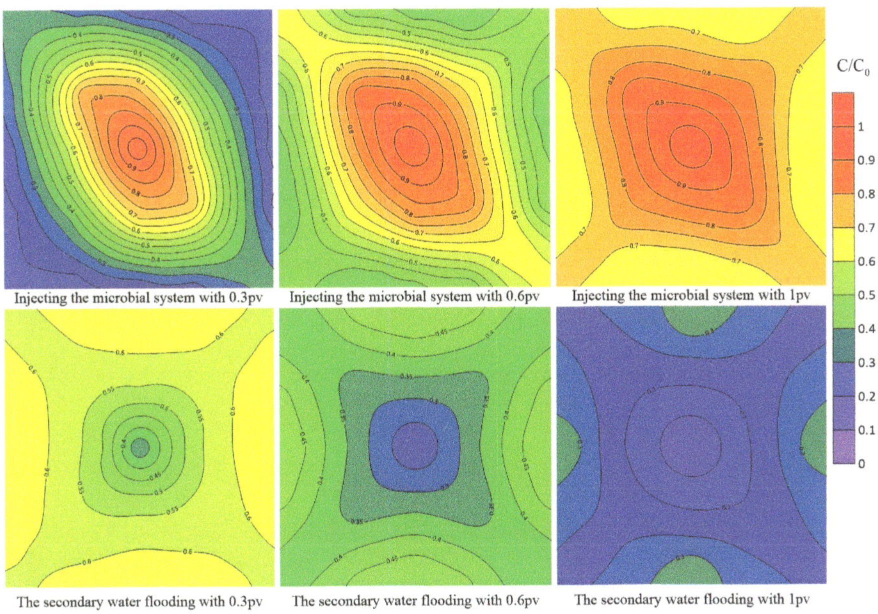

Figure 6. Distribution and variation in nutrients during the displacement process.

To some extent, the distribution and variation in nutrients are similar to those of microorganisms, as the distributions of both are affected by the injection methods and geological conditions of the reservoir. A comparison of the distributions and changes in the microorganisms and nutrient concentrations shown in Figures 5 and 6 reveals that cells migrate more slowly in the core than in nutrients. The reason for this phenomenon is that microorganisms, through electrostatic interactions, hydrophobic interactions, and other means, bind to the surfaces of porous media, resulting in an adsorption coefficient significantly higher than that of dissolved nutrients. Moreover, they are capable of forming biofilms on the surfaces of porous media, which enhances their attachment stability. The secreted extracellular polymeric substances can clog pore throats, thereby further restricting the migration of the microorganisms. This phenomenon is consistent with the conclusions drawn by Sun et al. [35]. The chromatographic separation between microorganisms and nutrients is not very pronounced, likely due to the relatively small spacing between the injection and production wells in the core.

In the early stage of microbial injection, the rapid accumulation of nutrients near the injection well (with concentrations reaching 90% of the injection concentration) is primarily dominated by advection. As the injection continues, mechanical dispersion and diffusion gradually come into play, causing the nutrients to gradually spread outward along the main flow path. By the end of the injection process, a concentration gradient of nutrients is formed along the main flow path (high-permeability channels), with concentrations exceeding 70% on the main flow path and ranging from 60% to 70% on either side.

According to Darcy's Law, the velocity of fluid flow in porous media is directly proportional to the pressure gradient. When other conditions (such as fluid viscosity and medium permeability) remain constant, the greater the pressure gradient, the faster the fluid velocity. As shown in Figure 3, the pressure difference between the injection and production wells is the largest. Near the injection well and along the main flow paths, the high velocity results in high shear stress, which strips the adsorbed nutrients and causes their concentration to decrease more rapidly. By the end of the secondary water flooding, the nutrient concentration near the injection well and along the main flow path drops to below 20% of the injection concentration, while in the low-velocity zones on either side of the main flow path, there are still 30% to 40% of the nutrients remaining, which can provide sufficient nutrition for the growth and reproduction of microorganisms.

3.4. Distribution and Variation in Surface Tension

Surface tension plays a crucial role in oil extraction. The reduction in surface tension can promote oil–water emulsification, making it easier for the oil droplets to detach from the rock surface, thereby more effectively releasing the crude oil that was originally trapped in the rock pores. This process can significantly improve the recovery rate of crude oil. The variation in surface tension at different locations within the model during the displacement process is shown in Figure 7.

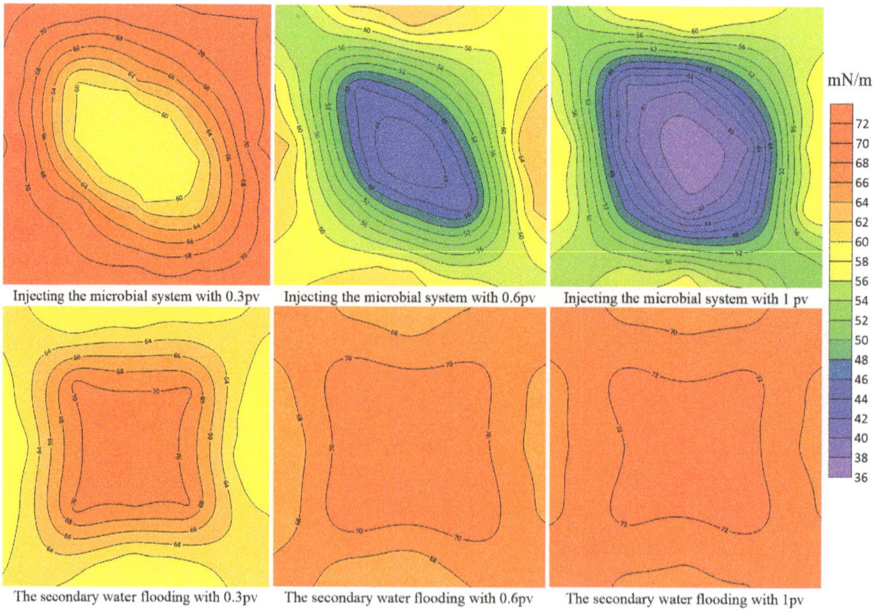

Figure 7. Distribution and variation in surface tension during the displacement process.

The reduction in surface tension is nonlinearly coupled with the microbial concentration, with greater reductions in surface tension occurring in areas of the model with higher microbial concentrations. When 0.3 PV of the microbial system was injected, the interfacial tension near the injection well decreased to 50–60 mN/m, indicating that the microorganisms produce biosurfactants. These biosurfactants have unique amphiphilic properties and can be adsorbed onto rock surfaces or oil–water interfaces, thereby reducing oil–water interfacial tension and promoting oil–water emulsification. With a further increase in the injection amount of the microbial system, the tension reduction effect gradually spreads along the main flow paths and both sides. When 0.6 PV of the microbial system was injected, the interfacial tension within the model had already achieved satisfactory results, with the tension near the injection well decreasing to 40–45 mN/m. By the end of the microbial flooding, the interfacial tension near the injection well further decreased to 36 mN/m, representing a reduction of 49.8%. The interfacial tension along the main flow path decreased to 50–55 mN/m, while on the sides of the main flow path, it decreased to 60 mN/m. These reductions in interfacial tension were effective in mobilizing the residual oil. Mulligan et al. [41] reported that a good surfactant can reduce the surface tension of water to as low as 35 mN/m. When the degree of surface tension reduction near the injection well is close to this value, it indicates that the injected microbial system has achieved a significant tension-reducing effect. On the sides of the main flow path, the microbial concentration is relatively low, and the amount of surfactant produced is also relatively small. Therefore, the tension-reducing effect is not very obvious.

However, when the secondary waterflooding reached 0.3 PV, the interfacial tension near the injection point and along the main flow path in the model rapidly increased to around 70 mN/m, while the tension on both sides of the main flow path also rose to approximately 60 mN/m. By the time the second waterflood reached 0.6 PV, the overall interfacial tension within the model had increased to around 70 mN/m. This was because of the dilution of the microbial system and the removal of biosurfactants by water flow. The increase in surface tension increases the difficulty of oil droplets detaching from the rock surface, which adversely affects the recovery efficiency. On the basis of these observations, it is possible to consider adopting a method of injecting the microbial system in multiple stages, which can help maintain a lower surface tension throughout the displacement process, thereby increasing the recovery rate of crude oil.

3.5. *Evaluation of the Microbial Flooding Effect*

The variations in oil recovery, water cut, and injection well pressure during the oil displacement process are shown in Figure 8. At the beginning of the primary waterflooding, the injection well pressure gradually increased due to the high amount of crude oil in the core. Water breakthrough occurred at 0.1 PV of water injection, and the injection well pressure rose to 0.96 MPa at 0.2 PV. As the crude oil was gradually displaced and preferential flow channels were formed during waterflooding, the injection well pressure began to decline. When the water injection reached 0.5 PV, the water cut reached 98%, and the pressure stabilized at 0.7 MPa. At the end of the primary waterflood, the oil recovery degree was 33.6%.

After the microbial system was injected, the surfactants and other substances produced by the microbes emulsified and reduced the viscosity of the crude oil. The crude oil was completely emulsified into water-in-oil or oil-in-water emulsions, reducing the viscosity from 5.6 mPa·s to 4.3 mPa·s. Meanwhile, the adsorption and retention of microbes in the core, along with the plugging of large pores by the produced polymers, gradually increased the pressure in the model, reaching a maximum of 1.43 MPa. This effectively expanded the sweep efficiency. As the displacement continued, the residual oil in the core gradually decreased with the amount of oil available for microbial action. At 0.4 PV of the microbial injection process, the water cut rose to 98% again. During this process, the water cut reached a minimum of 76%, and the oil recovery rate was increased by 4.2%.

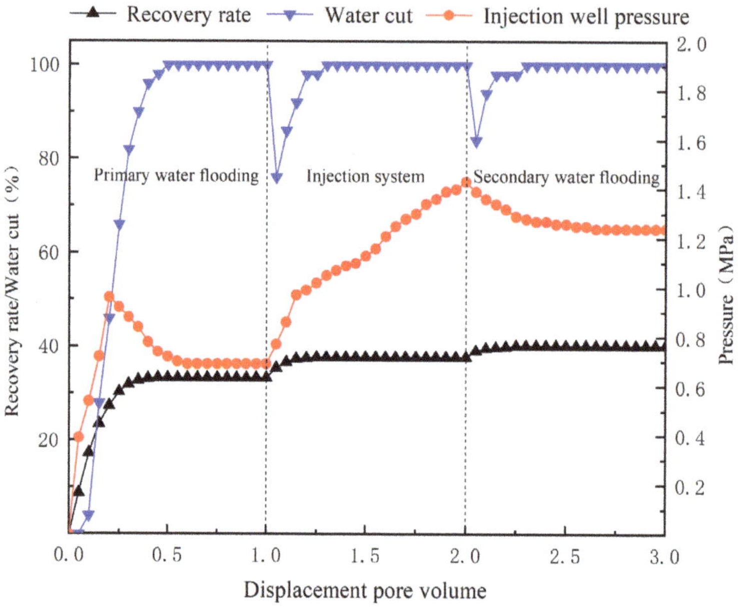

Figure 8. Curves of the oil recovery rate, water cut, and pressure change during the displacement process.

After a 7-day shut-in period for microbial growth and reproduction and their full interaction with the crude oil, a secondary waterflood was conducted. The crude oil viscosity further decreased to 3.4 mPa·s, which is 60.7% of the initial viscosity, and the water cut dropped to 84%. At 0.4 PV of the secondary water flooding, the water cut rose to 100% again. The secondary waterflooding increased the oil recovery rate by 2.3%, resulting in a final recovery rate of 40.1%. The total increase in the oil recovery rate due to microbial action was 6.5%. The increase in recovery efficiency achieved in this study was slightly lower than the 10% improvement reported by Sui et al. [42] following waterflooding and MEOR. This discrepancy is attributed to the increased number of production wells, which facilitates the formation of main flow paths in the porous medium during displacement. These channels restrict the effective range of microbial action, thereby reducing the efficiency of oil recovery. During the secondary waterflood, as the microbes and their metabolites were flushed out of the core by the water flow, the injection well pressure gradually decreased and stabilized at around 1.2 MPa. However, this was still higher than the pressure at the end of the primary waterflood. Although the blockages formed by the high concentration of microbial cells were partially relieved by the waterflood, the plugging effect caused by the microbes and their metabolites in the core did not completely disappear, leaving a certain residual resistance with a residual resistance coefficient of 1.71. The existence of this residual resistance further confirms the synergistic role of microbes and their metabolites in the oil displacement process. They not only changed the dynamic balance of oil and water flow through the physical plugging but also altered the properties of the oil–water interface through biochemical actions, thereby playing a positive role in enhancing crude oil mobility and improving oil recovery.

4. Conclusions

By establishing a two-dimensional physical model, the distribution changes in the pressure field, microbial field, nutrients, and surface tension during microbial flooding were studied, and the effects of microbial flooding were evaluated. The following conclusions were drawn:

1. During the process of injecting the microbial system, microorganisms tend to distribute preferentially near the injection well and then extend outward along the main flow paths. The concentration of microorganisms along the main flow paths is greater than that on either side of the paths. Owing to the adsorption and retention of microorganisms within the core, the overall pressure within the model increases, and the swept area expands.
2. The distribution and variation patterns of nutrients are somewhat similar to those of microorganisms, but the transport and diffusion rates of nutrients are faster than those of microorganisms, leading to an earlier breakthrough time. This is because microorganisms, with their relatively larger size and ability to form biofilms, exhibit stronger adsorption and retention effects in porous media.
3. In areas of the model where the microbial concentration is high, the reduction in surface tension is also significant, with the lowest value reaching 36 mN/m, a decrease of up to 49.8%, which can effectively mobilize residual oil.
4. During the injection of the microbial system, owing to the adsorption and retention of microorganisms and the plugging of large pores by the biopolymers they produce, the pressure within the model significantly increases. After secondary water flooding, the residual resistance coefficient is 1.71, indicating that the injection of the microbial fermentation mixture has a significant profile control effect, and that microbial flooding can increase the oil recovery rate by 6.5%.

This study has unveiled the coupling mechanism of transport and retention of microorganisms and nutrients in the MEOR process, providing a theoretical basis for optimizing injection schemes in MEOR. By understanding the distribution patterns of microorganisms and nutrients, as well as their impacts on the pressure field and interfacial tension, injection schemes can be better designed to enhance oil recovery efficiency.

This research is based on a two-dimensional physical model, which, despite its ability to effectively capture the primary dynamic changes in MEOR, still represents a simplification compared to the complex three-dimensional structures of actual reservoirs. Future research could consider developing more complex three-dimensional models to more accurately simulate the MEOR process in real reservoirs. Additionally, this study assumed a single microbial species, whereas microbial communities in actual reservoirs are likely to be more complex. The interactions among different microbial species and their impacts on oil recovery efficiency still require further investigation.

Author Contributions: Writing—original draft preparation, Y.Z.; writing—review and editing, J.X. and L.H.; supervision, L.Y. and Y.M. All authors have read and agreed to the published version of the manuscript.

Funding: This research received no external funding.

Data Availability Statement: The original contributions presented in this study are included in the article; further inquiries can be directed to the corresponding author.

Conflicts of Interest: The authors declare no conflict of interest.

References

1. Chen, X.; Hou, Q.; Liu, Y.; Liu, G.; Zhang, H.; Sun, H.; Zhu, Z.; Liu, W. Experimental Study on Surfactant–Polymer Flooding After Viscosity Reduction for Heavy Oil in Matured Reservoir. *Energies* **2025**, *18*, 756. [CrossRef]
2. Gómez-Delgado, J.L.; Gutierrez-Niño, N.; Carrillo-Moreno, L.F.; Martínez-López, R.A.; Santos-Santos, N.; Mejía-Ospino, E. Experimental Assessment of Magnetic Nanofluid Injection in High-Salinity and Heavy-Crude-Saturated Sandstone: Mitigation of Formation Damage. *Energies* **2025**, *18*, 212. [CrossRef]
3. Ke, C.Y.; Sun, R.; Wei, M.X.; Yuan, X.N.; Sun, W.J.; Wang, S.C.; Zhang, Q.Z.; Zhang, X.L. Microbial enhanced oil recovery (MEOR): Recent development and future perspectives. *Crit. Rev. Biotechnol.* **2024**, *44*, 1183–1202. [CrossRef]
4. Kandiel, Y.E.; Attia, G.; Metwalli, F.; Khalaf, R.; Mahmoud, O. Innovative Role of Magnesium Oxide Nanoparticles and Surfactant in Optimizing Interfacial Tension for Enhanced Oil Recovery. *Energies* **2025**, *18*, 249. [CrossRef]
5. Qiao, M.; Zhang, F.; Li, W. Rheological Properties of Crude Oil and Produced Emulsion from CO_2 Flooding. *Energies* **2025**, *18*, 739. [CrossRef]
6. Abolhasanzadeh, A.; Khaz'ali, A.R.; Hashemi, R.; Jazini, M. Experimental study of microbial enhanced oil recovery in oil-wet fractured porous media. *Oil Gas Sci. Technol.–Rev. D'ifp Energ. Nouv.* **2020**, *75*, 73. [CrossRef]
7. Sen, R. Biotechnology in petroleum recovery: The microbial EOR. *Prog. Energy Combust. Sci.* **2008**, *34*, 714–724. [CrossRef]
8. Lourdes, R.S.; Cheng, S.Y.; Chew, K.W.; Ma, Z.; Show, P.L. Prospects of microbial enhanced oil recovery: Mechanisms and environmental sustainability. *Sustain. Energy Technol. Assess.* **2022**, *53*, 102527. [CrossRef]
9. Patel, J.; Borgohain, S.; Kumar, M.; Rangarajan, V.; Somasundaran, P.; Sen, R. Recent developments in microbial enhanced oil recovery. *Renew. Sustain. Energy Rev.* **2015**, *52*, 1539–1558. [CrossRef]
10. Alkan, H.; Mukherjee, S.; Kögler, F. Reservoir engineering of in-situ MEOR; impact of microbial community. *J. Pet. Sci. Eng.* **2021**, *195*, 107928. [CrossRef]
11. Quraishi, M.; Bhatia, S.K.; Pandit, S.; Gupta, P.K.; Rangarajan, V.; Lahiri, D.; Varjani, S.; Mehariya, S.; Yang, Y.-H. Exploiting Microbes in the Petroleum Field: Analyzing the Credibility of Microbial Enhanced Oil Recovery (MEOR). *Energies* **2021**, *14*, 4684. [CrossRef]
12. Zhou, J.; Wang, X.; Sun, X.; Fan, Z. Insight into the Microscopic Interactions Among Steam, Non-Condensable Gases, and Heavy Oil in Steam and Gas Push Processes: A Molecular Dynamics Simulation Study. *Energies* **2025**, *18*, 125. [CrossRef]
13. He, L.; Ren, S.; Zhang, H. Fingering crossover and enhanced oil recovery mechanism of water-alternating-CO_2 injection. *Phys. Fluids* **2023**, *35*, 053322.
14. Yang, Y.; Liu, W.; Yu, J.; Liu, C.; Cao, Y.; Sun, M.; Li, M.; Meng, Z.; Yan, X. Technology Progress in High-Frequency Electromagnetic In Situ Thermal Recovery of Heavy Oil and Its Prospects in Low-Carbon Situations. *Energies* **2024**, *17*, 4715. [CrossRef]
15. Gao, C. Experiences of microbial enhanced oil recovery in Chinese oil fields. *J. Pet. Sci. Eng.* **2018**, *166*, 55–62. [CrossRef]
16. Niu, J.; Liu, Q.; Lv, J.; Peng, B. Review on microbial enhanced oil recovery: Mechanisms, modeling and field trials. *J. Pet. Sci. Eng.* **2020**, *192*, 107350. [CrossRef]
17. Wu, B.; Xiu, J.; Yu, L.; Huang, L.; Yi, L.; Ma, Y. Research advances of microbial enhanced oil recovery. *Heliyon* **2022**, *8*, e11424. [CrossRef]
18. Amani, H.; Sarrafzadeh, M.H.; Haghighi, M.; Mehrnia, M.R. Comparative study of biosurfactant producing bacteria in MEOR applications. *J. Pet. Sci. Eng.* **2010**, *75*, 209–214. [CrossRef]
19. Safdel, M.; Anbaz, M.A.; Daryasafar, A.; Jamialahmadi, M. Microbial enhanced oil recovery, a critical review on worldwide implemented field trials in different countries. *Renew. Sustain. Energy Rev.* **2017**, *74*, 159–172. [CrossRef]
20. Andrade, A.; Mehl, A.; Mach, E.; Couto, P.; Mansur, C.R.E. Application of biosurfactants in enhanced oil recovery ex-situ: A review. *Braz. J. Microbiol.* **2024**, *55*, 3117–3139. [CrossRef]
21. Zhang, J.; Gao, H.; Xue, Q. Potential applications of microbial enhanced oil recovery to heavy oil. *Crit. Rev. Biotechnol.* **2020**, *40*, 459–474. [CrossRef]
22. Jeong, M.S.; Lee, Y.W.; Lee, H.S.; Lee, K.S. Simulation-Based Optimization of Microbial Enhanced Oil Recovery with a Model Integrating Temperature, Pressure, and Salinity Effects. *Energies* **2021**, *14*, 1131. [CrossRef]
23. Saravanan, A.; Kumar, P.S.; Vardhan, K.H.; Jeevanantham, S.; Karishma, S.B.; Yaashikaa, P.R.; Vellaichamy, P. A review on systematic approach for microbial enhanced oil recovery technologies: Opportunities and challenges. *J. Clean. Prod.* **2020**, *258*, 120777. [CrossRef]
24. Yu, X.; Li, H.; Song, Z.; Zhu, W. Long-Term Pore-Scale Experiments on MEOR by Surfactant-Producing Microorganisms Reveal the Altering Dominant Mechanisms of Oil Recovery. *Energies* **2023**, *16*, 6854. [CrossRef]
25. Dopffel, N.; Kögler, F.; Hartmann, H.; Costea, P.I.; Mahler, E.; Herold, A.; Alkan, H. Microbial induced mineral precipitations caused by nitrate treatment for souring control during microbial enhanced oil recovery (MEOR). *Int. Biodeterior. Biodegrad.* **2018**, *135*, 71–79. [CrossRef]
26. Yue, M.; Zhu, W.; Song, Z.; Long, Y.; Song, H. Study on distribution of reservoir endogenous microbe and oil displacement mechanism. *Saudi J. Biol. Sci.* **2017**, *24*, 263–267. [CrossRef] [PubMed]

27. Lin, J.; Hao, B.; Cao, G.; Wang, J.; Feng, Y.; Tan, X.; Wang, W. A study on the microbial community structure in oil reservoirs developed by water flooding. *J. Pet. Sci. Eng.* **2014**, *122*, 354–359. [CrossRef]
28. Bi, Y.Q.; Yu, L.; Huang, L.X.; Ma, T.; Xiu, J.L.; Yi, L.N. Microscopic profile control mechanism and potential application of the biopolymer-producing strain FY-07 for microbial enhanced oil recovery. *Pet. Sci. Technol.* **2016**, *34*, 1952–1957. [CrossRef]
29. Yao, C.; Meng, X.; Qu, X.; Cheng, T.; Da, Q.A.; Zhang, K.; Lei, G. Kinetic model and numerical simulation of microbial growth, migration, and oil displacement in reservoir porous media. *ACS Omega* **2022**, *7*, 32549–32561. [CrossRef]
30. O'Bryan, O.D.; Ling, T.W.D. The effect of the bacteria Vibrio desulfuricans on the permeability of limestone cores. *Tex. J. Sci* **1949**, *1*, 117–128.
31. Jack, T.R.; Stehmeier, L.G.; Islam, M.R.; Ferris, F.G.C. F-6 Microbial Selective Plugging to Control Water Channeling. *Dev. Pet. Sci.* **1991**, *31*, 433–440.
32. Lei, G.L.; Li, X.M.; Chen, Y.M. The migration ability and rule of microbial in reservoir. *Pet. Explor. Dev.* **2001**, *28*, 75–78.
33. Bi, Y.; Yu, L.; Xiu, J.; Yi, L.; Huang, L.; Wang, T. Migration and retention mechanism of microorganisms for oil recovery in porous media. *Acta Pet. Sin.* **2017**, *38*, 91.
34. Guo, S.; Liu, T.; Guo, L. Oil displacement effect and growth distribution of microbial in cores with different permeability. *Oilfield Chem.* **2018**, *35*, 131–134.
35. Sun, G. Distribution of activator for endogenous microbe in large scale oil recovery experiments. *Chin. J. Bioprocess Eng.* **2016**, *14*, 1.
36. Ke, C.; Wu, G.; You, J. Study on Consumption, Adsorption and the Migration Laws of Nutrient Components of Oil Displacement Microbes. *Drill. Prod. Technol.* **2014**, *37*, 40–43.
37. Zhao, X.; Liao, Z.; Liu, T.; Cheng, W.; Gao, G.; Yang, M.; Li, G. Investigation of the transport and metabolic patterns of oil-displacing bacterium FY-07-G in the microcosm model using X-CT technology. *J. Appl. Microbiol.* **2023**, *134*, lxad281. [CrossRef] [PubMed]
38. Hu, J.; Guo, L.; Sun, G.; Wu, X.; Song, Y.; Lin, J.; Wang, Z. Relationship between the growth and metabolism of indigenous microorganisms and oil displacement efficiency during dynamic displacement. *Acta Pet. Sin.* **2020**, *41*, 1127.
39. Chen, Y.W.; Chiang, P.J. An automated approach for hemocytometer cell counting based on image-processing method. *Measurement* **2024**, *234*, 114894. [CrossRef]
40. Li, Y.; Yu, L.; Zhang, L.; Duan, L.; Wu, J.; Wang, Q. Experimental study of microbial flooding physical simulation for Baolige oilfield. *Oil Drill. Prod. Technol./Shiyou Zuancai Gongyi* **2013**, *35*, 95.
41. Mulligan, C.N. Environmental applications for biosurfactants. *Environ. Pollut.* **2005**, *133*, 183–198. [CrossRef] [PubMed]
42. Sui, J. Microbial EOR studies on the microorganisms using petroleum hydrocarbons as sole carbon source. *Acta Pet. Sin.* **2001**, *22*, 53.

energies

MDPI

Article

Research on the Fixation Strength of High-Temperature Geothermal Drilling Cone Bit Teeth

Yan Yang [1,2,3], Dongdong Song [4,*], Lian Chen [5], Yingxin Yang [2,5], Haitao Ren [5], Shunzuo Qiu [1,3] and Zequan Huang [6]

1 School of Mechanical and Electrical Engineering, Yibin University, Yibin 644000, China; belsionyy@yibinu.edu.cn (Y.Y.); 2020100006@yibinu.edu.cn (S.Q.)
2 State Key Laboratory of Oil and Gas Reservoir Geology and Exploitation, Southwest Petroleum University, Chengdu 610500, China; yangyingxin@swpu.edu.cn
3 Key Laboratory of Oil and Gas Equipment, Ministry of Education, Southwest Petroleum University, Chengdu 610500, China
4 CNPC Chuanqing Drilling Engineering Co., Ltd., Chuanxi Drilling Company, Chengdu 610051, China
5 College of Mechanical and Electrical Engineering, Southwest Petroleum University, Chengdu 610500, China; chenlian@swpu.edu.cn (L.C.); renhaitao@swpu.edu.cn (H.R.)
6 School of Electronic Information Engineering, Yibin University, Yibin 644000, China; 2022040002@yibinu.edu.cn
* Correspondence: sondd_sc@cnpc.com.cn

Abstract: During the drilling process of high-temperature geothermal wells, the high temperature at the bottom of the well and the complex lithology of the formation lead to poor tooth loss prevention in cone drill bits. This issue seriously affects the life and efficiency of geothermal drilling. The stability of the wellbore is one of the key issues in the drilling process of high-temperature geothermal wells, and the fixed-tooth strength of the roller drill bit directly affects the stability of the wellbore and drilling efficiency. The heat transfer effect of the wellbore will exacerbate the thermal expansion and performance degradation of the drill bit material in high-temperature environments, leading to a decrease in the strength of the fixed teeth. To address this, this study used a high-temperature experimental apparatus to systematically test the fixed-tooth strength of roller drill bits. By using five types of tooth spacing: 4, 6, 8, 10, and 12 mm, three types of tooth diameters: 12, 14, and 16 mm, and three types of interference fit: 0.075, 0.095, and 0.115 mm, the maximum fastening force of fixed teeth was measured under different conditions, and its variation pattern was analyzed. The experimental results show that the higher the temperature, the weaker the tooth-fixing strength. Under the same perforation distance, the maximum fastening force decreases with increasing temperature. Compared with normal temperature, the maximum fastening force decreases by about 49.6–64.5%. At the same temperature, the maximum fastening force is the largest when the perforation distance is 10 mm. When the temperature increases, the maximum fastening force increases with the tooth diameter; that is, the larger the tooth diameter, the better the tooth-fixing effect. At the same temperature, the maximum fastening force first increases and then decreases with increasing interference. The maximum fastening force is the largest when the interference is 0.095 mm. At 120 °C, 180 °C, and 240 °C, the maximum fastening force is reduced by 21.9%, 29.4%, and 56.6%, respectively, compared to normal temperature. The study reveals the variation law of tooth-fixing strength under high-temperature conditions and proposes tooth-fixing methods and suggestions suitable for high-temperature geothermal wells. This provides a scientific basis for solving the problem of tooth loss of roller bits in high-temperature geothermal drilling and has important theoretical and practical application value.

Keywords: high-temperature geothermal well; cone bit; well wall heat transfer; interference fit; well wall stability; fastening force

1. Introduction

The formation lithology of high-temperature geothermal drilling is complex, the bottom hole temperature is high, and there are many uncertain factors. In recent years, significant progress has been made in the geothermal energy extraction technology of high-temperature geothermal wells, especially in high-temperature drilling technology and efficient utilization of geothermal resources. The development of high-temperature geothermal wells faces challenges from high temperature, high pressure, and complex geological conditions, especially the wear and failure of drill bits during the drilling process. Research has shown that the performance of drill bit materials significantly decreases under high-temperature environments, leading to a decrease in drilling efficiency [1–5]. Related studies have investigated the crack propagation behavior of CO_2 fracturing fluid in unconventional low-permeability reservoirs, and found that the physical and chemical properties of the fracturing fluid and reservoir characteristics have a significant impact on crack propagation [6]. In addition, the wellhead stability during the development of combustible ice reservoirs in the northern South China Sea was analyzed, and it was pointed out that changes in formation pressure and wellhead structure design were key factors. Optimization suggestions were also proposed. These studies provide important references for the development of related fields [7]. In high-temperature geothermal drilling, the cone bit is mainly used. However, tooth loss has become the most serious failure form of the cone bit (as shown in Figure 1), which greatly shortens the bit's service life and increases drilling costs [8–12]. The Kadera bit, developed by Smith Company of the United States, was used in a geothermal well in Italy. The drilling time was 3–37% lower than that of the traditional cone bit, and the bit exhibited good tooth-fixing ability [13,14]. To improve the fixed-tooth strength of the cone bit, experimental research on the assembly stress of the cone bit was conducted. It was concluded that the stress on the tooth causes deformation of the tooth hole, the interference increases gradually, and the surface circumferential and radial stress of the tooth hole is transformed into tensile stress. This is an important factor leading to tooth-hole cracking [15,16]. Experimental analysis and numerical simulation were conducted to study the fixed-tooth strength of three-cone drill bits, and a personalized new product of three-cone drill bits was developed. The results showed that the fixed-tooth strength significantly decreased under high-temperature conditions [17–20]. Generally, chamfering the root of the teeth with different interference fits is beneficial for reducing stress concentration at the root of the tooth hole [21,22]. Numerical simulation of tooth interference fit assembly revealed that the gradient of compressive stress distribution on the hole wall increases during the assembly process, and the deformation around the hole increases. A shorter assembly distance can achieve a more reasonable stress distribution [23,24]. A shock-resistant drill bit has been developed for hard rock formations and high-temperature environments, effectively improving mechanical drilling speed and reducing drilling costs [25,26]. In addition, researchers have analyzed and calculated the contact stiffness and damping of the interference fit interface under different interference amounts through numerical simulation methods. They also calculated the frequency response function of the interference fit and verified the correctness of the contact stiffness and damping [27]. A new experimental testing method for tooth fixation ability has been proposed for the unique drilling technology of high-temperature geothermal wells [28–31].

Figure 1. Tooth loss phenomenon of tri-cone drill bits in high-temperature geothermal drilling.

Although scholars both domestically and internationally have conducted extensive simulation research on tooth fixation, various simulation models have oversimplified the tooth fixation process. Additionally, while there are experimental studies on tooth fixation, these experiments have not been designed to account for factors such as temperature and tooth stress. Therefore, this paper presents an experimental study on the tooth-fixing technology of cone bits in high-temperature environments, aiming to propose methods and suggestions for mitigating tooth loss in cone bits during high-temperature drilling operations.

2. Materials and Equipment

During the drilling process of high-temperature geothermal wells, the phenomenon of tooth loss in roller bits seriously affects drilling efficiency and equipment life. In order to reduce the failure rate and extend the service life of the equipment, this study proposes a new method for testing the strength of tooth fastening and conducts relevant experimental research. Through experiments, the influence of factors such as temperature, tooth diameter, and tooth-hole spacing on the tooth fixation strength of roller drill bits is analyzed in depth. The goal is to grasp their inherent laws and provide a scientific basis for optimizing the design of roller drill bits. The roller base material used in the experiment is 20CrMo, and the tooth material is YG16C. Both materials are provided by Henan Hard Alloy Company in China (Zhengzhou Hard Alloy Company, Henan Province, China), and the material specifications and performance indicators are determined according to experimental requirements. A procurement contract is signed with the supplier to clarify the quality standards and delivery time. Strict quality inspection shall be conducted before delivery to ensure that the materials meet the experimental requirements. The processed base and teeth, prepared according to the experimental requirements, are shown in Figure 2. The properties of the base and cutting tooth materials are listed in Table 1.

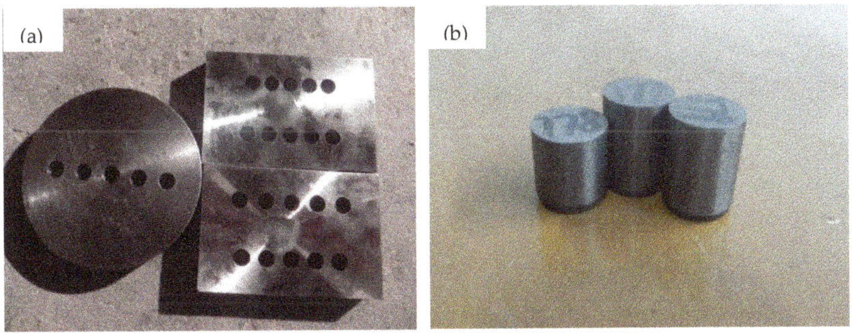

Figure 2. (**a**) Cone base. (**b**) Experimental tooth.

Table 1. The material properties of cone matrix and cutting teeth.

	Elastic Modulus (GPa)	Poisson's Ratio	Expansion Coefficient (K^{-1})	Yield Limit (MPa)	Strength Limit (MPa)
20CrMo	210	0.28	13.2×10^{-6}	626	885
YG16C	640	0.23	5.3×10^{-6}	1120	5460

In this experiment, a hydraulic universal testing machine (as shown in Figure 3) is used to complete the loading process. The hydraulic universal testing machine model AG-IS 100 kN, with a maximum test pressure of 100 kN and an accuracy level of 0.5, is suitable for testing the mechanical properties of various materials such as tension, compression, and bending. It is manufactured in Taizhou City, Jiangsu Province, China. A pressure sensor is threaded onto the indenter of the press, and the displacement of the indenter is monitored by a displacement sensor. Both sensors are connected to a computer with data acquisition capabilities, enabling real-time acquisition and storage of pressure and displacement data. The tooth and the experimental specimen (tooth wheel matrix) are placed on the workbench plane, with the indenter and tooth axis always aligned in a straight line. The indenter can press into and out of the tooth through the upward movement of the hydraulic cylinder.

Figure 3. Hydraulic universal testing machine.

3. Testing Process

This experiment was meticulously designed and executed in strict compliance with the relevant standards of the International Organization for Standardization (ISO), particularly in material testing, mechanical property evaluation, and high-temperature environment simulation. Specific reference was made to ISO 6892-1:2019 [32], "Tensile testing of metallic materials—Part 1: Room temperature test method". The tooth hole type employed in the fastening force test experiment is a through-hole. The experiment investigates the influence of different interference fits on the tooth-holding force at various temperatures. Specifically, the fastening force of the tooth hole is measured by applying axial loading to press the tooth out of the hole. In the experiment, tooth diameters of 16 mm, 14 mm, and 12 mm were used, with tooth hole distances of 4 mm, 6 mm, 8 mm, 10 mm, and 12 mm, respectively. Three interference fit ranges were tested: 0.065–0.085 mm, 0.085–0.105 mm, and 0.105–0.125 mm. Axial loading experiments were conducted at room temperature, 120 °C, 180 °C, and 240 °C. To eliminate experimental variability, each group of experiments was repeated five times. The temperature of the heated test piece was measured using an HT-8830 infrared thermometer, as illustrated in Figure 4.

In the fastening force experiment, the tooth-pressing process was carried out at different temperatures. During the pressing process, the press must be controlled to pressurize slowly. The tooth axis should always remain perpendicular to the working surface to avoid damaging the tooth hole surface by pressing the tooth askew. After the tooth reaches the specified depth in the tooth hole, it should be unloaded slowly to zero. When pressing out the teeth to test the fastening force, the press must be controlled to load slowly. Once the pressure curve stabilizes, it can be considered that the teeth have begun to move. The test is concluded when the tooth displacement reaches 1 mm. The experimental process is shown in Figure 5.

Figure 4. Temperature test specimen.

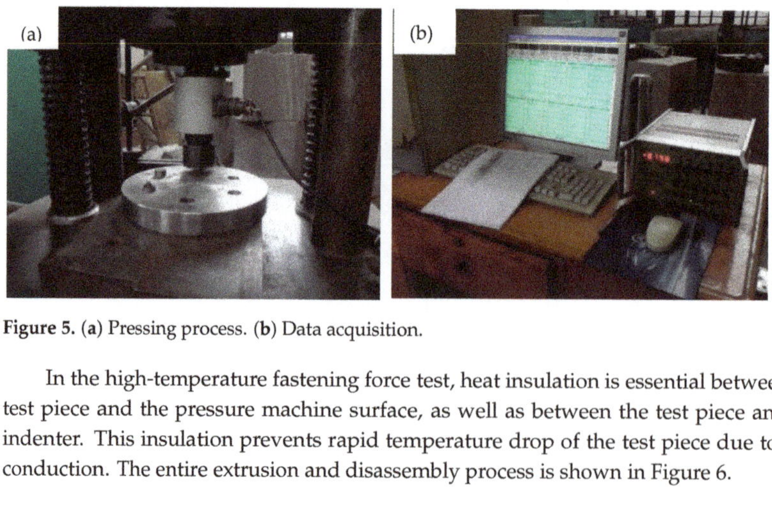

Figure 5. (**a**) Pressing process. (**b**) Data acquisition.

In the high-temperature fastening force test, heat insulation is essential between the test piece and the pressure machine surface, as well as between the test piece and the indenter. This insulation prevents rapid temperature drop of the test piece due to heat conduction. The entire extrusion and disassembly process is shown in Figure 6.

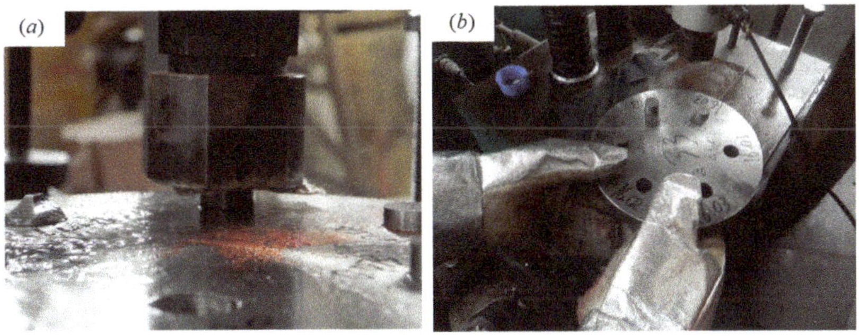

Figure 6. (**a**) Extrusion process. (**b**) Loading and unloading process.

In the experiment, to account for the influence of adjacent teeth on the tooth-tightening force, it is necessary to press all five teeth on the test piece into their respective holes before beginning the process of pressing out the teeth. The experimental process is shown in Figure 7.

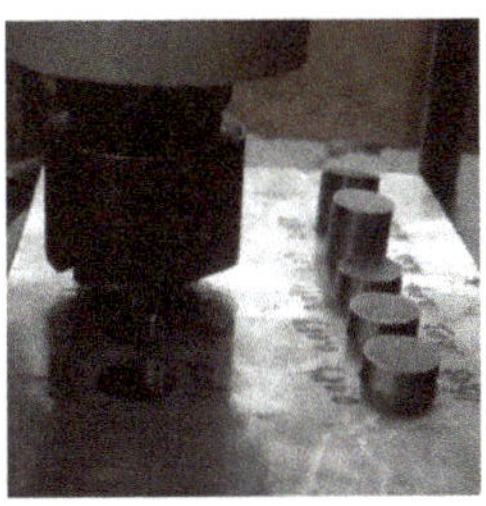

Figure 7. Pressing process.

4. Experimental Result and Numerical Simulation

4.1. Curve Analysis of Tooth Ballast Process

The displacement and drilling pressure curves of the teeth during pressing and extruding were obtained by the experimental acquisition system, as shown in Figure 8. During the tooth-pressing process, the maximum tooth insertion force occurs at point A, corresponding to the continuous increase in the load. In the extrusion process, the teeth initially move slightly to overcome the friction on the joint surface as the load continues to increase. Significant displacement movement begins once the maximum static friction, caused by the interference fit, is overcome. Since the contact area between the teeth and the perforations remains constant, the pressure curve stabilizes after the teeth start moving, and the load reaches the maximum fastening force at point B.

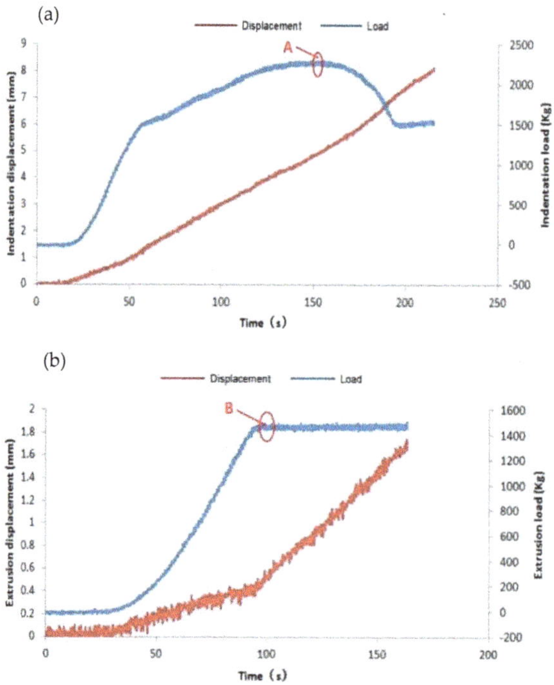

Figure 8. (**a**) Indentation curve. (**b**) Extrusion curve.

4.2. The Influence of Perforation Distance on Fastening Force

Establish a finite element three-dimensional model to analyze the effect of different tooth spacings on the fixed-tooth force. After the three-dimensional model is established, import it into Abaqus for calculation. The three-dimensional model is shown in Figure 9. The tooth spacings are set at 4 mm, 6 mm, 8 mm, 10 mm, and 12 mm, and their respective influences are analyzed. The stress cloud map is shown in Figure 10.

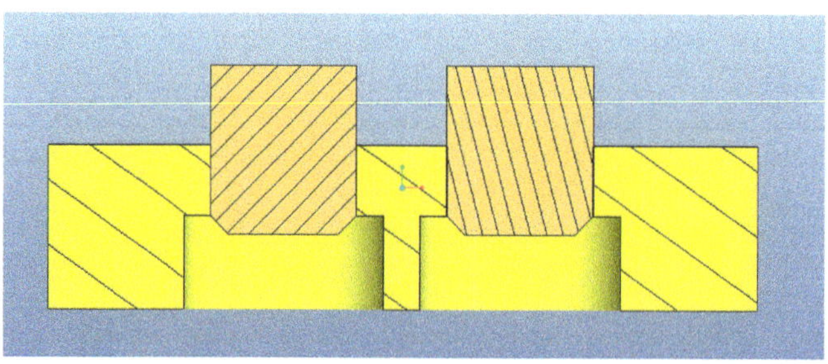

Figure 9. Finite element model.

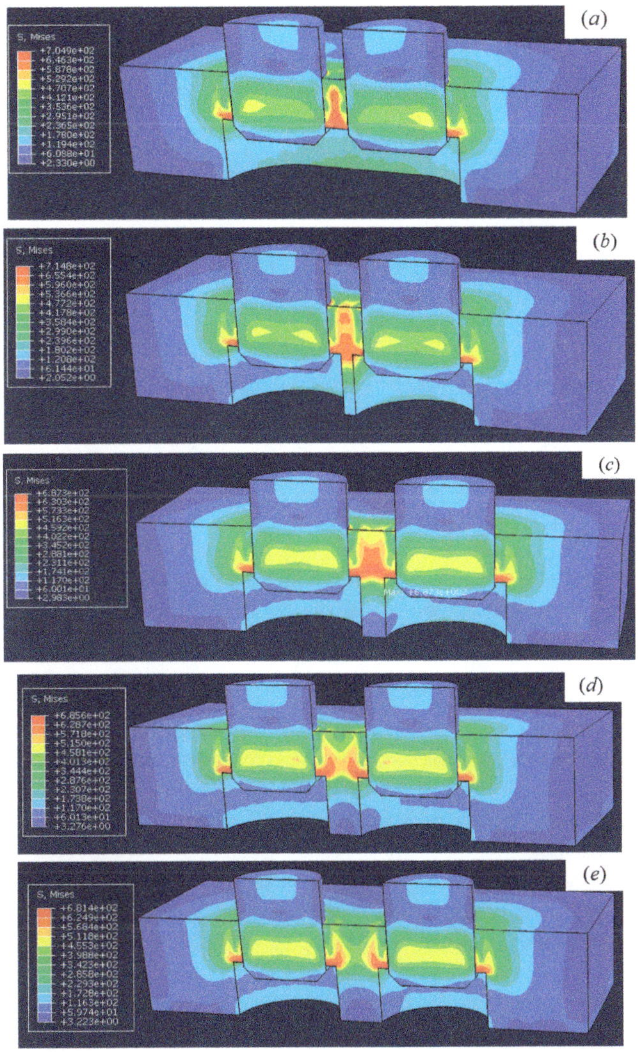

Figure 10. Stress cloud maps for different perforation distances: (**a**) 4 mm. (**b**) 6 mm. (**c**) 8 mm. (**d**) 10 mm. (**e**) 12 mm.

The relationship between the maximum fastening force during simulation and experimental testing, i.e., the pressure value during tooth pressing, and the perforation distance at different temperatures is shown in Figure 11. From the simulation data, it can be seen that the larger the distance between the teeth, the greater the maximum fastening force and the better the fastening effect. When the tooth pitch increases to 10 mm, the increase in surface contact pressure is most significant. This is because when the distance between adjacent teeth is too close, stress superposition during the tooth insertion process causes plastic deformation of the tooth hole material. The hole wall cannot maintain the fastening effect on the teeth, resulting in a decrease in fastening force. This is consistent with the experimental results.

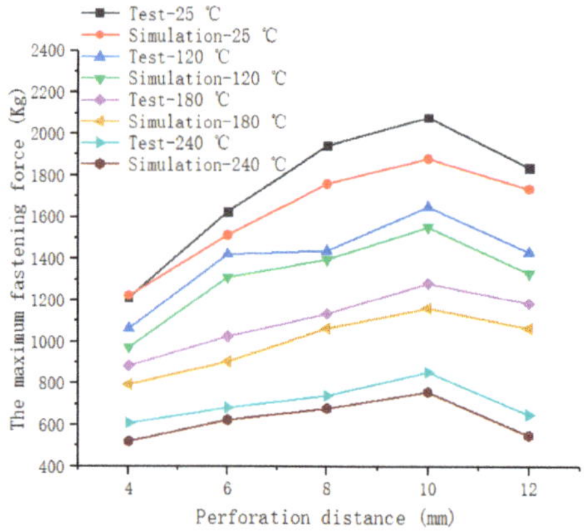

Figure 11. Relationship between perforation distance and maximum fastening force.

From the experimental data, it can be seen that at the same temperature, the maximum fastening force first increases and then decreases as the perforation distance increases. The maximum fastening force is highest when the perforation distance is 10 mm, with values of 2080.8 kg, 1652.1 kg, 1284 kg, and 856.9 kg at different temperatures, respectively. Conversely, the fastening force is lowest when the perforation distance is 4 mm, decreasing by approximately 28.5–41.6% compared to that at 10 mm. Therefore, under an interference of 0.115 mm, the tooth-fixing strength is optimal when the tooth perforation spacing is 10 mm. Under the same tooth spacing, the maximum fastening force decreases with increasing temperature, being highest at normal temperature. Compared to normal temperature, the maximum fastening force decreases by approximately 49.6–64.5% at elevated temperatures, indicating that higher temperatures significantly weaken the tooth-fixing strength of the cone bit. The main reason for this phenomenon is the influence of the thermal expansion effect. Under a high-temperature environment, the matrix material (20CrMo) and tooth material (YG16C) of the roller drill bit will undergo thermal expansion. When the distance between the perforations is 10 mm, the thermal expansion of the material may lead to a reduction in the interference fit between the perforations and the teeth. Due to the fact that interference fit is a key factor in the strength of fixed teeth, a decrease in interference fit directly leads to a decrease in fastening force. In addition, thermal expansion may also cause plastic deformation of the tooth hole wall, further weakening the fastening effect between the teeth and the tooth hole. The second is the change in material properties, where the strength and hardness of the material decrease at high temperatures, resulting in a weakened support capacity of the tooth hole wall. When the tooth spacing is 10 mm, the decrease in the performance of this material may be more significant, making the tooth hole wall unable to effectively resist the pressure of the teeth, resulting in a rapid decrease in fastening force. The third is caused by stress concentration. When the spacing between teeth is small, the stress between adjacent teeth may overlap, which may lead to stress concentration. At high temperatures, this stress concentration will further intensify the deformation of the tooth hole wall, reducing its fastening ability to the teeth. At 120 °C, the main reason for the continued rebound after a slight decrease in tooth spacing is due to the nonlinear characteristics of thermal expansion. The thermal expansion of the material is not completely linear, especially near the yield point of the material. At 120 °C, the thermal expansion of the material may cause a slight decrease in the tooth spacing, but as the temperature further increases, the plastic deformation of the material begins to take effect, causing the tooth spacing to increase again. This nonlinear thermal expansion characteristic may lead to a brief decrease in tooth spacing and subsequent rebound. The second is the elastic recovery of the material. At 120 °C, although the elastic modulus of the material decreases, it still has a certain degree of elastic recovery ability. When the distance between teeth decreases slightly due to thermal expansion, the elastic recovery of the material may cause the distance between teeth to increase again, leading to a rebound in fastening force.

4.3. Effect of Tooth Diameter on Fastening Force

The fixation strength of different tooth diameters was analyzed under an interference fit of 0.115 mm. Figure 12 illustrates the relationship between tooth diameter and maximum fastening force at various temperatures. The fastening force consistently decreases as the temperature rises. When the tooth diameter is held constant, the tooth fixation effect is poorest at 240 °C, with the maximum fastening force decreasing by approximately 58.7–73.9% compared to that at room temperature. At room temperature, the maximum fastening force initially decreases and then increases with increasing tooth diameter, reaching a peak of 1392.8 kg when the tooth diameter is 16 mm. As the temperature increases, the maximum fastening force also increases with larger tooth diameters, peaking at 16 mm. In other words, at higher temperatures, larger tooth diameters result in better tooth fixation effects. For a tooth diameter of 16 mm, the maximum fastening forces at elevated temperatures are 1053.2 kg at 120 °C, 895 kg at 180 °C, and 549.2 kg at 240 °C, respectively.

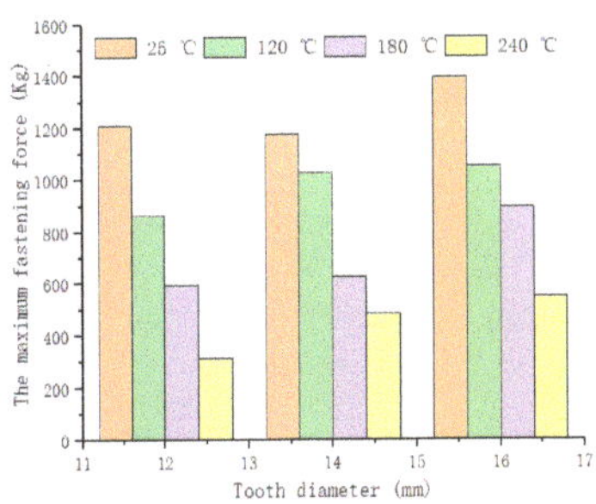

Figure 12. Relationship between tooth diameter and maximum fastening force.

Establish different tooth diameter models for finite element analysis, and their models and stress cloud maps are shown in Figure 13. It can be seen that the larger the diameter of the teeth, the better the fastening effect. This is because appropriately increasing the diameter of the teeth can significantly increase the contact area between the teeth and the surface of the tooth cavity, thereby effectively improving the clamping force of the tooth cavity on the teeth. This design optimization can effectively prevent the teeth from falling off during use and improve the stability and reliability of the structure.

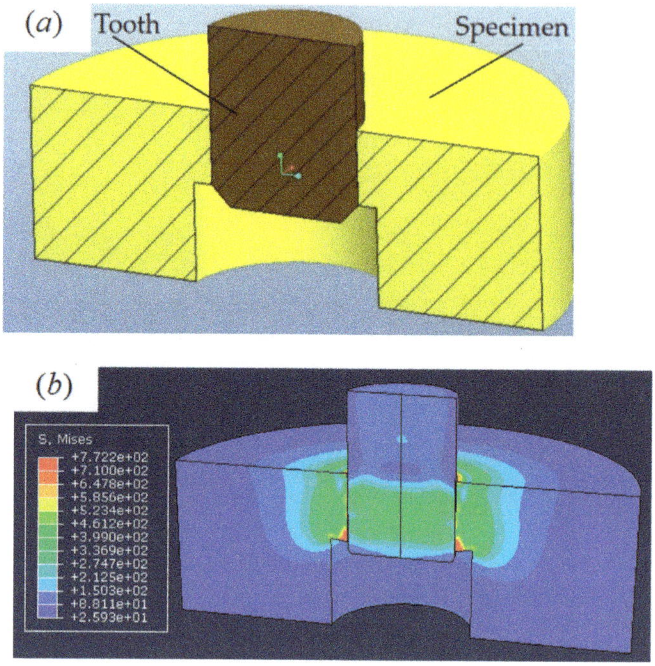

Figure 13. (**a**) Finite element model and (**b**) stress cloud map.

Extracting the maximum fastening force from the stress cloud map, as shown in Figure 14, the simulation results are highly consistent with the experimental data. This not only verifies the accuracy and reliability of the finite element analysis but also provides strong theoretical support for subsequent structural optimization design.

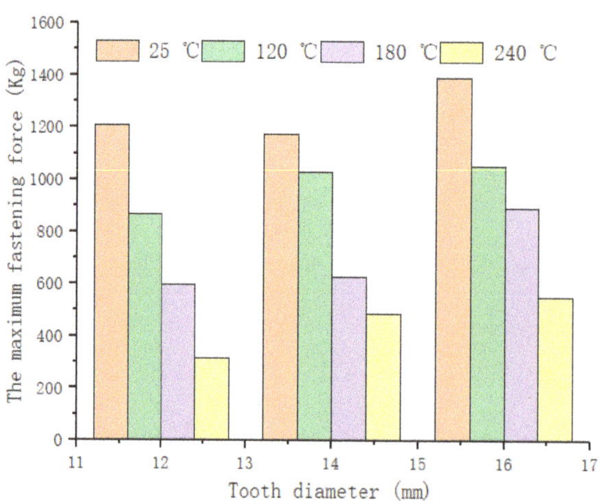

Figure 14. Simulation results.

4.4. The Effect of Interference on the Fastening Force

As shown in Figure 15, which illustrates the relationship between the interference amount and the maximum fastening force of 16 mm teeth at different temperatures, it can be seen that the maximum tooth-fixing force decreases with increasing temperature under the same interference amount. Specifically, when the interference amount is 0.075 mm, the maximum fastening force at 120 °C, 180 °C, and 240 °C decreases by approximately 25.9–64.4% compared to normal temperature. When the interference is 0.095 mm and 0.115 mm, the maximum fastening force is 56.6% and 60.6% lower than that at normal temperature, respectively. The maximum fastening force decreases with increasing temperature mainly due to the degradation of material properties. As the temperature increases, the mechanical properties of the base material (20CrMo) and tooth material (YG16C) of the roller drill bit significantly decrease. The second is the thermal expansion effect. At high temperatures, the thermal expansion coefficients of the matrix material and the tooth material are different, resulting in a decrease in the interference fit between the tooth hole and the tooth. Thirdly, stress relaxation occurs in materials at high temperatures, where the stress gradually decreases over time under constant strain. This phenomenon will gradually weaken the fastening force between the tooth hole wall and the teeth.

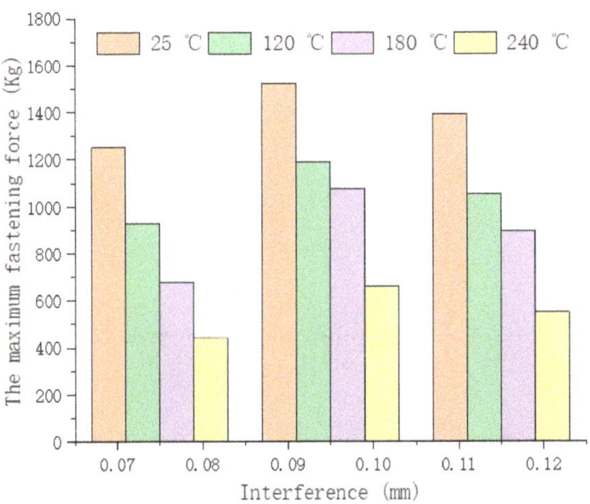

Figure 15. Relationship between interference and maximum fastening force.

Simulation analysis was conducted on different interference amounts, and the stress cloud maps and extracted maximum stresses are shown in Figures 16 and 17. It can be seen that the tooth fastening force decreases with increasing temperature, and the larger the interference amount, the smaller the maximum fastening force. This is because when the interference amount is too large, the tooth hole wall undergoes plastic deformation. The hole wall cannot maintain the fastening effect on the teeth, resulting in a decrease in the strength of the fixed teeth. This is consistent with the experimental results. The main reason for the maximum fastening force, when the interference fit is 0.095 mm, is that the contact pressure between the tooth hole wall and the teeth is moderate, providing sufficient friction and support to ensure maximum fastening force. The second is the optimization of stress distribution. An appropriate interference fit can optimize the stress distribution between the tooth hole wall and the teeth, avoiding stress concentration. A 0.095 mm interference fit can ensure more uniform contact between the tooth hole wall and the teeth, thereby improving the fastening force. The third is the matching of material properties. In high-temperature environments, an appropriate interference fit can better match the thermal expansion coefficients of the matrix material and the tooth material, thereby reducing the decrease in fastening force caused by differences in thermal expansion.

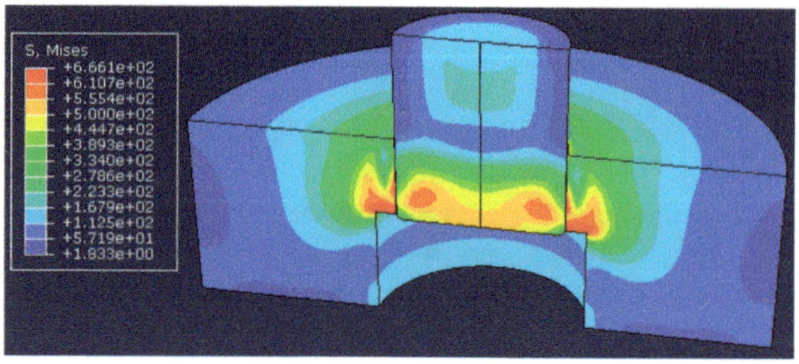

Figure 16. Stress cloud map.

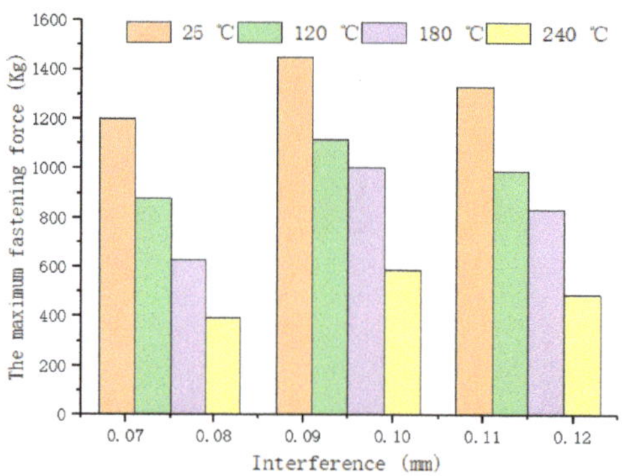

Figure 17. Simulation results.

5. Conclusions and Discuss

5.1. Conclusions

This study proposes an experimental testing method for the tooth fixation ability of roller drill bits in high-temperature geothermal wells, and conducts systematic tooth fixation experiments in high-temperature environments. Through finite element analysis and indoor experiments, the influence of tooth diameter, tooth pitch, interference fit, and temperature on the strength of fixed teeth was studied, aiming to provide a theoretical basis for the design optimization of roller drill bits. The experimental results indicate that:

(1) The higher the temperature, the weaker the strength of the fixed teeth. At the same tooth spacing, the maximum fastening force decreases continuously with increasing temperature, and compared to room temperature, the maximum fastening force decreases by about 49.6–64.5%. This indicates that high temperatures significantly weaken the tooth fixation strength of the roller drill bit.

(2) At the same temperature, the maximum fastening force is highest when the tooth spacing is 10 mm, with maximum values of 2080.8 kg (room temperature), 1652.1 kg (120 °C), 1284 kg (180 °C), and 856.9 kg (240 °C), respectively. When the interference fit is 0.115 mm and the tooth spacing is 10 mm, the best tooth-fixing effect is achieved.

(3) When the tooth diameter is 16 mm, the maximum fastening force is the highest, with a maximum value of 1392.8 kg. As the temperature increases, the maximum fastening force increases with the increase in tooth diameter. When the interference fit is 0.115 mm, choosing a tooth diameter of 16 mm is more conducive to improving the strength of the fixed teeth.

(4) At the same temperature, as the interference fit increases, the maximum fixed-tooth force first increases and then decreases, and the maximum fixed-tooth force is highest when the interference fit is 0.095 mm. At 120 °C, 180 °C, and 240 °C, the maximum fastening force decreased by 21.9%, 29.4%, and 56.6%, respectively, compared to room temperature.

5.2. Discuss

This study systematically analyzed the variation law of fixed-tooth strength of roller drill bits under a high-temperature environment through a combination of experiments and numerical simulations. The research results provide an important theoretical basis for optimizing the design of roller drill bits. The following is a further discussion of the research results:

(1) The impact of high temperature on material properties. Under high-temperature conditions, the mechanical properties of the matrix material (20CrMo) and tooth material (YG16C) of the roller drill bit significantly decrease. The strength limit and hardness of the material decrease, and the difference in thermal expansion coefficient leads to a reduction in interference fit, thereby weakening the fastening force between the teeth and the tooth hole.

(2) Optimization of tooth spacing. The experimental results show that the fixed-tooth strength is the highest when the tooth spacing is 10 mm. This may be because appropriate tooth spacing can better balance the stress on teeth and the thermal expansion of materials. Too small a tooth spacing can lead to stress concentration, while too large a tooth spacing can reduce the support effect of the teeth. This discovery provides an important reference for the design of roller drill bits.

(3) Selection of tooth diameter. In high-temperature environments, larger tooth diameters (such as 16 mm) can provide stronger resistance to thermal expansion and structural stability. Therefore, when drilling high-temperature geothermal wells, it is recommended to prioritize selecting teeth with larger diameters to improve the strength of the fixed teeth.

(4) Optimization of interference fit. The experimental results show that the fixed-tooth strength is the highest when the interference fit is 0.095 mm. The optimization of interference fit is crucial for improving the strength of fixed teeth. Excessive interference fit may lead to material fatigue and stress concentration, and instead reduce the strength of fixed teeth.

This study provides a scientific basis for the design and optimization of roller bits for drilling high-temperature geothermal wells. In practical applications, it is recommended to choose the appropriate tooth spacing, tooth diameter, and interference fit based on the specific temperature and geological conditions of the geothermal well. In addition, future research can further explore new materials and fixed-tooth technologies to better address the challenges in high-temperature geothermal drilling.

Author Contributions: Y.Y. (Yan Yang): Writing—review and editing, Writing—original draft, Methodology, Investigation, Formal analysis. D.S.: Writing—review and editing, Writing—original draft, Visualization, Validation, Methodology, Investigation, Formal analysis. L.C. and H.R.: Writing—review and editing, Writing—original draft, Visualization, Methodology, Formal analysis. Y.Y. (Yingxin Yang) and S.Q.: Writing—review and editing, Supervision, Methodology, Formal analysis. Z.H.: Writing—review and editing, Supervision, Methodology, Formal analysis, Experiment. All authors have read and agreed to the published version of the manuscript.

Funding: This study was supported by the Open Fund (PLN202426) of State Key Laboratory of Oil and Gas Reservoir Geology and Exploitation (Southwest Petroleum University), Open Fund (OGE202303-19) of Key Laboratory of Oil & Gas Equipment, Ministry of Education (Southwest Petroleum University), and the "Sailing" Program for High-level Talents of Yibin University (2022QH19).

Data Availability Statement: The original contributions presented in this study are included in the article.

Conflicts of Interest: D.S. was employed by CNPC Chuanqing Drilling Engineering Co., Ltd. and Chuanxi Drilling Company. The remaining authors declare that the research was conducted in the absence of any commercial or financial relationships that could be construed as a potential conflict of interest.

References

1. Yang, Y.X.; Song, D.D.; Ren, H.T.; Huang, K.; Zuo, L. Study of a new impregnated diamond bit for drilling in complex, highly abrasive formation. *J. Pet. Sci. Eng.* **2019**, *187*, 106831. [CrossRef]
2. Xu, G.; Xiao, F.J.; Cui, X.M.; Yu, D.K.; He, C.; Ye, X.T. Arsenic and fluorine concentrations and sources of Yangbajing high temperature geothermal water in Xizang. *Rock Miner. Test.* **2024**, *43*, 487–500.
3. Shulyupin, A.N. Steamwater flow instability in geothermal wells. *Thermo-Phys. Aeromechanics* **2015**, *22*, 475–480. [CrossRef]
4. Simisinov, D.I.; Gorodilov, L.V.; Simisinov, A.D. Parameters of dimension chains of roller cone drill bits. *J. Min. Sci.* **2024**, *60*, 629–638. [CrossRef]
5. Sporin, G.V.Z. The characterization of wear in roller cone drill bit by rock material-Sandstone. *J. Pet. Sci. Eng.* **2019**, *173*, 1355–1367. [CrossRef]
6. Li, Q.; Li, Q.; Cao, H.; Wu, J.; Wang, F.; Wang, Y. The crack propagation behaviour of CO_2 fracturing fluid in unconventional low permeability reservoirs: Factor analysis and mechanism revelation. *Processes* **2025**, *13*, 159. [CrossRef]
7. Cheng, Y. Wellhead stability during development process of hydrate reservoir in the Northern South China Sea: Evolution and mechanism. *Processes* **2024**, *13*, 40. [CrossRef]
8. Saksala, T.; Gomon, D.; Hokka, M.; Kuokkala, V.-T. Numerical and experimental study of percussive drilling with a triple-button bit on Kuru granite. *Int. J. Impact Eng.* **2014**, *72*, 56–66. [CrossRef]
9. Iman, R.; Babak, A.; Brian, E. Analysis of rock cutting process with a blunt PDC cutter under different wear flat inclination angles. *J. Pet. Sci. Eng.* **2018**, *171*, 771–783.
10. Song, D.D.; Ren, Z.P.; Yang, Y.X.; Chen, Y.; Nie, G.; Tan, L.; Peng, H.; Li, Z.; Chen, X.; Li, M.; et al. Drilling performance analysis of impregnated micro bit. *Mech. Sci.* **2022**, *13*, 867–875. [CrossRef]
11. Ji, W.S.; Cao, Z.; Zhong, L.; Wu, X.; Leng, X. Failure analysis of roller bearings in KX732-250 mining roller drill bit. *Min. Equip.* **2024**, 150–155.
12. Li, F.F.; Jiang, X.; Zhang, Y. Application of force analysis in failure analysis of roller drill bits. *J. Liaoning Univ. Sci. Technol.* **2024**, *26*, 32–34+60.
13. Li, F.F.; Liu, B.; Shi, Z.W.; Zhang, Y. Failure analysis of KX84 series roller bit claw large runway. *Fail. Anal. Prev.* **2023**, *18*, 55–61.
14. Bill, R.; Abraham, S.; Paul, S. Sccessfully applying micronized cellulose to minimize lost circulation on the PUNA geothermal venture wells. Geothermal Resource Council. In Proceedings of the 34th Annual Meeting, Reno, NV, USA, 19–22 April 2010; pp. 1–15.
15. Majd, H.M.; Hassani, B. Improvement of roller cone drill bit design by using finite element method and experimental study. *Int. J. Oil Gas Coal Technol.* **2022**, *31*, 382–405. [CrossRef]
16. Yan, B.Y. Research on the mechanism and improvement technology of drill bit wear. *West. Explor. Eng.* **2023**, *35*, 58–60.
17. Ren, H.T.; Tian, H.P.; Yu, H.B.; Zhou, C.; Yang, Y.; Liu, Q.; Song, D.; Chen, L.; Chen, X. Development and application of three cone drill bit for high temperature geothermal wells. *Nat. Gas Ind.* **2021**, *41*, 95–100.
18. Ravina, K.; Yang, N.; Brocoum, S.; Pasco-Anderson, J.; Walker, R.L.; Khan, M.; Cabodi, M.; Holsapple, J. Conical drill bit for optimized external ventricular drain placement: A proof-of-concept study. *J. Neurosurg.* **2023**, *139*, 881–891. [CrossRef]
19. Hu, H.; Qi, L.; Chao, X. Physics-informed Neural Networks (PINN) for computational solid mechanics: Numerical frameworks and applications. *Thin-Walled Struct.* **2024**, *205*, 112495. [CrossRef]
20. Yu, H.; Wang, H.; Lian, Z. An assessment of seal ability of tubing threaded connections: A hybrid empirical-numerical method. *J. Energy Resour. Technol.* **2023**, *145*, 052902. [CrossRef]
21. Smolyanitsky, B.N.; Popelyukh, A.I.; Karpov, V.N.; Alekseev, S.E.; Timonin, V.V. Downhole high-pressure air hammers for open pit mining. *J. Min. Sci.* **2014**, *50*, 929–993.

22. Shephard, M.S.; Yehlan, A. Computational strategies for nonlinear and fracturemechanics problem. *Comput. Struct.* **1985**, *20*, 211–223. [CrossRef]
23. Huang, K.L.; Ai, Z.J.; Yang, Y.X. The improved rock breaking efficiency of an annular-groove PDC bit. *J. Pet. Sci. Eng.* **2018**, *172*, 425–435. [CrossRef]
24. Kuang, Y.C.; He, P.; He, J.B.; Wei, Q.; Zhou, S.W.; Dong, Z.Z. Research on sealing test technology of roller Drill bit and performance evaluation of sealing ring. *J. Eng. Des.* **2024**, *31*, 383–392.
25. Wang, Y.J.; Nie, D.J.; Yuan, B.H.; Zhao, C.L.; Zhang, T.; Liu, Z.X. Research and application of PDC drill bits for high-temperature hard rock geothermal drilling in Tianyang Basin. *Drill. Eng.* **2024**, *51*, 125–131.
26. Schmitz, T.; Powell, K. Shrink fit tool holder connection stiffness/damping modeling for frequency response prediction in milling. *Int. J. Mach. Tools Manuf.* **2007**, *47*, 1368–1380. [CrossRef]
27. Pryhorovska, T.A.; Chaplinskiy, S.S. Finite element modeling of rock mass cutting by cutters for PDC drill bits. *Oil Ind. J.* **2015**, *42*, 38–41. [CrossRef]
28. Simone, O.; Enel, I.; Regillion, K. New HT/HP Technology for geothermal application significantly increases on-bottom drilling hours. In Proceedings of the IADC/SPE Drilling Conference and Exhibition, San Diego, CA, USA, 6–8 March 2012; pp. 1–20.
29. Burkhart, D.I.C.; Schollmayer, D.I.T.; Vorst, B.V.D.; Sansalone, M.; Thielen, S.; Sauer, B. Development of an online-wear-measurement for elastomer materials in a tribologically equivalent system for radial shaft seals. *Wear* **2021**, *476*, 203671. [CrossRef]
30. Wang, H.X. Strait development and application of deep sea shallow jet three cone drill bit. *Chin. Pet. Chem. Stand. Qual.* **2024**, *44*, 115–117.
31. Slipchuk, A.M.; Jakym, R.S.; Korendiy, V.M.; Lytvyniak, Y.M. Design and technological aspects of functionally oriented technology of manufacturing the three-cone drill bits. *IOP Conf. Ser. Mater. Sci. Eng.* **2023**, *1277*, 012015. [CrossRef]
32. *ISO 6892-1:2019*; Tensile testing of metallic materials—Part 1: Room temperature test method. International Organization for Standardization: Geneva, Switzerland, 2019.

energies

Article

Design Optimization of Valve Assemblies in Downhole Rod Pumps to Enhance Operational Reliability in Oil Production

Seitzhan Zaurbekov [1,*], Kadyrzhan Zaurbekov [1], Doszhan Balgayev [1], Galina Boiko [1], Ertis Aksholakov [1], Roman V. Klyuev [2] and Nikita V. Martyushev [3,*]

[1] Institute of Metallurgy and Ore Beneficiation, Satbayev University, Almaty 050010, Kazakhstan; k.zaurbekov@satbayev.university (K.Z.); dos_mjk@mail.ru (D.B.); g.boiko@satbayev.university (G.B.); taitik@mail.ru (E.A.)

[2] Department of Technique and Technology of Mining and Oil and Gas Production, Moscow Polytechnic University, 107023 Moscow, Russia; kluev-roman@rambler.ru

[3] Department of Information Technologies, Tomsk Polytechnic University, 634050 Tomsk, Russia

* Correspondence: s.zaurbekov@satbayev.university (S.Z.); martjushev@tpu.ru (N.V.M.)

Abstract: This study focuses on the optimization of valve assemblies in downhole rod pumping units (DRPUs), which remain the predominant artificial lift technology in oil production worldwide. The research addresses the critical issue of premature failures in DRPUs caused by leakage in valve pairs, i.e., a problem that accounts for approximately 15% of all failures, as identified in a statistical analysis of the 2022 operational data from the Uzen oilfield in Kazakhstan. The leakage is primarily attributed to the accumulation of mechanical impurities and paraffin deposits between the valve ball and seat, leading to concentrated surface wear and compromised sealing. To mitigate this issue, a novel valve assembly design was developed featuring a flow turbulizer positioned beneath the valve seat. The turbulizer generates controlled vortex motion in the fluid flow, which increases the rotational frequency of the valve ball during operation. This motion promotes more uniform wear across the contact surfaces and reduces the risk of localized degradation. The turbulizers were manufactured using additive FDM technology, and several design variants were tested in a full-scale laboratory setup simulating downhole conditions. Experimental results revealed that the most effective configuration was a spiral plate turbulizer with a 7.5 mm width, installed without axis deviation from the vertical, which achieved the highest ball rotation frequency and enhanced lapping effect between the ball and the seat. Subsequent field trials using valves with duralumin-based turbulizers demonstrated increased operational lifespans compared to standard valves, confirming the viability of the proposed solution. However, cases of abrasive wear were observed under conditions of high mechanical impurity concentration, indicating the need for more durable materials. To address this, the study recommends transitioning to 316 L stainless steel for turbulizer fabrication due to its superior tensile strength, corrosion resistance, and wear resistance. Implementing this design improvement can significantly reduce maintenance intervals, improve pump reliability, and lower operating costs in mature oilfields with high water cut and solid content. The findings of this research contribute to the broader efforts in petroleum engineering to enhance the longevity and performance of artificial lift systems through targeted mechanical design improvements and material innovation.

Keywords: downhole rod pumping units (DRPU); valve assembly optimization; flow turbulizer; shut-off element rotation; oilfield equipment

1. Introduction

Despite decarbonization and the transition to renewable energy sources, oil is still one of the most important energy resources today. In addition, oil is a raw material for the production of plastics and other mass-market materials. Therefore, research in the field of increasing the energy efficiency of oil production is relevant today. Most of the oil produced is extracted using pumps. The global oil and gas pumps market is projected to reach USD 10.80 billion by 2028 at a compound annual growth rate of 6.4% from USD 6.6 billion in 2020 [1–3]. More than half of the world's active well stock is equipped with downhole rod pumping units (DRPU). Therefore, in particular, in the USA, 85% of the entire well stock (more than 470 thousand), about 53% in Russia (about 76 thousand), and about 60% in Kazakhstan are operated by this method. The wide application of DRPU is conditioned by their high reliability and service life, constructive simplicity, non-deficiency, and cheapness of materials used in their manufacture, as well as unpretentiousness in maintenance. These factors have ensured the conservativeness of the DRPU design, which has not been changed for a long period of time [4–8].

The operation of downhole rod pumping units, despite their relative simplicity and prevalence, is subject to a number of problems affecting their reliability and efficiency. One of the key factors reducing the reliability of DRPUs is the mechanical wear and tear of components. Constant friction between the string rod and tubing, as well as between the ram and cylinder, leads to gradual wear of these elements. This is especially true for wells with curved wellbores or when running heavy and abrasive fluids. Corrosion is also a serious problem, as water and the aggressive components of the produced oil contribute to the deterioration of the metal parts of the pumping unit. Improper equipment selection, such as an inadequate pump or rod of inappropriate strength, can lead to premature failures and reduced well production rates. DRPU reliability also depends on the operating environment. Operating at high temperatures and pressures, especially in deep wells, places increased demands on pump materials and design. Gas plugs, often encountered during production, can cause unstable pump operation and hydraulic shocks that damage valves and other components. Well watering, when the produced fluid contains large amounts of water, can lead to salt and paraffin deposits, making it difficult for the pump to operate and reducing its efficiency.

The problem of improving DRPU reliability can be addressed through various solutions. These can be both solutions to change the design and material of pump components and the optimization of the operating modes of pumping units and accompanying equipment.

Improving the reliability of downhole rod pumping units (DRPU) is an important challenge in oil production. There are several approaches to solving this problem. Optimization of operating regimes, including controlling the dynamic fluid levels and reducing the pumping rates, can reduce stresses on the equipment and extend its life. However, this can reduce the well production rate. The use of modern materials for rods, pumps, and tubing, such as high-strength steels [5] and composites [6], significantly increases their resistance to corrosion and wear [7], but requires substantial capital investment. The use of centralizers on rods can reduce friction and wear, especially in curved wells, but may be ineffective in wells with high solids content. Application of advanced methods of corrosion protection [8], such as inhibitors and coatings [9], reduces the intensity of corrosion processes, but requires constant monitoring and maintenance of reagent concentrations. An effective system of monitoring and diagnostics of equipment condition [10,11] allows timely detection and elimination of faults, preventing accidents, but requires implementation of sophisticated software and qualified personnel. The use of downhole rod pumps with extended overhaul intervals, such as pumps with carbide friction pairs [12] or pumps with elastomeric seals [13,14], can significantly reduce the number of repairs, but their cost is usually higher. Improvements in the design of equipment components also contribute to reliability, such as the use of internal bypasses in valves [15] and optimization of sealing elements [16,17]. The calculation and selection of equipment parameters [18,19], consideration of temperature factors [20,21], and analyzing the causes of defects and damage [22–24] are also important aspects. The aim of this study is to optimize the design of the valve assembly in downhole rod pumping units (DRPU) in order to improve the operational reliability of oil production systems and to increase the time interval between equipment overhauls [25–28].

To achieve this aim, the following objectives were pursued:

- To collect and analyze statistical data on the failure modes of DRPUs operating in actual field conditions, with a focus on valve-related issues;
- To develop a modified valve assembly design incorporating a flow turbulizer intended to reduce localized wear and to improve sealing performance;
- To perform the laboratory testing of the proposed valve design under simulated downhole conditions in order to evaluate its mechanical behaviour and functional efficiency;
- To conduct pilot field trials of the redesigned valves in real oil wells and to assess their durability and performance in comparison with standard designs.

2. Methods and Materials

The whole work consisted of several stages. At the first stage, the statistics of DRPU failures at oil-producing wells located in the Republic of Kazakhstan were collected. Based on the results of the analysis, the causes and the mechanism of DRPU failure were determined. Necessary changes were made to the DRPU design to eliminate its failure. DRPUs of the modified design were tested with the help of the laboratory bench, and then their industrial testing was carried out.

2.1. Collection of Oilfield Equipment Failure Statistics

Failure statistics were collected at oil production wells of the Uzen field of Ozenmunaigas JSC (Republic of Kazakhstan). The field is located in the Mangistau region of the Republic of Kazakhstan, on the Mangyshlak peninsula (north-eastern coast of the Caspian Sea). A total of 927 wells are operated at the Uzen field, of which 917 are operated by sucker rod well pumping units, 19 are operated by electric centrifugal pump units, and 1 is an operated electric screw pump unit. Hence, almost 98% of the wells are DRPU [14]. The statistics were collected during 2022.

2.2. Design of the DRPU Structure and Manufacture of DRPU Elements

The design and development of technical documentation, including engineering drawings and 3D models for the fabrication of prototype turbulizers and modified valve assemblies, were performed using the KOMPAS-3D V18 software package. This CAD system was selected due to its capabilities in parametric modeling and convenient export of geometry to formats compatible with additive manufacturing equipment.

The work was limited to geometric and structural design; no numerical simulation (e.g., CFD or FEA) was conducted at this stage. The focus was on the physical fabrication and the experimental testing of the proposed design solutions.

The turbulizer prototypes were developed for valves used in downhole rod pumps of the HH2Б-70-35-12-1 type with a nominal diameter of 70 mm. For enhancement of the DRPU valve assembly, an additional flow element (referred to as a "turbulizer") was installed beneath the valve seat. Using additive FDM technology, these components were manufactured from PLA plastic on a Creality CR-5 H 3D printer (Shenzhen Creality 3D Technology Co., Ltd., Shenzhen, China).

2.3. Laboratory Tests of the Developed DRPU

To conduct experimental bench studies on the DRP valve unit operation, a model of the experimental bench pumping unit (EBPU) was created, as shown in Figure 1.

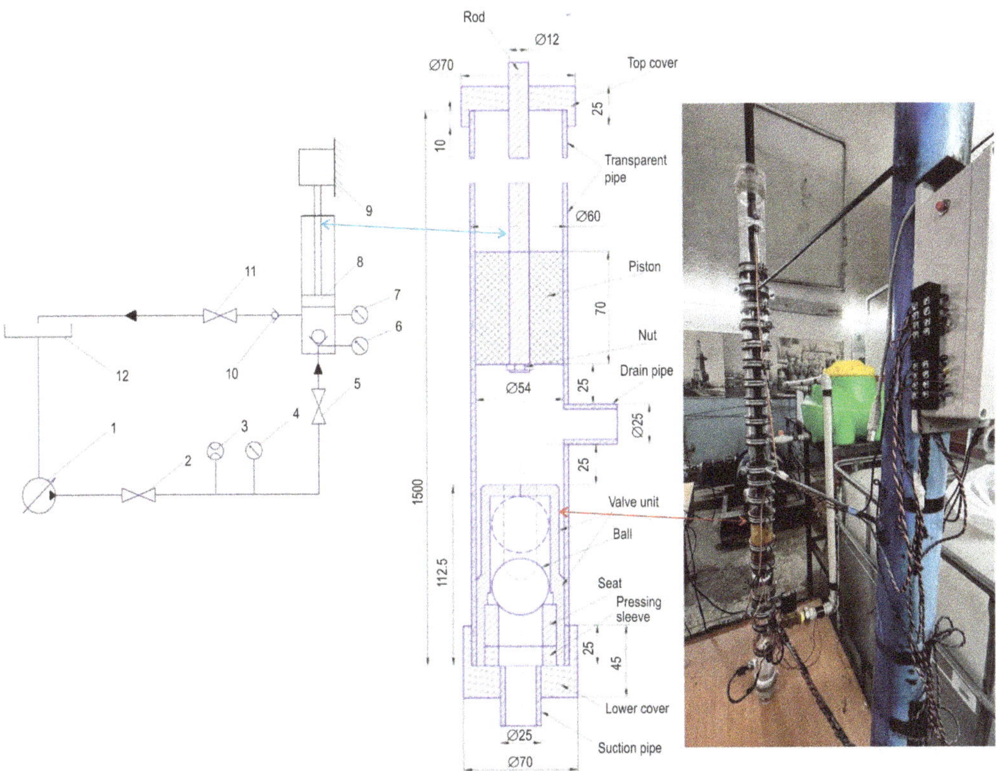

Figure 1. The scheme of the experimental bench pumping unit (EBPU): 1—pump (the Pedrollo DG-PED3 pump station with the inverter); 2, 5, 11—cut-off valves; 3—liquid flow meter; 4—pressure gauge No. 1 (electronic pressure sensor); 6—pressure gauge No. 2 (electronic pressure sensor); 7—pressure gauge No. 3 (electronic pressure sensor); 8—plunger; 9—rod drive; 10—electromagnetic valve; 12—tank.

The dimensions of the EBPU allow for the life-size modeling of the DRP valve assembly, thus eliminating scaling errors. The seat is made of organic glass, and the turbulizer is made of plastic used in 3D printers. In the experimental setup, a natural sample shut-off element (a valve ball made of steel) was used. Although water was used as the working fluid in the laboratory tests, and its physical properties, particularly viscosity, differ significantly from those of crude oil, this choice was justified by the need for safe handling, transparency for visual observation, and stable flow behaviour in a controlled environment. Importantly, water exhibits a lower viscosity compared to most types of oil, especially high-wax-content crude. Therefore, the fact that stable and sustained rotation of the valve ball was achieved in water implies that similar or even more pronounced rotational effects are to be expected under actual oilfield conditions, where the fluid medium exerts greater resistance and shear forces on the valve elements. To validate this hypothesis and to verify the operability of the proposed turbulizer design under realistic conditions, a series of full-scale field tests were conducted in active oil wells. These field experiments confirmed the effectiveness of the turbulizer configuration in inducing ball rotation and improving valve longevity in high viscosity, high water-cut production environments.

High-speed video recording (up to 4800 frames per second) was used to provide the accurate counting of the ball rotation frequency during the valve opening within 7 s (ball lifting and its landing on the seat with rotation).

The "EVERCAM high speed rapid camera shooting" was used to video the movement of the valve shut-off element, which is an HD video of excellent 720 p quality at up to 4800 frames per second [21]. The camera has small dimensions (100 × 90 × 100 mm) and is lightweight (1 kg without optics). The ultimate frame rate, depending on the resolution, is in the range of 1000−22,500 fps.

To assess the reliability of measuring the ball rotation rate and time to failure, standard deviations and 95% confidence intervals were calculated. The data were processed in Python 3.0 using the scipy.stats library.

Electronic sensors with a range of 0–10 bar were used to measure the pressure in the system, which ensured the reliable recording of pulsations and pressure drops in various valve operating modes. The fluid flow rate was measured by an impeller flow meter with a pulse output and an operating range of 0.2–10.0 L/s, which significantly exceeds the actual values of the measured flow rate (0.27–1.34 L/s) and eliminates the risk of signal saturation. The data were processed using an Arduino controller and Python-based software.

2.4. Industrial Testing of the Developed DRPU

A small pilot batch of 10 "turbulizers" made of Al-Si-Mg aluminum alloy was manufactured for pilot field tests using additive manufacturing technologies to create a physical object from an electronic model on a 3D printer using the layer-by-layer FDM (fused deposition modeling) method. Installation of turbulizers required the lengthening of the lower part of the valve body (cage). The valve with the turbulizer had 5 parts in contrast to the series produced with 4 parts. A small pilot batch (10 pcs) of turbulizers was produced, which were installed on 5 DRP of the HN2B-70-35-12-1 type with a nominal diameter of 70 mm (5 pcs on suction and 5 pcs on discharge valves).

At the production service premises, "OzenMunayGas" valves of new design were installed on 3 old pumps. On each pump, there were new suction and discharge valves, totalling 6, and 2 new pumps had 4 valves in total with a "turbulizer". The valves were assembled, pressurized, and fitted to the pumps as shown in Figure 1.

The pumps, equipped with the new design valves, were delivered to the oil production area and run into the wells. The pumps, equipped with the new suction and discharge valves, were run into active operated wells.

The depth of the pumps was between 400–912 m; all wells were oil and vertical; pump diameters were 70 mm. The stroke lengths were 3.0 and 3.5 m, the number of strokes is from 4.93 to 6.2 per min. The dynamic level was 138–420 m; the theoretical (planned) liquid flow rate was within 40–85 m^3/day. The actual flow rate was 33–74 m^3/day; in the case of high water cut (planned 90%, actual 95–98.3%), the theoretical flow rate was 2.52–5.88 t/day. The actual oil flow rate was 0.88–1.35 t/day.

3. Results and Discussion

3.1. Analysis of Failure Statistics and Refinement of the DRP Valve Assembly Design

Currently, 99% of production wells in the Uzen field are operated by DRPU, and they account for about 97–98% of the entire oil produced [29–33]. At the same time, the number of repairs of wells equipped with DRPU during 2022 was 2794, i.e., each well was repaired on average 3 times during the year. The duration of repairs depends on the type of well repair. The reasons for wells' repair are in Table 1.

Table 1. Gradation by causes of field equipment failure.

		Gradation by Causes of Field Equipment Failure								
Production and Service Enterprise	**Number of Well Repairs**	**Mechanical Impurities Paraffin-Salting Salt-Salting Pump Valve Leakage Leakage of Pump Valves Oil Leaks**	**Paraffin-Salting**	**Salt-Salting**	**Leakage of Pump Valves**	**Oil Leaks**	**Tubing Leakage**	**Rods Breakage**	**Slag, Scale in Gasoil Equipment**	**Accident with Field Equipment**
Oil and Gas Production Division-2	2764	596	118	10	388	487	810	240	0	14

The analysis of DRPU failure causes showed that the total share of failures due to leakage of pump valve pairs was 14–15% of the total number. The loss of tightness of failed valves was caused by high content of mechanical impurities and paraffin, the ingress of which, when the shut-off element was seated in the same place on the seat, led to rapid wear of the ball, because the area of impact of the shut-off element on the seat was heavily overloaded. Therefore, the DRP valve assembly required more attention to improve the performance of the pumping equipment [34].

The commercially available DRP valves consist of a body that houses the shut-off element (ball) and seat, which are held in place by a seat holder that is threaded to the body. There are a number of patents for improvements in the design of the DRP valve assembly [35–39]. In order to improve the service life of the DRP valve assembly, the authors proposed an improved design providing for the installation of an additional element "turbulizer" of the flow under the valve seat. The proposed design of the valve assembly is shown in Figure 2.

The improved design of the DRP valve assembly includes the installation of an additional flow "turbulizer2 element (Figure 2). Within the framework of this research work, the turbulizer with the valve axis deviation from the vertical by 5°, 10°, 15° (Figure 3) and with the spiral plate swirler of 5, 7.5, 10 mm in width (Figure 4) was manufactured. "Turbulizers" were manufactured using additive technologies by an electronic model using a 3D printer model of Creality CR-5 H by the fused deposition modeling (FDM) method.

"Turbulizers" with the valve axis deviation from the vertical by 5°, 10°, and 15°, and the spiral plate of the flow swirler were made with certain geometrical relations. The thickness of the turbulizer is equal to two thicknesses of a standard valve seat. The outer diameter is equal to the diameter of a standard valve seat. These dimensions are due to the choice of DRP with a nominal diameter of 70 mm, which is most widely used in the field "Uzen" as the object of research.

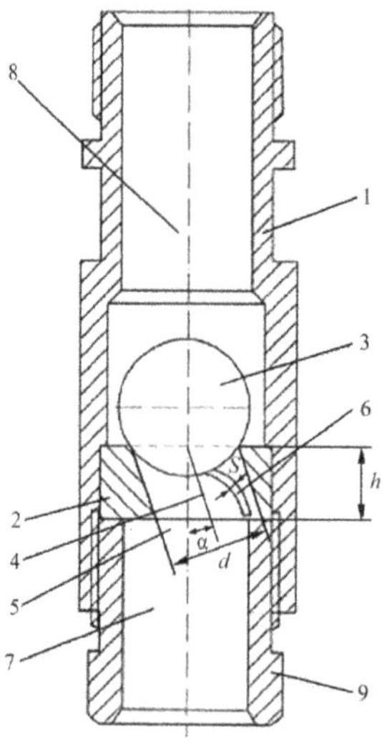

Figure 2. The ball valve of the modified design. 1—body; 2—seat; 3—ball; 4—shifted centreline of the central hole; 5—vertical axis of valve symmetry; 6—360° screw plate; 7—sub-valve cavity; 8—supra-valve cavity; 9—seat holder.

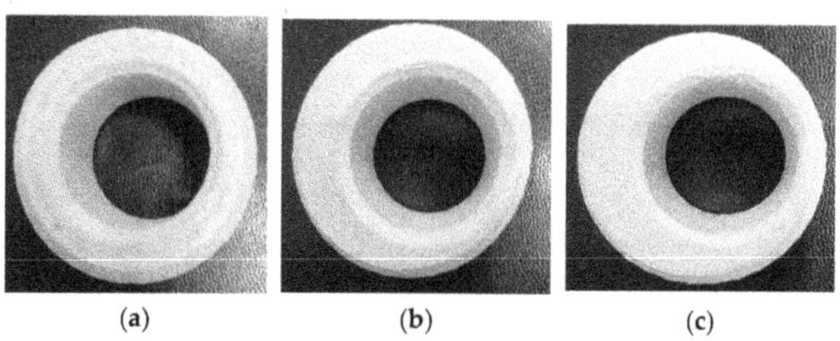

Figure 3. "Turbulizers" with axial deviation of the passage opening (PLA prototypes used in laboratory testing): (**a**)—5°; (**b**)—10°; (**c**)—15°.

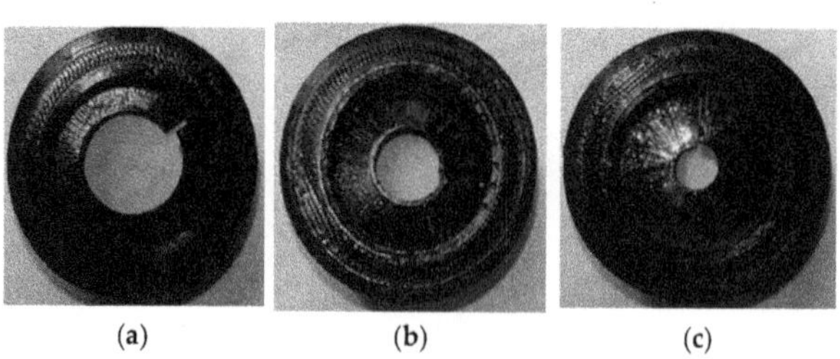

Figure 4. "Turbulizers" with screw plate width (PLA prototypes used in laboratory testing): (**a**)—5 mm; (**b**)—7.5 mm; (**c**)—10 mm.

"Turbulizers" are installed directly under each suction and discharge valve seat and require the fabrication and installation of an extended valve body structure.

3.2. Conducting Bench Experimental Tests

In the bench-scale experimental studies, readings were taken from the flow meter (3), electronic pressure gauges (4, 6, and 7) (Figure 1). In this case, pressure gauges (6 and 7) show the pressure before and after the valve with the "turbulizer". The readings of the sensors in increments of 3 s were processed in the "Arduino" programme and presented in the form of tables and graphs.

During the bench studies, high-speed photography of the valve shut-off element movement during its opening and closing at angles of the valve axis deviation from the vertical of 5°, 10°, and 15° was used (Figure 5). Larger angles were not considered due to the non-technological manufacturing of such "turbulizers". Each "turbulizer" with the angles of deviation of the valve axis from the vertical of 5°, 10°, and 15° was studied using a spiral plate swirler with a width of 5, 7.5, and 10 mm.

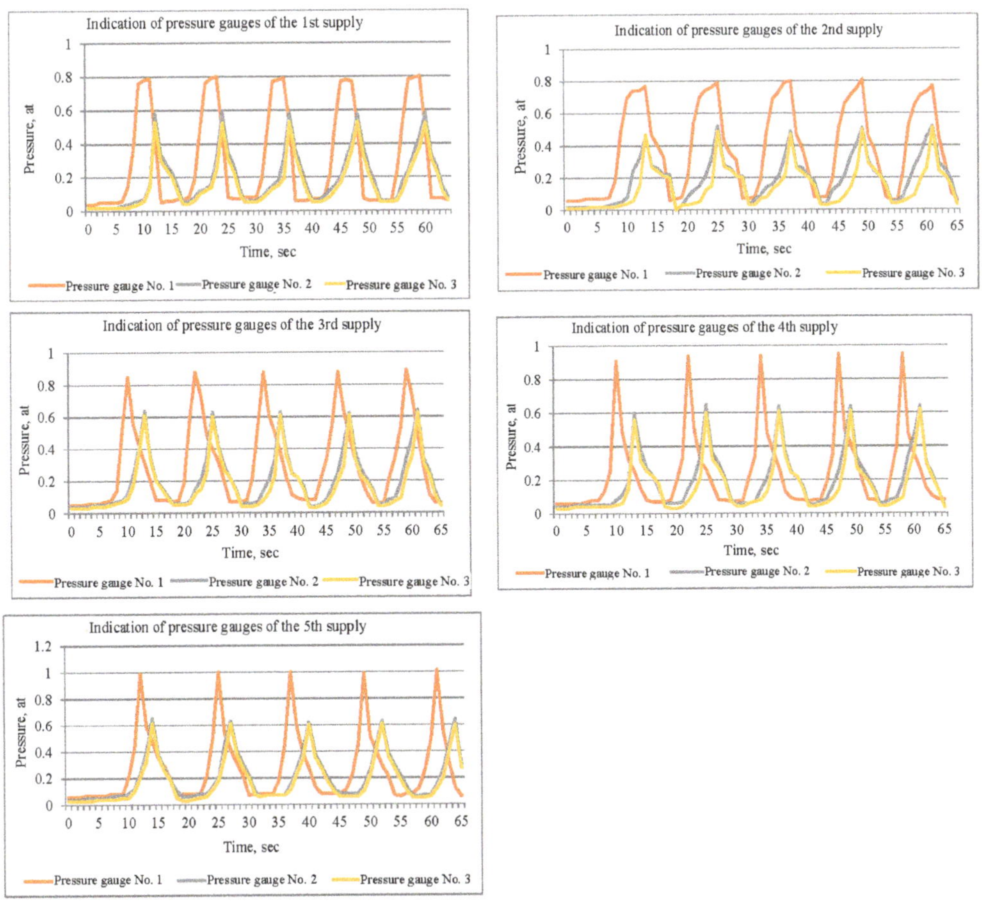

Figure 5. Pressure change readings in different pump operation modes: mode 1—pump flow is 0.27 L/s; mode 2—pump flow is 0.54 L/s; mode 3—pump flow is 0.81 L/s; mode 4—pump flow is 1.08 L/s; mode 5—pump flow is 1.34 L/s.

Figure 5 shows a series of graphs illustrating the change in pressure at different operating modes of the pump with the turbulizer installed. The measurements were taken on an experimental setup simulating the operation of a deep rod pump at five different flow rates: 0.27, 0.54, 0.81, 1.08, and 1.34 L/s. As the flow rate increased, there was a regular increase in the average pressure, an increase in the amplitude of pulsations, and an increase in the frequency of oscillations. These changes were not random but regular in nature and were related to the physical nature of the fluid flow and the functioning of the valve assembly.

At low flow rates (mode 1), the pressure was stable and the changes on the graph were practically smoothed out. This was because at low flow rates, the valve ball moves smoothly, without pronounced water hammer, and the flow resistance remains minimal. Accordingly, pressure fluctuations had a small amplitude, and pulsations had a low frequency. In mode 2 (0.54 L/s), there was an increase in both the average pressure and the amplitude of fluctuations, which indicated an increase in flow energy and more active operation of the valve mechanism. The graph became less smooth, and the pulsations become more pronounced. With a further increase in the flow rate of up to 0.81 and 1.08 L/s, more abrupt pressure drops occurred, caused by the rapid opening and closing of the valve, as well as an increase in vortex formation in the turbulizer zone. In these modes, the valve opens at a higher speed, and the ball abruptly breaks away from the seat. And when closing, it collides with the seat with residual rotation and significant momentum, which causes hydraulic shocks and pressure surges. In mode 5 (1.34 L/s), the pressure graph took on an almost sinusoidal shape with sharp fronts of growth and decline, with the amplitude of pulsations reaching its maximum. This is explained by the intense turbulence of the flow and the resonant operation of the valve, where each phase of the cycle (opening/closing) is accompanied by clearly pronounced pressure changes.

The shape of the curves reflected the combined influence of many factors: the fluid flow rate, the ball mass and kinematics, the turbulizer geometry, and the nature of the flow before and after the valve. As the flow rate increased, the turbulizer more actively forms vortex structures, which increase the rotation speed of the shut-off element. This increases the efficiency of wear equalisation, but at the same time creates conditions for peak loads on the seating surfaces. The most stable and efficient operation of the valve was achieved in the range of modes 3 and 4, where a sufficiently high ball rotation frequency and an acceptable level of pulsation were ensured. At the same time, according to other sections of the work, a configuration with a spiral plate width of 7.5 mm provided an optimal balance among the flow intensity, the pressure level, and the ball rotation speed.

In view of this, the differences between the curves in Figure 5 are due to the increase in flow energy, changes in the nature of the interaction between the ball and the seat, and the degree of vortex development initiated by the turbulizer. The results obtained confirm that an increase in the flow rate significantly affected the pressure dynamics and operation of the valve assembly, and the shape of the curves directly reflected the physical processes occurring in the system in each specific mode.

Figure 6 illustrates the change in the liquid flow rate in five operating modes of the pump with a valve unit equipped with a turbulizer: 0.27, 0.54, 0.81, 1.08, and 1.34 L/s. In all the modes, the graphs had a pulsating shape, typical of reciprocating pumps, where oscillations were caused by the plunger movement, the valve operation, and the hydrodynamic resistance of the system.

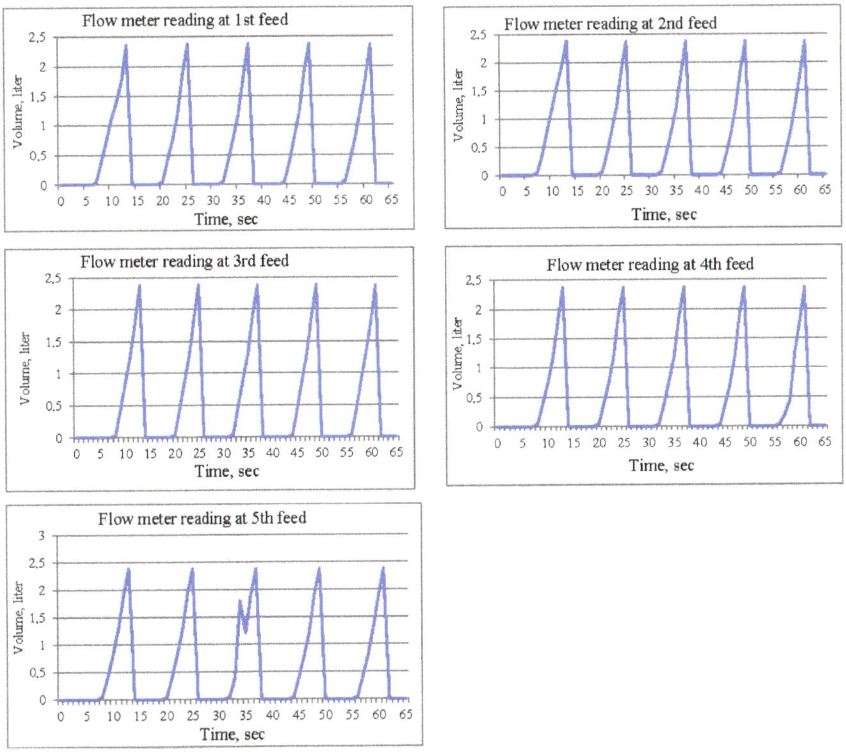

Figure 6. Flow rate variation readings in different pump operating modes: mode 1—pump flow rate is 0.27 L/s; mode 2—pump flow rate is 0.54 L/s; mode 3—pump flow rate is 0.81 L/s; mode 4—pump flow rate is 1.08 L/s; mode 5—pump flow rate is 1.34 L/s.

At a flow rate of 0.27 L/s, the oscillations were minimal and symmetrical, indicating stable and laminar operation of the system. As the flow rate increased up to 0.54 and 0.81 L/s, the amplitude and unevenness of the oscillations increased due to increased turbulence and inertial effects. The turbulizer enhanced vortex flows, promoting the ball rotation, and it also caused short-term flow disturbances that were represented by sharp peaks and dips.

In the modes of 1.08 and 1.34 L/s, the graphs become more dynamic: the amplitude and frequency of pulsations increase, the hydraulic shocks and the inertial effects intensify. Sharp changes in the flow occur at the moment of ball separation and landing, which is clearly displayed on the curves.

The turbulizer configuration also influences the shape of the graphs. With a screw plate width of 5 mm, the resistance was low, and the oscillations were smoothed out. At 7.5 mm, an optimal balance was achieved between the resistance and the efficiency of ball rotation, but pulsations increased. At 10 mm, excess resistance occurs, which decelerates the flow and made the oscillations sharper.

Therefore, the differences in Figure 6 are explained by both the change in the flow rate and the complex hydrodynamics of the interaction of the turbulizer, the ball and the valve channel. The shape of the curves allows determining the optimal parameters for a stable flow and uniform wear.

The results of the experiments presented in Figure 7 allow for a detailed comparative analysis of the rotation dynamics of the valve assembly's shut-off element under various turbulizer design configurations. For each turbulizer configuration, 50 repeated measurements of the ball rotation frequency were made. The mean value for the configuration with a vertical axis and a 7.5 mm wide spiral plate was 4.29 rpm (SD = 0.31), 95% CI: (4.19; 4.39). For the configuration with a 10 mm width, it was 3.11 rpm (SD = 0.44), 95% CI: (2.98; 3.24), and it was 3.68 rpm (SD = 0.27), 95% CI: (3.59; 3.77) for 5 mm. The main focus is on the influence of two parameters: the angle of deviation of the valve axis from the vertical (5°, 10°, and 15°) and the width of the spiral swirling plate of the turbulizer (5 mm, 7.5 mm, and 10 mm). They influence the rotation frequency of the ball during valve opening. Each turbulizer variant was tested under conditions simulating the operation of a pump with a nominal diameter of 70 mm at a pressure of 6 bar and a flow rate of 60 litres per minute. For each case, 50 repeated measurements were performed, which provided a stable sample for statistical and physical analysis.

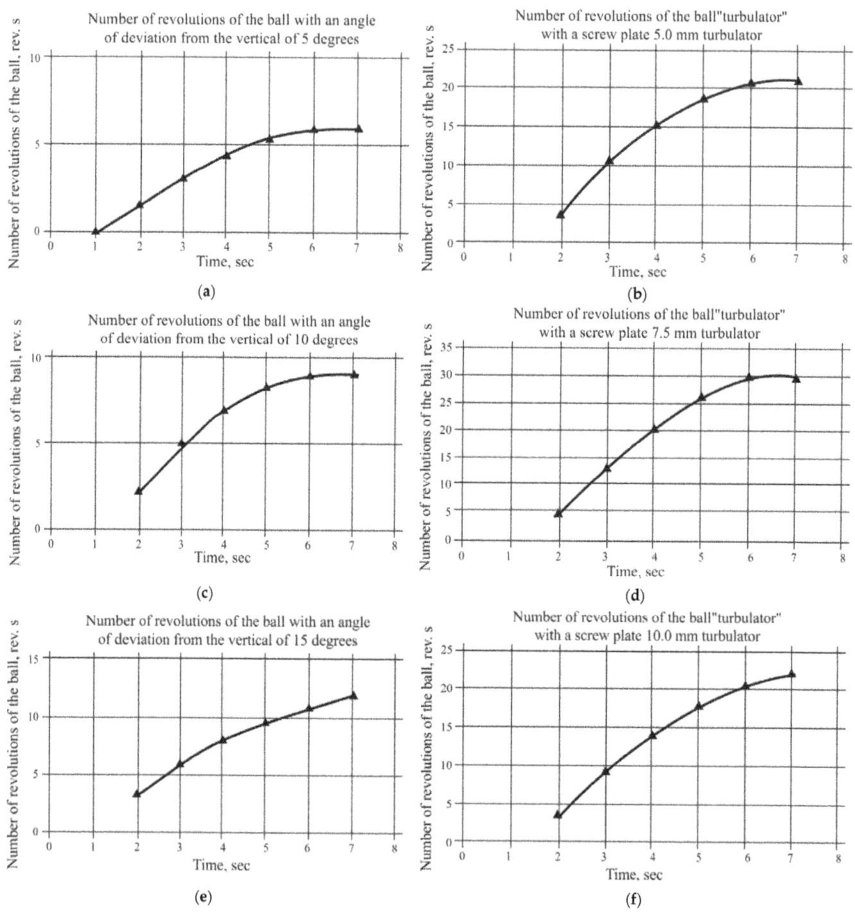

Figure 7. Experimental results of the valve shut-off element speed during 5° (**a**), 10° (**c**), and 15° (**e**) change of the seat axis from the vertical and "turbulization" with a screw plate of 5 (**b**), 7.5 (**d**), and 10 (**f**) mm in thickness.

The graphs in Figure 7 clearly show that the highest rotation speed of the shut-off element is achieved when using a turbulizer with a 7.5 mm wide screw plate, installed strictly vertically and generating a full rotation (360°) inside the flow channel. In this case, the ball made about 30 revolutions in 7 s of lifting. Such intensity was conditioned by the optimal ratio between the created vortex flow and the resistance of the turbulizer. With a smaller plate width (5 mm), the vortex was not sufficiently pronounced, and the torque transmitted to the ball was weak. When the width was increased to 10 mm, the flow section was blocked excessively, the flow decelerated, and the rotation speed drops.

The behaviour of the ball was additionally influenced by the angle of deviation of the valve axis from the vertical. When the angle increased up to 10° and 15°, the efficiency of vortex motion transmission decreased due to the appearance of asymmetric flows and stagnant flow zones. This was confirmed by the decrease in the rotation frequency of the ball in the corresponding configurations in Figure 7.

Hence, the maximum rotation speed was achieved with a vertical valve axis and a spiral width of 7.5 mm. In this case, symmetrical and stable vortex flows were formed, ensuring rotation of the ball not only during lifting, but also during landing on the seat. Residual rotation contributed to more uniform running-in, reduced local wear, and increased the reliability of the unit.

The obtained data confirm that the geometry of the turbulizer significantly affected the dynamics of the ball. The most effective configuration was that with a vertical axis and a 7.5 mm plate, while other options provide less rotation stability and worse wear resistance conditions. The differences in the shapes of the graphs in Figure 7 are explained by the degree of vortex development, flow symmetry and the resistance created by the design of the valve channel.

Figure 8 shows the results of the visual video and separate footage from the EVERCAM-1000-8-M (Evercam Ltd., Dublin, Ireland) camera of the valve ball rotation for clarity purposes.

Figure 8. Visual video results and selected EVERCAM-1000-8-M footage of the valve ball rotation.

The best performance of the "turbulizer" with a screw inner plate width of 7.5 mm, in our opinion, is due to the rational value of liquid flow overlap at the entrance to the valve seat. The plate width of 5.0 mm overlaps a smaller cross-section of the fluid flow and, consequently, the ball has less pressure of the flow on the shut-off element of the valve, forcing it to rotate. The plate width of 10 mm blocks most of the fluid flow entering the valve seat, creates additional resistance, and reduces the flow velocity and, hence, the rotation frequency of the shut-off element of the valve (the ball).

3.3. Conducting Pilot Field Tests

Field tests were conducted at oil-producing wells of the 'Uzen' field equipped with downhole NNAB-70-35-12-1 (Shengji Petroleum, Dongying, China) rod pumps. The work was performed at five wells.

A small pilot batch of 10 "turbulizers" made of the Al-Si-Mg aluminum alloy with application of additive technologies was manufactured for pilot field tests. Mechanical characteristics of the used alloy are given in Table 2.

Table 2. Properties of the Al-Si-Mg aluminum alloy used for turbulizer fabrication.

No.	Characteristics	Units of Measurement	Al-Si-Mg Aluminum Alloy D16T (Hardened and Aged)
1	Yield strength (min)	MPa	180
2	Tensile strength (min)	MPa	300
3	Hardness (max)	HB	105
4	Fatigue strength (min)	N/mm^2	100
5	Relative elongation (min)	%	8–10
6	Density	g/cm^3	2.8
7	Heat capacity	J/kg K	922

Installation of turbulizers required the lengthening of the lower part of the valve body (cage) (Figure 9b). The valve with the turbulizer had five parts in contrast to the series produced with four parts (Figure 9c). A small pilot batch (10 pcs) of turbulizers has been produced, which are planned to be installed on five SSN of the HH2Б-70-35-12-1 type with the conditional diameter of 70 mm (five pcs on suction and five pcs on discharge valves).

(a) (b) (c)

Figure 9. The newly designed valve for pilot field tests: (**a**)—duraluminum turbulizers; (**b**)—serial and new valve body; (**c**)—valve of new design.

At the production service premises "OzenMunaiGas", valves of the new design were installed on three old pumps. On each pump, there were new suction and discharge valves (six in total), and four valves with the "turbulizer" in total on two new pumps. The valves were assembled, pressurized, and fitted to the pumps, as shown in Figure 10.

Figure 10. Installation of newly designed valves on DRP pumps.

The pumps equipped with the newly designed valves were delivered to the oil production site and run into the wells. Initial inspections after short-term operation revealed that wear of the turbulizer components was primarily due to abrasive interaction with mechanical particles. This indicates that, in addition to structural optimization, the material properties of the turbulizer significantly influence the operational lifespan of the valve assembly. The depth of the pumps was between 400 and 912 m; all wells were vertical oil wells. Pump diameters were 70 mm, stroke lengths were 3.0 and 3.5 m, stroke rates were from 4.93 to 6.2 per min, and the dynamic level was 138–420 m. The theoretical (planned) liquid flow rate was 40–85 m^3/day, the actual flow rate was 33–74 m^3/day, with the high water cut of 90% (planned) and 95–98.3% (actual). The theoretical flow rate was 2.52–5.88 tonnes/day, and the actual oil flow rate was 0.88–1.35 tonnes/day. The data on the wells where the pumps with the newly designed valves were run are given in Table 3.

To objectively assess the efficiency of the turbulizers implemented in the design of the valve unit, a comparative testing was conducted using data on the operation of standard pumps that were not modified. As a control group, 15 deep-well sucker rod pumps of the NN2B-70-35-12-1 type, operated in 2022–2023 at the same horizons of the Uzen field, where field tests of the modernized designs were made, were considered.

When forming the control sample, key parameters were taken into account to ensure the comparability of operating conditions. In particular, the pump running depth ranged from 400 to 912 m, which corresponds to the depths of the wells in which valves with turbulizers were used. The degree of water cut of the product was also close and varied from 95 to 98%, while in all cases, an increased content of mechanical impurities was noted: rust, sand, and solid inclusions. Therefore, the analysis was carried out under conditions that were as close as possible to those in which the prototypes were used.

The design of the pumps in the control group completely coincided with the upgraded versions: the diameter was 70 mm, the plunger stroke was from 3.0 to 3.5 m, and the pumping frequency ranged from 4.9 to 6.2 pumping cycles per minute. This made it possible to exclude the influence of design differences on the operating results and focus on the influence of the new turbulent elements.

Earlier, the work already indicated that one of the key causes of sucker rod pump failures in the Uzen field was the loss of tightness of valve pairs, which accounted for about 14–15% of the total number of failures. The main factor causing wear of the shut-off elements is the high concentration of mechanical impurities and paraffin, precipitating in the contact zone of the ball with the seat. The repeated impact of the ball on the same section of the seat in the absence of rotation leads to local wear and subsequent leakage.

Table 3. Pumped well data from the pilot field test programme.

Data on Discharged Pumps					
Date of running	28 July 23	14 August 23	15 August 23	17 August 23	17 August 23
Well number	No. 6356	No. 4702	No. 8509	No. 2093	No. 3093
Pump number	302,036	303,169	303,149	308,086/11	307,932/10
Well data					
Well type	Vertical	Vertical	Vertical	Vertical	Vertical
Field	Uzen	Uzen	Uzen	Uzen	Uzen
Horizon number	15	15	13	17	14
Well purpose	Oil	Oil	Oil	Oil	Oil
Commissioning date	30 June 2014	31 July 2012	24 April 1993	21 September 1974	25 May 1979
Diameter of production casing/completed production casing, mm	168	168	146	168/114	168
Pump running depth, m	696	680	912	400	640
Stroke length, m	3.0	3.0	3.0	3.0	3.5
Number of swings, rpm	4.93	6.15	6.2	6.15	6.2
Pump diameter, mm	70	70	70	70	70
Dynamic level	138	420	415	326	113
Static level	0	44	265	0	237
Rpl	113.7	122.1	128	108	100.7
Rzab	106.2	91.4	71.9	120.2	101.9
Actual bottom hole, m	1215	1324	1170	950	1116
Depth of tubing, run	696	680	912	400	640
Liquid flow rate, m^3/day, (mode/actual)	70.0/33.0	70.0/64.0	40.0/42.0	85.0/74.0	60.0/62.0
Water cut, % (mode/fact)	90.0/95.4	90.0/95.0	90.0/96.0	95.0/97.9	95.0/98.3
Oil flow rate, tonnes per day (mode/fact)	5.88/1.27	5.8/1.35	3.4/1.32	3.57/1.29	2.52/0.88

A typical example is the pump installed in well No. 6356. After 39 days of operation, it was lifted, and upon inspection, the turbulizer was found to be damaged because of wear of the screw plate. The cause was the intense abrasive action of the contaminated environment containing rust, sand, and solid inclusions (Figure 11). At the same time, the turbulizers on the discharge valves remained in satisfactory condition. Following the discussions with representatives of the underground repair service, it was established that the failure was caused by a combination of an increased content of mechanical impurities in the pumped liquid and insufficient wear resistance of the turbulizer material (Table 4 and Figure 11).

(a) (b) (c) (d)

Figure 11. The condition of the turbulizer and valve components after retrieval of the pump from well No. 6356 (operation time is 39 days). **(a)** The general view of the valve assembly with the removed turbulizer showing mechanical damage. **(b)** Close-up of the turbulizer surface with traces of abrasive wear on the screw plate. **(c)** The inner surface of the turbulizer illustrates thinning and edge erosion caused by solid particles. **(d)** Accumulated deposits and contamination in the valve seat area.

Table 4. Test results of the new valve design (the turbulizer was manufactured from the AL-Si-Mg aluminum alloy).

Data on Discharged Pumps						
Well number	6356	4702	8509	2093	3093	9846 (repeat descent)
Pump number	302,036	303,169	303,149	308,086/11	307,932/10	30,3962
Run date	28 July 2023	14 August 2023	15 August 2023	17 August 2023	17 August 2023	3 September 2023
Lift date	6 September 2023	23 November 2023	1 November 2023	10 November 2023	31 August 2023	19 November 2023
Reason for lifting the pump	Wear of the turbulizer of the intake valve	Clamping of the pump	Wear of the turbulizer of the intake valve	Wear of the turbulizer of the intake valve	Mechanical impurities from pipes	Wear of the turbulizer of the intake valve
Pump operating time with newly designed valves, days	37	101	78	85	15	77

Therefore, the comparative assessment using the control group confirms the practical efficiency of the turbulizers under heavy-duty conditions. At the same time, cases of premature wear in aggressive environments emphasize the need to use more durable materials, such as the 316 L steel, in the design of the vortex element.

Four other pumps with the newly designed valves were used in wells No. 4702 for one hundred and one days, No. 2093 for eighty-five days, No. 8509 for seventy-eight days, and No. 9846 for seventy-seven days. The mean operating time was 75.3 days (SD = 9.6), 95% confidence interval: (66.7; 83.9). For standard valves, the mean operating time was 52.3 days (SD = 10.2), 95% CI: (42.1; 62.5). The difference between the mean values is statistically significant ($t = 3.41$, $p = 0.008$, $n = 5$). This, respectively, exceeded the average working life of ordinary pumps not reworked in this oil-producing section of the "OzenMunaiGas" JSC when they were lifted due to valve failure by 44%, 21%, 12%, and 11%. The main reason of suction valves failure is also abrasive wear, because of the low resistance of the material of the "turbulizers" of the valves, being made of the AL-Si-Mg aluminum alloy, having essentially less resistance to abrasive wear in comparison with steel. These findings highlight the need to treat the selection of construction materials with equal priority as geometrical design. The aluminum alloy, while easy to process and lightweight, demonstrated insufficient resistance to prolonged exposure to sand and rust particles. This led to accelerated erosion and a loss of effectiveness of the turbulizer's vortex function.

A proposal was made to produce newly designed valves made of 316 L stainless steel, which has increased strength and anti-corrosion resistance, in connection with which steel "turbulizers" can be used in highly aggressive environments. For "turbulizers" (newly designed valve elements), the main indicator of resistance is the material strength, which determines the longevity of the turbulizer. For the 316 L steel, the minimum material tensile strength is more than 1.5 times higher than that for D16T duraluminum. Consequently, a significant increase in the operating life of the newly designed valves can be expected before they are replaced.

4. Discussion

The results obtained during the experimental and field studies aimed at improving the reliability of the valve assembly of deep rod pumps through the introduction of turbulizers allow drawing a number of important conclusions that reveal the physical laws governing the processes occurring inside the pump system under various operating conditions. The analysis shows that a significant proportion of submersible pump failures is caused by a loss of tightness in the valve pairs, which is caused by both mechanical impurities and paraffin deposits entering the gap between the ball and the valve seat. This leads to localized wear of the working surfaces, disruption of the running-in process, the uneven seating of the ball, and, ultimately, leaks. As an engineering solution, a variant of the unit modernisation was proposed and implemented: the installation of a turbulizer under the valve seat, the task of which is to form a vortex motion of the fluid to force the ball to rotate and ensure uniform wear of its surface.

Laboratory tests conducted on the test bench with visualization of the ball's movement using high-speed video recording showed a stable relationship between the geometric parameters of the turbulizer and the rotation speed of the shut-off element. The results obtained in laboratory and field conditions were statistically verified. The values of the rotation frequency of the shut-off element and the service life between repairs had narrow confidence intervals and low values of standard deviation, which confirms the high reproducibility of the effect. The statistical significance of the differences between the standard and modernized design was confirmed by the Student test ($p < 0.01$).

The maximum rotation frequency (about 30 revolutions in 7 s) was achieved when using a turbulizer with a screw plate 7.5 mm wide, made without deviating the axis from the vertical. This was confirmed by rotation graphs and analysis of high-speed video frames recording the residual rotation of the ball when it landed on the seat. This configuration creates the most favourable hydrodynamic flow structure: the fluid flow passing through the spiral plate evenly flows around the ball, spinning it with high stability. Increasing or decreasing the width of the plate leads to a deterioration in the rotation conditions: in the first case, due to excessive hydraulic resistance and a decrease in flow velocity, and in the second, due to insufficient vortex formation.

In addition, it was found that the angle of deviation of the valve axis from the vertical also had a significant effect on the rotation dynamics. With deviations of 10° and 15°, the flow loses symmetry, secondary circulation zones are formed, and the efficiency of torque transmission decreases. This is reflected in a decrease in the ball rotation frequency, which can lead to uneven wear of the valve working surfaces. Hence, it can be argued that the configuration of the turbulizer with an axial arrangement and a 7.5 mm spiral plate is optimal in terms of creating stable vortex motion.

Field tests conducted at wells in the Uzen field confirmed the applicability of the developed design in real conditions. In most cases, the new valve configuration showed an increase in the inter-repair life compared to standard solutions. However, in some cases, increased wear of the aluminum alloy turbulizers themselves was observed. This indicates the need to select more wear-resistant materials when operating in conditions with a high content of mechanical inclusions. The proposal to replace aluminum alloy with the 316 L stainless steel seems reasonable and logical based on the observed data, since this material has higher strength, corrosion resistance, and resistance to abrasive wear [39]. In [39–43], the authors stated that 316 L retains its mechanical strength and microstructural stability after a long-term exposure to brine solutions at 80–120 °C under conditions comparable to those at the Uzen field. Moreover, comparative tribological studies showed that 316 L exhibits lower friction coefficients and higher wear resistance when sliding on nitrided steels or ceramics, indicating its suitability for ball-to-seat interactions in pump valves. However, some limitations are also noted in the literature. The 316 L steel can suffer from mechanical fatigue when exposed to vibration stresses in combination with erosion by solid particles. This makes it critical to consider not only the static properties but also the interaction of dynamic loads, which is typical of turbulizer operation during cyclic pumping. Despite these concerns, the overall balance of corrosion resistance, manufacturability (including laser powder melting), and mechanical reliability makes the 316 L steel a preferred material over aluminum alloys, especially in high-abrasive loading conditions.

A comparative analysis with other researchers confirms the relevance and effectiveness of the approach used. For example, Fakher [44,45], in the review of the effectiveness of the rod pumps, cited data that showed that improvements in valve design and diagnostics reduced the failure rate by 15–20%. However, specific values for the ball rotation speed were not provided, which made our results (30 revolutions per 7 s) a unique indicator of the effectiveness of turbulent valve control.

The work of AC2T Research and UniLeoben [46] on the dynamics of valve ball movement led to the discovery of the phenomenon of "mid-cycle closing" and noted significant accelerations of up to 5–7 m/s^2 during closing. In our case, the stable rotation of the turbulizer removes some of these impact accelerations, which probably reduces the amplitude of water hammer and reduces excessive loads on the seat and ball. The rotation frequencies in our study demonstrate a controllable and predictable operating mode, whereas in the study described, the movement was chaotic and led to premature wear.

The frequency-elastic pump speed control technology described in Palka and Czyz [47] showed a 23% reduction in maximum impulse loads and an increase in overall efficiency from 24% to 38% at a speed of ~five strokes per minute (≈0.083 Hz). Our rotation rates (4.3 rpm) are significantly higher than the stroke frequency, indicating active vortex action within each stroke, creating additional lapping cycles and thus potentially increasing efficiency and reducing wear even in the absence of controlled frequency control.

The authors of [48,49] proposed upgrading valve assemblies to increase pump life. It mentioned the same approach of modifying the seat geometry and using turbulence, but without numerical data on ball rotation frequency. Our results complement this approach with specific figures that allow for estimating the potential resource efficiency.

In summary, a comparative analysis shows that our development is not just the addition of a vortex element, but a controlled hydrodynamic optimization that ensures surface lapping and slows valve wear. Numerical indicators of rotation frequency (4.3 rpm), reduction of flow asymmetry at the vertical axis, and relatively high resources in field conditions confirm the reliability and practical significance of the proposed design in comparison with existing approaches, which mainly operate with qualitative improvements without quantitative reference to valve dynamics. Moreover, further development of the valve assembly modeling may benefit from poroelastic formulations that account for nonhydrostatic in situ stress conditions and material anisotropy. Such an approach is particularly relevant when evaluating stress distribution and deformation behaviour in valve components operating under complex loading in subsurface environments. Fan et al. [50] presented a representative study, where poroelastic solutions were derived for a semipermeable borehole within transversely isotropic media under anisotropic stress conditions, providing valuable insights into stress transfer mechanisms in engineered subsurface systems.

The proposed valve assembly design with a turbulizer ensures uniform wear of the contact surfaces and extends the service life of the valves, and is technologically compatible with existing pumps. This makes the solution particularly attractive for use in mature oil fields with high water cut and the presence of mechanical impurities. Moreover, the experimental results clearly demonstrate that the durability of the turbulizer is materially dependent. The aluminum AL-Si-Mg alloy used in pilot tests could not withstand prolonged operation under high-solid-content flow conditions. The inner spiral plates suffered erosion, reducing the intended vortex effect. Switching to the 316 L stainless steel, with its superior hardness and corrosion resistance, is expected to mitigate such failures and enhance operational reliability. The results obtained allow recommending this design for further industrial application, provided that the durability of the turbulizers is increased through the use of more durable materials.

5. Conclusions

The results obtained during the work allow establishing a number of stable patterns demonstrating the influence of the turbulizer design on the rotation dynamics of the shut-off element and, as a consequence, on the wear resistance of the valve assembly. The most effective design in terms of forming a stable rotational movement of the ball was a turbulizer with a 7.5 mm wide spiral plate installed strictly along the axis of the valve channel. In this case, during the 7 s of valve opening, the ball made an average of 30 revolutions, which corresponds to a frequency of 4.29 rpm. Such rotation ensures a regular change in the contact area between the ball and the seat, which reduces the likelihood of local damage and contributes to more uniform wear of the surfaces. In other configurations with a spiral width of 5 mm and 10 mm, a decrease in the rotation frequency of up to 18–21 and 14–17 revolutions, respectively, was observed, which is explained by either insufficient or excessive impact on the flow. A 10° deviation of the channel axis from the vertical reduced the ball rotation by 22–25%, and by 35–40% at 15°, which indicates the high sensitivity of the vortex mechanism to flow symmetry.

Significant improvements were also recorded during field tests. The average service life of valves with a turbulizer was 75.3 days (SD = 9.6), which is 28% longer than that of standard designs (52.3 days). The rotation frequency (4.29 ± 0.31 rpm) and service life extension (75.3 ± 9.6 days) were statistically verified, and the improvements were found to be significant at $p < 0.01$. In some cases, the service life reached 101 days without repair, while the average service life of standard valves did not exceed 70 days. This confirms not only the reproducibility of the effect in the laboratory but also its stability under real operating conditions. Additional observation showed that the material of the turbulizer (AL-Si-Mg aluminum alloy) is not sufficiently resistant to abrasive effects in the presence of rust and sand in the liquid. This was the basis for switching to a more durable material, i.e., the 316 L stainless steel, which has a tensile strength of at least 485 MPa and a hardness of up to 217 HB. This transition may lead to a further increase in the service life of the unit and a reduction in the frequency of repairs.

The results obtained are of high practical significance. The design of the turbulizer is easily adaptable to serial pumps and can be implemented without changing the overall layout of the equipment. In conditions of mass application in a field of 900 wells, even an increase in the interval between repairs by 15–20 days can have a significant economic effect. Additional advantages include stabilisation of pump operation, reduction of accidents, reduction of reagent treatment volume, and reduction of unscheduled downtime.

The performed work reveals broad prospects for further research, including numerical modeling (CFD) of hydrodynamics in the internal volume of the valve, the development of new geometric solutions for turbulizers adapted to specific well operating conditions, as well as the use of wear-resistant alloys and composite materials. Particular attention should be paid to the numerical modeling of flows inside the modified valve design, since this will allow a deeper understanding of the nature of the interaction of the formed vortices with the shut-off element and optimizing the conditions of its rotation. In addition, the development of new-design solutions for turbulizers that take into account the specifics of wells (including a high level of mechanical impurities, temperature, and hydraulic conditions) will allow adapting the proposed solution to a wider range of operating situations.

One of the promising areas is the integration of digital twin technology of valve assemblies with pump monitoring systems. This will allow the implementation of predictive diagnostics of the technical condition of equipment based on the analysis of the rotation dynamics of the shut-off ball, comparing it with the accumulated wear statistics. This approach will increase reliability and allow for the advanced planning of maintenance. In addition, the proposed method for generating a controlled vortex flow inside the valve assembly demonstrates high efficiency and stability, making it a reliable tool for increasing the service life of equipment and reducing operating costs in oil production.

It is worth noting that the approach with inducing rotation of the shut-off element is applicable not only to sucker rod pumps, but can also be adapted to other types of lifting systems, such as screw and electric centrifugal pumps, especially in conditions of increased content of mechanical inclusions in the liquid. An additional area of development is the study of new materials for the manufacture of turbulizers, in particular, the use of hybrid and composite solutions with increased resistance to abrasive wear, which is especially important in aggressive environments. This will significantly increase the resource of the valve assembly with minimal costs for its improvement and implementation.

Scaling the proposed solution to the level of the entire field, where hundreds and even thousands of wells are operated, opens up significant opportunities for increasing energy efficiency, reducing unscheduled repairs, and reducing maintenance costs. This also helps to reduce the environmental impact by reducing the frequency of technological interventions and increasing the reliability of equipment, which is especially important in the context of the global transition to sustainable energy. Therefore, the results of this work not only demonstrate the practical efficiency of the proposed technical approach but also set the vector for its further scientific and industrial development.

Author Contributions: Conceptualization, S.Z. and K.Z.; methodology, D.B., G.B. and E.A.; validation, D.B., G.B. and E.A.; formal analysis, R.V.K.; investigation, R.V.K.; data curation, D.B., G.B. and E.A.; writing—original draft preparation, S.Z., K.Z. and N.V.M.; writing—review and editing, S.Z., K.Z. and N.V.M.; supervision S.Z. and K.Z.; project administration, N.V.M.; funding acquisition, R.V.K. All authors have read and agreed to the published version of the manuscript.

Funding: This research is funded by the Committee of Science of the Ministry of Science and Higher Education of the Republic of Kazakhstan (Grant No. BR24992868).

Institutional Review Board Statement: Not applicable.

Informed Consent Statement: Not applicable.

Data Availability Statement: The data presented in this study are available from the corresponding authors upon reasonable request.

Conflicts of Interest: The authors declare no conflicts of interest.

References

1. Exactitude Consultancy. *Oil & Gas Pumps Market (2019–2028); Forecast to Reach USD 10.80 Billion by 2028 at CAGR 6.4%*; Global Report; GlobeNewswire: Luton, UK, 2022.
2. Extractive Industries Transparency Initiative. Kazakhstan Country Facts. EITI. 2022. Available online: https://eiti.org/countries/kazakhstan (accessed on 1 July 2025).
3. Jurasz, J.; Canalesc, F.A.; Kiesd, A.; Guezgouz, M.; Beluco, A. A review on the complementarity of renewable energy sources: Concept, metrics, application and future research directions. *Sol. Energy* **2020**, *195*, 703–724. [CrossRef]
4. A.T. Consulting. Market Overview of Submersible Pumps for Oil Production in Russia. 2023. Available online: https://atconsult.ru/oilpump.html (accessed on 1 March 2025).
5. Bakhtizin, R.N.; Urazakov, K.R.; Topolnikov, A.S.; Azizov, A.M.; Komkov, A.G.; Ishmukhametov, B.H. *Oil Production by Rod Installations in Complicated Conditions*; UGNTU Publishing House: Ufa, Russian, 2016.

6. Stosiak, M.; Lubecki, M.; Karpenko, M. Designing a Composite Hydraulic Cylinder Using Genetic Algorithms. *Actuators* **2025**, *14*, 77. [CrossRef]
7. Molchanov, A.V. Stankov-kachalki: Problems and Prospects of Improvement. *Promyshlennye Vedom.* **2007**, *10*, 46–50.
8. Adonin, A.N. *Oil Extraction by Rod Pumps*; Nedra: Moscow, Russian, 1979.
9. Andrew, S.P.; Eiselstein, L.E.; Caligiuri, R.D.; Kim Parnell, T. Evaluation of a failure in a chlorine production facility. In Proceedings of the ASME International Mechanical Engineering Congress and Exposition, New York, NY, USA, 11–16 November 2001; Volume 2, pp. 771–781.
10. Belogolov, Y.I.; Pogodin, V.K.; Gozbenko, V.E.; Kargapoltsev, S.K.; Olentsevich, V.A.; Gladkih, A.M. Method for determining the sealing forces in a shut-off valve. *IOP Conf. Ser. Mater. Sci. Eng.* **2021**, *1064*, 012036. [CrossRef]
11. Elagina, O.Y.; Buklakov, A.G.; Dubinov, Y.S.; Dedok, D.V. Gas-abrasive wear of shut-off valves and process piping of compressor and gas distribution stations. *Mater. Phys. Mech.* **2023**, *51*, 84–92. [CrossRef]
12. Syrkin, V.V.; Balakin, P.D.; Treyer, V.A. Study on hydraulic direct-acting relief valve. *J. Phys. Conf. Ser.* **2017**, *858*, 012035. [CrossRef]
13. Bartenev, D.A.; Boiko, A.Y.; Matyushin, E.G.; Regush, L.A.; Fedorov, O.I.; Grebkov, A.A. Prestressed polymeric sealing elements of pipeline shut-off valves. *Chem. Pet. Eng.* **1986**, *22*, 265–267. [CrossRef]
14. Alekhnovich, V.; Syasko, V.; Umanskii, A. Multi-Parameter Complex Control of Metal Coatings on Ball Plugs of Pipeline Shut-Off Valves. *Inventions* **2024**, *9*, 78. [CrossRef]
15. Alexandrov, I.A.; Muranov, A.N.; Mikhailov, M.S. Development of an Algorithm for Automated Evaluation of the Operability of Structural Elements of Shut-off Valves. In Proceedings of the 2021 IEEE International Conference "Quality Management, Transport and Information Security, Information Technologies" T and QM and IS 2021, Yaroslavl, Russian, 6–10 September 2021; pp. 257–261. [CrossRef]
16. Arbuzov, N.S. Protection of the marine oil terminal against hydraulic shock using the fast shut-off valve. *Neft. Khozyaystvo—Oil Ind.* **2012**, *2*, 106–108.
17. Soboleva, J.; Kerimov, A.; Lampezhev, A. Consideration of Temperature Factors when Designing Butterfly Check Valves for Hazardous Production Facilities. *Civ. Eng. J.* **2023**, *9*, 263–274. [CrossRef]
18. Hu, C.-S.; Qin, H.; Zhu, J.; Zhao, Z.-X. Analysis of the causes of valve stem nut fracture in shut-off valves. *Dongbei Daxue Xuebao/J. Northeast. Univ.* **2012**, *33* (Suppl. 2), 76–78.
19. Chinyaev, I.R.; Fominykh, A.V.; Tel'minov, A.V. *Gate Valve with Internal Bypass*; Lecture Notes in Mechanical Engineering; Springer: Cham, Switzerland, 2022; pp. 746–751. [CrossRef]
20. Hasanov, I.I.; Guliyev, A.A. Development of effective sealing units for oil and gas equipment in the oil and gas production system [Opracowanie skutecznych zespołów uszczelniających dla urządzeń naftowych w systemie wydobywczym ropy i gazu]. *Nafta—Gaz* **2023**, *79*, 464–472. [CrossRef]
21. Gan, X.; Akin, J.E. Low-cycle fatigue life estimation of a subsurface safety valve. In Proceedings of the Materials Science and Technology Conference and Exhibition 2010, MS and T'10, Houston, TX, USA, 17–21 October 2010; Volume 2, pp. 1486–1494.
22. Vasin, S.A.; Plakhotnikova, E.V. Electric drive setting torque calculation method in electric drive shut-off valve system with rectilinea displacement of the valve closure. *J. Min. Inst.* **2018**, *232*, 407–412. [CrossRef]
23. Turbygin, V.; Klabukov, A.G.; Cheshlya, R.R.; Vikulin, Y.A. Wear resistance of metal-rubber sealing pairs in shut-off and regulating valves. *Sov. Eng. Res.* **1991**, *11*, 42–45.
24. Zhukov, D.V.; Melnikov, A.A.; Konovalov, S.V.; Afanasyev, A.V. Analysis of manufacturing internal pressure induced defects of the pipe wall with modeling of stresses. *Chernye Met.* **2021**, *10*, 49–55. [CrossRef]
25. Pogodin, V.K.; Belogolov, Y.I.; Gozbenko, V.E.; Kargapoltsev, S.K.; Olentsevich, V.A.; Gladkih, A.M. Calculation of sealing pressures of shut-off valves. *IOP Conf. Ser. Mater. Sci. Eng.* **2021**, *1064*, 012035. [CrossRef]
26. Soboleva, J.A.; Kerimov, A.M.; Anudinov, S.S.; Kocharov, M.A. Operation Simulation for a Check Valve Used in High-Performance Systems. *Civ. Eng. J.* **2023**, *9*, 3202–3217. [CrossRef]
27. Ivanov, V.V.; Oleinikov, A.V.; Denisevich, D.S.; Pektimirov, B.G. Emergency Shut-Off Valve for Pipelines. *Chem. Pet. Eng.* **2013**, *49*, 400–402. [CrossRef]
28. Afinogentov, A.A.; Ivanova, N.I.; Fedotova, I.A. The Method of Shut-off Valves Optimal Placing on Trunk Oil Pipeline. *IOP Conf. Ser. Earth Environ. Sci.* **2019**, *272*, 022038. [CrossRef]
29. Kislyakov, Y.P. Strategy for further development of the Uzen oil and gas field. In *Collection of Proceedings Geology, Development, Drilling, Oil and Gas Production, Ecology, Training and Professional Development*; JSC 'NIPIneftegaz': Ufa, Russian, 2014; Volume 1.
30. Nurshakhanova, L.K. Analysis of Methods of Impact and Assessment of Changes in Oil Properties in the Process of Development of the Uzen Field. Ph.D. Thesis, Gubkin Russian State University, Moscow, Russian, 2005.

31. Shinyakov, Y.; Sukhorukov, M.; Torgaeva, D.; Soldatov, A. Methods and Facilities for Monitoring the Operation of a Sucker Rod Pump. *Int. J. Mech. Eng. Technol.* **2018**, *9*, 1224–1231. Available online: https://iaeme.com/MasterAdmin/Journal_uploads/IJMET/VOLUME_9_ISSUE_12/IJMET_09_12_124.pdf (accessed on 1 July 2025).
32. Fakher-Sefat, S.; Khlaifat, A.; Hossain, M.E.; Nameer, H. A Comprehensive Review of Sucker Rod Pumps' Components, Diagnostics, Mathematical Models, and Common Failures and Mitigations. *J. Pet. Explor. Prod. Technol.* **2021**, *11*, 3815–3839. [CrossRef]
33. Togasheva, A. Well Operation by Plunger Rod Pumps in Difficult Conditions: A Case Study of the Uzen Field. *Nauk. Visnyk Natsionalnoho Hirnychoho Universytetu* **2025**, *1*, 22–30. [CrossRef]
34. Strekalovskaya, D.A.; Davydov, A.D.; Lyashenko, D.V.; Tleshev, M. Failure Analysis of Plunger Rod and Barrel of Sucker Rod Pumps. *Int. J. Mech. Prod. Eng. Res. Dev.* **2020**, *10*, 14835–14844. [CrossRef]
35. Dolov, T.R. Research of Work of Valve Units of Downhole Rod Pumping Units. Ph.D. Thesis, Gubkin Russian State University, Moscow, Russian, 2017.
36. Alper, R. Ball and Seat-Type Check Valve for Down-Hole Rod Pump. U.S. Patent 5,297,579 A, 22 March 1994.
37. Khaliullin, F.H. Valve of a Downhole Rod Pump. RU Patent RU2059885C1 MPK F04B 47/02, 10 May 1996.
38. Basos, G.Y.; Valovskiy, K.V. Suction Valve of a Depth Pump. RU Patent RU2487271C1 MPC F04B 53/10, 10 July 2013.
39. Wang, Y.; Chen, C.; Ren, R.; Xue, Z.; Wang, H.; Zhang, Y.; Wang, J.; Wang, J.; Chen, L.; Mu, W. Ferrite Formation and Decomposition in 316H Austenitic Stainless Steel Electro Slag Remelting Ingot for Nuclear Power Applications. *Mater. Charact.* **2024**, *218*, 114581. [CrossRef]
40. Ivey, R.K. Sucker Rod Pump with Improved Ball and Seat. U.S. Patent 8061381B2, 22 November 2011. Available online: https://patents.google.com/patent/US8061381B2 (accessed on 18 July 2025).
41. Zaurbekov, S.A.; Zaurbekov, K.S.; Balgaev, D.E.; Kadyrov, J.N. Ball Valve of Deep-Seated Pump. RK Patent No. 34,582, bulletin No. 37, 18 September 2020.
42. Wang, Y.; Chen, C.; Yang, X.Y. Solidification modes and delta-ferrite of two types of 316L stainless steels: A combination of as-cast microstructure and HT-CLSM research. *J. Iron Steel Res. Int.* **2025**, *32*, 426–436. [CrossRef]
43. Chen, L.; Wang, Y.; Li, Y.; Zhang, Z.; Xue, Z.; Ban, X.; Hu, C.; Li, H.; Tian, J.; Mu, W.; et al. Effect of Nickel Content and Cooling Rate on the Microstructure of as Cast 316 Stainless Steels. *Crystals* **2025**, *15*, 168. [CrossRef]
44. Fakher, S.; Khlaifat, A.; Hossain, M.E.; Nameer, H. Rigorous Review of Electrical Submersible Pump Failure Mechanisms and Their Mitigation Measures. *J. Pet. Explor. Prod. Technol.* **2021**, *11*, 3799–3814. [CrossRef]
45. Fakher, S.; Khlaifat, A.; Hossain, M.E.; Nameer, H. Improving Electric Submersible Pump Efficiency and Mean Time Between Failure Using Permanent Magnet Motor. *Upstream Oil Gas Technol.* **2022**, *9*, 100074. [CrossRef]
46. Peroutka, J. Sucker Rod Pump Standing Valve Movement Dynamics Investigation. Master's Thesis, Montanuniversität Leoben, Leoben, Austria, 2022.
47. Palka, M.; Czyż, A. Frequency-Elastic Control for Sucker-Rod Pumping Systems: Reducing Impulse Load and Increasing Efficiency. *Energy Technol.* **2021**, *9*, 1050–1060.
48. Cheng, H.; Yu, H.; Zeng, P.; Osipov, E.; Li, S.; Vyatkin, V. Automatic Recognition of Sucker-Rod Pumping System Working Conditions Using Dynamometer Cards with Transfer Learning and SVM. *Sensors* **2020**, *20*, 5659. [CrossRef] [PubMed]
49. Zhao, W.; Zhou, B.; Wang, Y.; Liu, W. Semi-Supervised Class-Incremental Sucker-Rod Pumping Well Operating Condition Recognition Based on Multi-Source Data Distillation. *Sensors* **2025**, *25*, 2372. [CrossRef] [PubMed]
50. Fan, Z.; Song, X.; Wang, D.; Ayasrah, M.; Li, S. Poroelastic Solutions of a Semipermeable Borehole under Nonhydrostatic In Situ Stresses within Transversely Isotropic Media. *Int. J. Geomech.* **2025**, *25*, 04024342. [CrossRef]

MDPI AG
Grosspeteranlage 5
4052 Basel
Switzerland
Tel.: +41 61 683 77 34

Energies Editorial Office
E-mail: energies@mdpi.com
www.mdpi.com/journal/energies

www.ingramcontent.com/pod-product-compliance
Lightning Source LLC
Chambersburg PA
CBHW040827250926
51850CB00007BA/1276
* 9 7 8 3 7 2 5 8 5 0 9 3 8 *